# Omega-3 Fatty Acids and the DHA Principle

# Omega-3 Fatty Acids and the DHA Principle

Raymond C. Valentine
David L. Valentine

CRC Press
Taylor & Francis Group
Boca Raton London New York

CRC Press is an imprint of the
Taylor & Francis Group, an **informa** business

CRC Press
Taylor & Francis Group
6000 Broken Sound Parkway NW, Suite 300
Boca Raton, FL 33487-2742

First issued in paperback 2019



ISBN-13: 978-1-4398-1299-0 (hbk)
ISBN-13: 978-1-138-37419-5 (pbk)



**Library of Congress Cataloging-in-Publication Data**

Valentine, R. C. (Raymond Carlyle), 1936-
Omega-3 fatty acids and the DHA principle / Raymond C. Valentine, David L. Valentine.
p. cm.
Includes bibliographical references and index.
ISBN 978-1-4398-1299-0 (hardcover : alk. paper)
1. Docosahexaenoic acid--Health aspects. 2. Omega-3 fatty acids--Health aspects. I. Valentine, David L. II. Title.

QP752.D63V35 2010
612.3'97--dc22 2009040278

**Visit the Taylor & Francis Web site at**
**http://www.taylorandfrancis.com**

**and the CRC Press Web site at**
**http://www.crcpress.com**

# *Dedication*

To the women in our lives: Isabella, Sienna, Carla, Annalisa, Lori, Rebecca, and Cindy, for understanding that omega-3 fatty acids are essential for any family activity.

# Contents

# *SECTION III General Properties of Omega-3s and Other Membrane Lipids*

## SECTION IV Cellular Biology of Omega-3s and Other Membrane Lipids

# SECTION V Lessons and Applications

# Preface

The physical–chemical properties of the omega-3 fatty acid, DHA (docosahexaenoic acid, 22:6), place it atop a hierarchy of lipids in its capacity to facilitate biochemical processes in the membrane. This DHA effect enables growth of bacteria at low temperatures, rapid energy generation to feed the wing beats of hummingbirds, photosynthesis in cold ocean water, vision in the human eye, the whiplike motion of swimming sperm, and the rapid impulses of the human brain. But DHA carries risks, too. Incorporation of DHA in membranes reduces their proton barrier properties and increases the rates of lipid oxidation, both of which can result in cellular death or disease.

The clear balancing of benefits and risks for DHA in membranes brings us to propose a general principle for all cellular membranes and lipid types. This *DHA principle* states that the blending of lipids to form cellular membranes is evolutionarily honed to maximize benefit while minimizing risk, and that a complex blending code involving conformational dynamics, energy stress, energy yield, and chemical stability underlies all cellular membranes. It is our challenge to understand this code.

# Acknowledgments

This book was approximately 16 years in the making, and would not have been possible without the encouragement, advice, facilities, equipment, supplies, and research of numerous individuals and institutions. RCV is grateful to Calgene (later Monsanto, Inc.) for sponsoring the early research on DHA and EPA in bacteria, and wishes to acknowledge Norman Goldfarb, Roger Salquist, Vic Knauf, Daniel Facciotti, Jim Metz, Bill Hiatt, Marilyn York, Tim Hickman, and Tom Hayes for their important contributions. The Sagami Chemical Research Center provided EPA plasmids and their deletions, and Dr. K. Yazawa coordinated research efforts. Facilities, equipment, and supplies at the University of California Davis were courtesy of Bruce German, Gary Smith, Glenn and Briana Young, and John Ramsey. Luke Hillyard and Laura Gillies, graduate students with Bruce German, carried out numerous experiments and helped keep this project alive. DLV is grateful to colleagues and students who have provided stimulating discussions and collaborated on this topic, notably Doug Bartlett, Eric Allen, and Alex Sessions. This research has been enabled in DLV's laboratory in the Department of Earth Science and Marine Science Institute at the University of California Santa Barbara through a series of awards from the National Science Foundation.

We are indebted to Scott Feller at Wabash College and Toby Allen at UC Davis for introducing us to the wonders of DHA conformations and generously providing the illustrations that appear throughout this book. Cindy Anders, Ellen McBride, and Evelyn Newey spent untold hours on preparation of the manuscript, and Stan Noteboom contributed original artwork. Bill Crandall encouraged us to include a chapter on neurodegenerative diseases and provided numerous research articles. We are greatly indebted to Richard Billings and Suzanne Arrington for voyages of exploration of the ecosystems of southeast Alaska on the *Creole* and the *Surveyor*. Note that the large salmon shown in Figure 10.6A is a result of this scientific research. Hilary Rowe encouraged, sponsored, and guided the book through the editing process at Taylor & Francis.

Thanks to Elliot Anders for encouraging RCV to purchase his first computer and for teaching him to "Google." Finally, special thanks to George Fareed for his faith in taking the senior author's scientific case, through proper channels of course, to the attention of the man himself (RMN).

# The Authors

**Raymond C. Valentine** is currently professor emeritus at the University of California Davis (Davis, California) and visiting scholar in the Marine Science Institute at the University of California, Santa Barbara (Santa Barbara, California). He was also the scientific founder of Calgene, Inc. (Davis, California), now a campus of Monsanto, Inc. The author's scientific interests involve the use of reductionism to address problems of fundamental scientific and societal importance, such as agricultural productivity and aging. Some of his scientific accomplishments include the discovery of ferredoxin, the identification and naming of the nitrogen fixation (*nif*) genes, and the development of Roundup®-resistance in crops. Dr. Valentine holds B.S. and Ph.D. degrees from the University of Illinois at Urbana-Champaign.

**David L. Valentine** is currently an associate professor of earth science with affiliations in ecology, evolution, and marine biology, as well as the Marine Science Institute, at the University of California, Santa Barbara. The author's scientific interests involve the use of a systems-based approach to investigate the interaction between microbes and the earth, particularly in the subsurface and oceanic realms. Dr. Valentine is best known for his research on the biogeochemical cycling of methane and other hydrocarbons, and for his works on archaeal metabolism and ecology. In contrast with the lead author, he asserts that reductionism is the tunnel at the end of the light. Dr. Valentine holds B.S. and M.S. degrees from the University of California at San Diego and M.S. and Ph.D. degrees from the University of California at Irvine.

# Section 1

## Introduction

The quandary that attracted us in 1993 when we started research on omega-3 fatty acids centered on why DHA is needed in neurosensory cells such as neurons. The natural counter to this question is why DHA is missing as a major membrane structural lipid in most of our cells. As we pondered this quandary we seized on several direct and indirect opportunities to investigate DHA in reference to other membrane lipids, such as archaeal isoprenoids. One particular system attracted special attention—DHA/EPA-producing bacteria. Thus, we restated our quandary in terms of specific questions: "What are the benefits and risks of DHA in deep-sea bacteria?" and "Do these benefits and risks translate to other organisms, including humans?" A brief history of this field is as follows:

- 1986—Discovery of DHA in deep-sea bacteria
- Late 1990s—DHA genes sequenced
- Late 1990s—EPA recombinant available
- Late 1990s—First bizarre growth properties of EPA recombinant recorded
- 2001—Polyketide pathway for DHA/EPA biosynthesis published in *Science*
- 1990s to early 2000s—Advances in marine molecular microbiology and ecology
- Early 2000s—Dynamic conformational model of DHA developed
- 2004—Unified concept of DHA published
- 2008—Genome sequence of *Moritella* completed
- 2008—Scientific marriage between conformational model and ecological perspective of DHA structure-function gains momentum

# 1 Molecular Biology of Omega-3 Chains as Structural Lipids: *Many Central Questions Remain Unanswered*

Hypothesis: Relationships between DHA structure and function are emerging from a range of disciplinary studies in physical chemistry, ecology, deep-sea microbiology, and molecular genetics, and these relationships translate to core biochemical and biophysical principles that dictate the occurrence and distribution of DHA in nature.

All known living organisms segregate their vital functions of life from the outside world with a lipid membrane. While life as we know it is not possible without lipid-based membranes, we know relatively little about how this essential protective and biochemically active layer operates at the molecular level. Deciphering relationships between membrane function and fatty acid structure has proved to be a difficult task because the activity of lipid chains is tied to their dynamic conformations. This explains why no generally accepted, unified concept of membrane lipid structure-function has yet been developed. Now, interdisciplinary teams of scientists are reporting on the extraordinary conformational dynamics of DHA chains, findings that illuminate what may be equally extraordinary functions. In this volume we integrate data across many fields of scientific inquiry in order to better understand the roles of the omega-3 fatty acids, docosahexaenoic acid (DHA) and eicosapentaenoic acid (EPA), in bacteria, plants and animals. Lessons from the investigation of DHA/EPA are also applied more generally to understand the essential roles played by other fatty acids.

The essential need for polyunsaturated fatty acids such as DHA in the animal diet was discovered almost 80 years ago during nutritional studies, and mysteries surrounding the molecular biology of "essential oils" continue today. Indeed, a structure-function relationship for membrane lipid chains is one of the last "codes" involving a major class of cellular biopolymers yet to be cracked. Mitchell's discovery of the essential role of proton ($H^+$) electrochemical gradients (i.e., low inside, high outside) in energizing cellular reactions (Figure 1.1) opened up a new era in understanding the membranes' contributions toward production and conservation of cellular energy. His Nobel Prize–winning research focused the attention of membrane biologists on how lipid chains must be melded together to form a tight barrier or molecular architecture to prevent uncoupling of critical electrochemical gradients

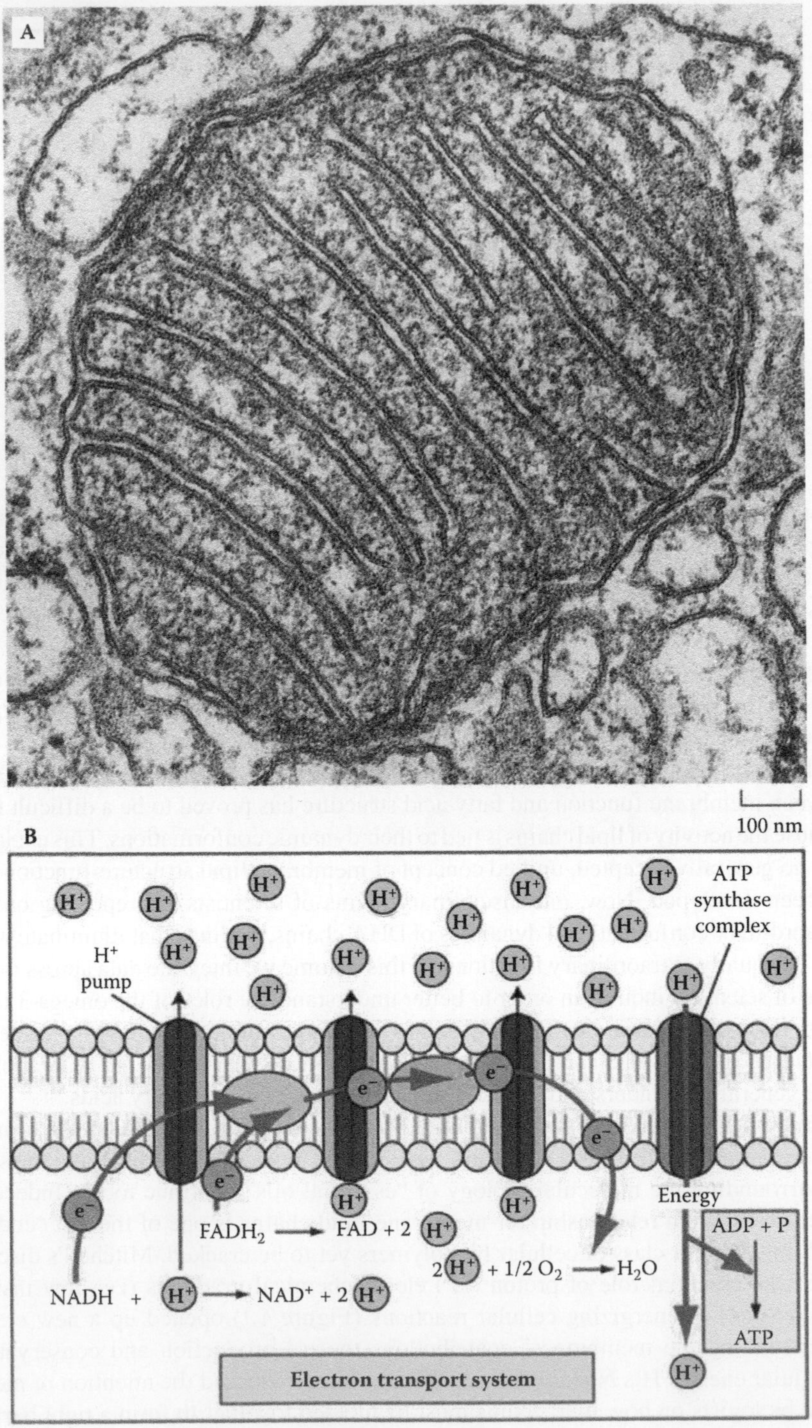
A
100 nm
B
ATP
synthase
complex
H+
pump
e−
Energy
ADP + P
ATP
FADH2 → FAD + 2 H+
2 H+ + 1/2 O2 → H2O
NADH + H+ → NAD+ + 2 H+
Electron transport system

of protons and other bioenergetically important cations. These chemiosmotic gradients, also known as the proton motive force, represent the primary energy form for powering the cell and require that cells evolve membranes that serve as a permeability barrier or defense against proton leakage.

In the early 1980s, a controversy arose when it was reported that membrane vesicles composed only of chemically defined lipids and devoid of membrane proteins are leaky for protons. This implicated the lipid portion of the bilayer as a dynamic factor in bioenergetics at the level of cellular energy conservation in that tightening membrane architecture could act as a permeability barrier against uncoupling of proton gradients—or vice versa. Proton leakiness is explained by a mechanism involving rapid tunneling of protons along chains of water forming in the membrane. As discussed in more detail in Chapter 8, molecular threads of water forming spontaneously, but infrequently, across the lipid portion of the membrane are now proposed to conduct protons at amazing speed, thus breaching the critical proton permeability barrier of the cell. Evidence is accumulating that lipid chains themselves can create or defend against proton leakage and thus contribute in a fundamental way to cellular bioenergetics.

Historically, the first successful attempts to decipher membrane fatty acid structure-function relationships at the physiological level used bacteria, including *Escherichia coli*, as research tools and involved another fundamental property of membranes called fluidity. The classic experiment involves explaining the need for "fluidizing fatty acid chains" (i.e., defined as the inverse of viscosity) in membranes. These studies almost 50 years ago established that bent or kinked fatty acid chains are required for growth of single-celled organisms such as *E. coli*. Thus, insertion of one double bond near the center of a long saturated fatty acid chain generates a kinked structure required for growth. Adding extra bulk, such as methyl branching or increasing the numbers of double bonds per chain beyond one, also satisfied this growth requirement for what initially was defined as membrane fluidity. Chain shortening also increases fluidity, but comes with serious drawbacks concerning permeability, as discussed in Chapter 16. Fluidizing chains as initially defined on the basis of growth assays are believed to work at the molecular level by preventing close packing among fatty acid chains, resulting in a liquid state of the membrane. Thus, fatty acids as phospholipids can be classified on the basis of their antibonding properties, which reduce weak chemical interactions that hold long lipid chains together. It was established that this property

**FIGURE 1.1 (opposite)** Membrane lipids play a central role in cellular energy production and conservation. (A) An electron micrograph of a mitochondrion, highlighting an extensive membrane structure. These complex membranes are needed to house components of the respiratory chain, which generate energy from oxidation of foodstuffs. Similar membranes are essential for photosynthesis. (Courtesy of Daniel S. Friend.) (B) Peter Mitchell won the Nobel Prize for his discovery that proton electrochemical gradients, high on one side of the membraine and low on the other, are critical for cellular energy generation and conservation. This mechanism has since proven to be the primary mode of energy conservation in the biosphere. In this diagram a proton energy gradient generated by the electron transport chain of a mitochondrion is used to synthesize ATP. Membrane lipid architecture houses the components of the electron transport chain and also forms a tight molecular permeability barrier to block spontaneous leakage of protons. This property of lipid chains is now recognized as essential for energy conservation. (Courtesy of Gary Kaiser.)

resides in the structures of fatty acid chains and is seen as essential for maintaining the bilayer in a physiological functional state. In the decades since the original work was published, this generalized concept of fluidity as originally defined using bacteria has faced criticism, not because of any doubts about the essential roles of kinked chains for growth of cells such as *E. coli* and yeast, but rather because of changes in the interpretation of the biochemical meaning of "physiologically functional state." The source of disagreement has now been at least partially explained.

About 10 years ago the concept of fluidity (i.e., the inverse of viscosity) in bacteria was essentially redefined, splitting out two terms, proton permeability and a new fluidity function. In our laboratory we use the term motion to replace fluidity, as discussed next. The studies linking proton permeability and lipid structure are convincing, but it is difficult to sort out any satisfactory biochemical mechanism to explain fluidity. To provide a way out of this dilemma we turn to a well-studied essential membrane whose function apparently requires rapid motion, but not permeability, rhodopsin disk membranes of rod cells. These membranes are disconnected from the plasma membrane and are believed to play solely a structural role. This specialized membrane, discussed in detail later, provides a tool for studying motion in its purest form, separated from permeability considerations. Studies of these membranes help define a second important biochemical function, lateral motion in the membrane, or more specifically in the case of rhodopsin disks, collisions between visual rhodopsin and its G protein, transducin. As described in Chapter 2, these protein-protein collisions occur in the membrane and time or trigger the visual cascade. The picture outlined above is about as far as the literature takes us toward a comprehensive biochemical definition of motion and does not go far enough to explain the molecular biology of the most complex fatty acid structures such as DHA chains in membranes. A new way of thinking about DHA structure-function comes from an unexpected source, ecology.

## 1.1 MEMBRANE LIPIDS: CONTRIBUTION TO ECOLOGY

Membrane lipids surrounding all cells and organelles play cardinal roles in allowing organisms to colonize an amazing diversity of environmental niches throughout the biosphere. Thus, there is an intimate linkage between membrane lipid structures and ecology (Figure 1.2). Molecular ecologists are interested in understanding how functions contributed by specific membrane lipids are harnessed for the benefit of one organism over another, whereas thermodynamic ecologists consider the membrane as a crucial component allowing cells to compete in acquiring and conserving energy from the environment. The concept that membrane lipids are in fact pivotal or decisive players in explaining ecological balance among life forms has gained momentum over the last decade or so. Perhaps the clearest example involves competition between the two prokaryotic life forms, Archaea versus Bacteria. There is convincing evidence that archaea have evolved robust, specialized membrane lipids that allow these cells to dominate many niches. For example, it is clear that at extremely high temperatures—for example, in the range of 100°C to 121°C—bacteria simply cannot compete with archaea. This is attributed to the robust, isoprenoid-based membrane architecture of archaea, which has evolved to maintain a tight permeability barrier against uncoupling of proton/sodium electrochemical gradients under

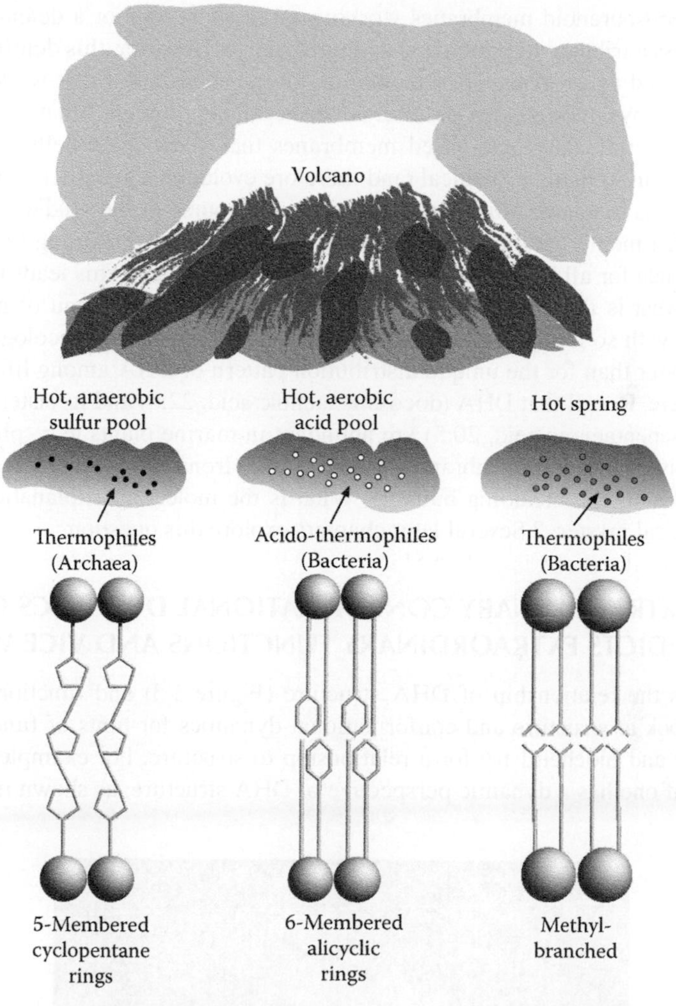


**FIGURE 1.2** Bacteria and Archaea adapt their membrane lipid structures to accommodate their environment. Membrane lipids adapted for life in three extreme environments are depicted in a artists' sketch highlighting their variable structures. Note that lipids of this kind are not necessary or found in human or most other cells, which have evolved membranes specific for their own needs.

such extreme conditions. It is a safe bet that the proton permeability barrier in bacteria essential for energizing cellular functions would be breached long before temperatures reach 121°C (the current record held by archaea!).

Similar membrane lipid-based scenarios have been developed to explain the apparent dominance of archaea over bacteria in other niches, such as extreme salinity, acidity, or more generally in "energy stressed" environments such as the deep subsurface. Indeed, archaea as a separate life form can be defined on the basis of their unique isoprenoid membranes working as an adaptation or a defense against energy stress attributed to proton and sodium leakage. However, this dominance can also be viewed as an example of membrane overspecialization that works against archaea in many environments, resulting in bacterial domination. What are the properties of bacterial, fatty acid-based membranes that allow these cells to outcompete archaea in so many ecological (and therefore evolutionary) battles? One goal of our research is to answer this question. During the course of our studies, it became apparent that membrane lipids are far more important in determining the ultimate ecological fate for all life forms than is currently appreciated. This leads to another question: what is the ecological significance of having a vast menu of membrane fatty acids with so many different structures? Seldom can a macroecological question be clearer than for the unique distribution pattern of DHA among life forms in the biosphere. Recall that DHA (docosahexaenoic acid, 22:6) and its sister molecule EPA (eicosapentaenoic acid, 20:5) are abundant in marine plants (e.g., phytoplankton) but missing among membranes of land plants. Ironically, they are required by many land animals, including humans. What is the molecular explanation behind this ecological mystery? Several later chapters explore this question.

## 1.2 EXTRAORDINARY CONFORMATIONAL DYNAMICS OF DHA PREDICTS EXTRAORDINARY FUNCTIONS AND VICE VERSA

To address the relationship of DHA structure (Figure 1.3) and function we have opted to look at structure and conformational dynamics for hints of function, and to ecology and biochemistry for a relationship to structure. For example, it seems clear that if one has a dynamic perspective of DHA structure, as shown in the lipid

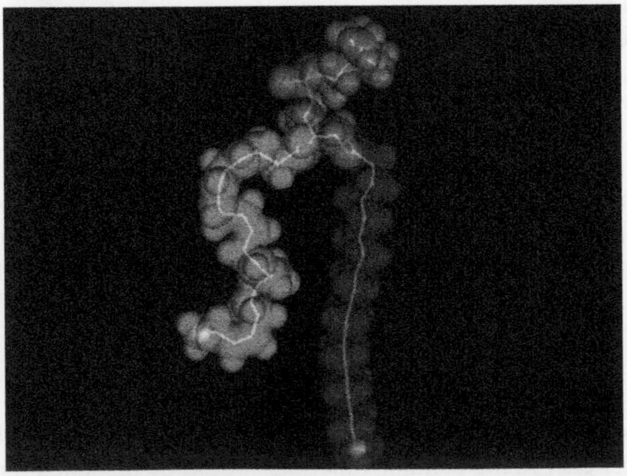

**FIGURE 1.3** Snapshot of a membrane phospholipid containing DHA and stearic acid (18:0), from a chemical simulation model. Despite having more carbons in its chain, DHA (shown at left) tends toward shorter and bulkier conformations compared to saturated chains such as 18:0 (shown at right). DHA also tends to display rapid changes between numerous conformations, presumably defining many properties of DHA membranes. (Courtesy of Scott Feller.)

animation of DHA by Dr. Scott Feller of the Wabash College Chemistry Department, specifically 500 ps dynamics of a stearoyl docosohexaenoyl PC lipid molecule (14 MB), then biological functions will become more clear. Conversely, as bioactivity of DHA-rich membranes becomes clear, the underlying conformational dynamics can be inferred. A structure-function relationship for DHA is being approached here from both perspectives, and a linkup now seems possible. It is our experience that solutions to tough questions like this are often closer at hand than originally thought. For example, the bridge across scientific divides can often be built with existing information held by different scientific disciplines that simply need to communicate with one another. This might be true of DHA. A case in point is that chemists can now visualize DHA chains as having extraordinary conformational dynamics when compared to more common fatty acid chains found in membranes (compare DHA and stearic acid). Thus, as a result of chemical analysis involving nuclear magnetic resonance (NMR) imaging, X-ray crystallography, computer simulation, and other methods, the following properties can be used to describe the extraordinary conformational dynamics of DHA (comments in parentheses relate to possible bioactivity discussed in later chapters):

- Great flexibility (antibonding? increases motion?)
- Diverse set of conformations (space-filling? adaptable?)
- Rapid transition times between states, that is, nanosecond to picosecond time scales (easily perturbed by environmental signals?)
- Elasticity (applied later to sperm tails, which may be stretched by mechanical stress?)
- $Na^+$ modulation of surface area (might be important in marine bacteria?)
- Sharp bends at the methyl end [note that this feature separates omega-3 chains like DHA from omega-6 chains such as arachidonic acid (20:4n6); bulking mechanism against cations?]
- Elbow conformation forming near the middle carbons of the chain (i.e., seems to punch its 18:0 partner converting this straight chain into a transiently bent conformation; antibonding?)
- Compact/compressed conformation (high packing density of carbon atoms; might plug holes against leakage of protons or sodium?)
- Space-filling or self-sealing properties (i.e., may allow chains to quickly fill or seal microspaces or momentary membrane defects?)
- Relatively hydrophilic (more water wires might form, weakening proton permeability barrier?)
- Extremely low phase transition temperatures (membrane antifreeze? disrupter of lipid rafts?)
- Rapid flip-flop (might uncouple proton gradients preferentially over sodium gradients?)

The conformational dynamics of DHA chains described above contrast with many previous perceptions about bioactivity of fatty acids in natural membranes. However, these conformational dynamics contrast less with the ecological perception of DHA because DHA and EPA are linked to equally bizarre extremophilic lifestyles, such as that of deep-sea bacteria. Such an extreme lifestyle seems to beg for membrane structural components such as DHA that are capable of delivering function under extraordinary conditions. Obviously, the diversity of functions of DHA such as seen in the microbial world is not readily apparent when considering DHA-enriched membranes of humans. The point is that the extraordinary conformational dynamics of DHA chains envisioned by chemists seem to solve the structure-bioactivity relationships of bacteria. Admittedly it requires a measure of faith to accept that this membrane code will apply to humans. However, that faith is tempered by a powerful rule of biochemistry as described next.

## 1.3 REDUCTIONIST STRATEGY FOR DHA RESEARCH

The major question that originally motivated our research involves a quandary that has puzzled membrane biologists for years. Why do certain cells in the human body, for example, neurons, sperm, and rod cells, enrich their bilayers with DHA? This becomes even more puzzling in view of the fact that most human cells incorporate only traces of DHA into their bilayers as though these cells seek to avoid DHA chains altogether. In ruminating about these questions, it soon became clear that here was an opportunity for "thinking out of the box," especially regarding the type of research tools employed. We chose a multifaceted approach, including a reductionist strategy as discussed next.

Reductionism in science involves the use of simple systems or models to answer complex questions, such as "what is the structure of the simplest atom?" or "what is heredity or the simplest form of heredity?" In answering the latter question, scientists such as Max Delbruck, Salvador Luria, and James Watson focused on using bacterial viruses and the bacterium *E. coli* as representing the simplest research tools for studying genetics. Reductionism concerning heredity in bacteria and their viruses had its "golden era" in the 1960s and 1970s, but the subsequent rapid advances in deciphering of plant, animal, and human genomes has diminished the need for such simple model systems as bacteria for the studies of animal and plant genes. However, there are still a few areas of core biology where reductionism can be applied fruitfully. Defining how DHA chains work in membranes of life forms ranging from deep-sea bacteria to human neurons is one of the remaining areas with implications for understanding a more general membrane code.

Omega-3 fatty acids—sometimes called fish oils—as a class have been popularized because of their importance in human health and nutrition. It is safe to say that this linkage to human health is the major driver for research in this area, but there is also increasing interest in the "molecular or chemical ecology" of these molecules, as discussed in Chapter 17. In addition to their structural roles in membranes as emphasized here, the study of eicosanoid hormones derived from omega-3 fatty acids is an important, well-established, and ongoing area of research. Omega-3 fatty acids are also widely used as energy storage molecules in many organisms, such as marine fishes that fuel their long upstream migrations with storage forms of these fatty acids. Understanding of the structural roles of omega-3 chains in the bilayer is less advanced and emerges as a potential target for a reductionist strategy. The senior author's journey down this path began about 20 years ago during strategy sessions at Calgene, Inc., on how to marry molecular agriculture with public health (e.g., preventive medicine) and nutrition with an eye toward the twenty-first century. Being avid Alaskan fishermen, we were well aware of the fact that most of our dietary supply of omega-3 fatty acids comes from marine fish, not land plants. The practical objective at the time involved transfer of DHA genes from bacteria and algae into oil crops, such as corn, canola, or soybean (see Chapter 18), with the ultimate goal of a new land-based system for production of marine oils whose supplies are not keeping up with demand. In the early 1990s, Vic Knauf, director of research, organized two teams to pursue this project in the industrial setting of Calgene, Inc., a small plant biotechnology company at Davis, California, of which the senior author is the scientific founder. Calgene is now owned by Monsanto, Inc., a well-known global plant biotechnology company with continued interest in oil seeds. The two teams were divided based on gene sources, one devoted to DHA/EPA genes from marine bacteria and the other focusing on genes from eukaryotic sources, such as algae, fungi, and plants.

The bacterial project captured our imagination and turned out to be more attractive from a basic perspective. For example, membrane fatty acid compositional data published for deep-sea bacteria showed a similarity in DHA content compared to neurons of animals. Exposure to data obtained by research on an EPA recombinant of *E. coli* (Figure 1.4) triggered a wave of reductionist thinking involving the similarities between membranes in deep-sea bacteria versus neurons. Collection of experimental data forming the basis of this book began in 1993 at Calgene and continues to the present time in laboratories at the University of California Davis and UC Santa Barbara. The idea began to take hold that DHA/EPA-producing bacteria and their

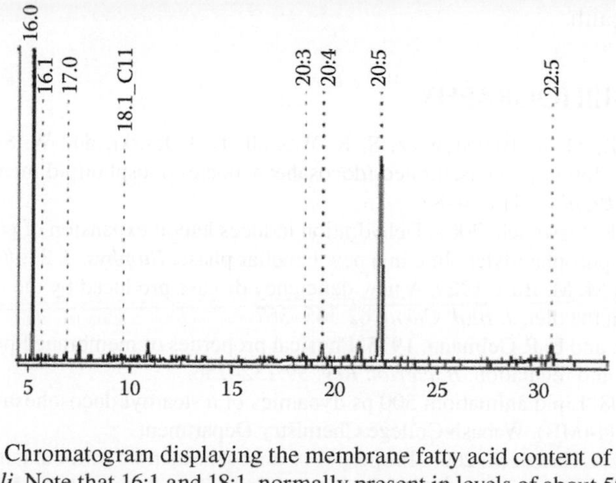


**FIGURE 1.4** Chromatogram displaying the membrane fatty acid content of an EPA recombinant of *E. coli*. Note that 16:1 and 18:1, normally present in levels of about 50% total chains in membranes of *E. coli*, are virtually absent, being replaced by EPA, which often pairs in phospholipids with 16:0. These cells display unusual growth properties more characteristic of marine bacteria than the *E. coli* recipient used in this cross. We suggest that these unexpected growth properties can be explained by the presence and functions of EPA in their membranes. Confronted with explaining the novel growth properties of EPA recombinant cells, we began exploring a new unified concept of DHA/EPA in membranes across life forms.

recombinants offered advantages as research tools for understanding the importance of omega-3 oils not only in ecology, but in human health as well. In undertaking initial bacterial studies of DHA with our partners at the Sagami Chemical Research Institute in Japan, it was assumed that a basic biochemical unity likely exists regarding DHA structure-function across life forms. Having been trained around reductionists, this concept did not seem to be an unusual leap of faith. It is important to point out that reductionism is nothing new in the membrane lipid field, where the following hierarchy of research tools has been established through the years:

Mycoplasmas → *E. coli* (Bacillus, Archaea) → Yeast and Algae → Plants → *C. elegans* → Mice and Rats → Humans

As already mentioned, a long-range goal of our research is to decipher what might be called the membrane lipid "code" for explaining why cells incorporate different lipid chains into their bilayers. We chose omega-3 fatty acids such as DHA/EPA as a focus because these chains stand out in terms of their extraordinary conformational dynamics, low melting points, and rapid rates of chemical oxidation. For example, DHA is the most highly unsaturated and oxidation-prone chain found in natural membranes. The reasoning was that if reductionism is not a fruitful approach toward understanding these chemically unique chains, then it is difficult to imagine much success in deciphering the roles of other common fatty acids with less distinct structures. The idea of a continuum or hierarchy of functions across all unsaturated fatty acid structures emerged early in our thinking, raising the possibility that solving structure-function relationships of omega-3 fatty acids might have applications toward understanding how membrane lipids in general contribute to ecology, agriculture, and health.

## SELECTED BIBLIOGRAPHY

Armstrong, V. T., M. R. Brzustowicz, S. R. Wassall, L. J. Jenski, and W. Stillwell. 2003. Rapid flip-flop in polyunsaturated (docosahexaenoate) phospholipid membranes. *Arch. Biochem. Biophys.* 414:74–82.

Binder, H., and K. Gawrisch. 2001. Dehydration induces lateral expansion of polyunsaturated 18:0-22:6 phosphatidylcholine in a new lamellar phase. *Biophys. J.* 81:969–982.

Burr, G. O., and M. M. Burr. 1929. A new deficiency disease produced by the rigid exclusion of fat from the diet. *J. Biol. Chem.* 82:345–367.

Cronan, J. E., Jr. and E. P. Gelmann. 1975. Physical properties of membrane lipids: Biological relevance and regulation. *Bacteriol. Rev.* 39:232–256.

Feller, S. E. 2008. Lipid animation: 500 ps dynamics of a stearoyl docosohexaenoyl PC lipid molecule (14MB). Wabash College Chemistry Department.

Feller, S. E. and K. Gawrisch. 2005. Properties of docosahexaenoic acid-containing lipids and their influence on the function of rhodopsin. *Curr. Opin. Struct. Biol.* 15:416–422.

Feller, S. E., K. Gawrisch, and A.D. MacKerell, Jr. 2002. Polyunsaturated fatty acids in lipid bilayers: Intrinsic and environmental contributions to their unique physical properties. *J. Am. Chem. Soc.* 124:318–326.

Holte, L. L., F. Separovic, and K. Gawrisch. 1996. Nuclear magnetic resonance investigation of hydrocarbon chain packing in bilayers of polyunsaturated phospholipids. *Lipids* 31:S199–S203.

Huber, T., K. Rajamoorthi, V. F. Kurze, K. Beyer, and M. F. Brown. 2002. Structure of docosahexaenoic acid-containing phospholipid bilayers as studied by $^{2}$H NMR and molecular dynamics simulations. *J. Am. Chem. Soc.* 124:298–309.

Huster, D., J. J. Albert, K. Arnold, and K. Gawrisch. 1997. Water permeability of polyunsaturated lipid membranes measured by $^{17}$O NMR. *Biophys. J.* 73:856–864.

Metz, J. G., P. Roessler, D. Facciotti et al. 2001. Production of polyunsaturated fatty acids by polyketide synthases in both prokaryotes and eukaryotes. *Science* 293:290–293.

Mitchell, P. 1961. Coupling of phosphorylation to electron and hydrogen transfer by a chemiosmotic type of mechanism. *Nature* 191:144–148.

Nichols, J., and D. Deamer. 1980. Net proton-hydroxyl permeability of large unilamellar liposomes measured by an acid-base titration technique. *Proc. Natl. Acad. Sci. U.S.A.* 70:2038–2042.

Rabinovich, A. L., and P. O. Ripatti. 1991. On the conformational, physical properties and functions of polyunsaturated acyl chains. *Biochim. Biophys. Acta* 1085:53–62.

Rajamoorthi, K., H. I. Petrache, T. J. McIntosh, and M. F. Brown. 2005. Packing and viscoelasticity of polyunsaturated omega-3 and omega-6 lipid bilayers as seen by $^{2}$H NMR and X-ray diffraction. *J. Am. Chem. Soc.* 127:1576–1588.

Rawicz, W., K. C. Olbrich, T. McIntosh, D. Needham, and E. Evans. 2000. Effect of chain length and unsaturation on elasticity of lipid bilayers. *Biophys. J.* 79:328–339.

Saiz, L., and M. L. Klein. 2001. Structural properties of a highly polyunsaturated lipid bilayer from molecular dynamics simulations. *Biophys. J.* 81:204–216.

van de Vossenberg, J. L. C. M., A. J. M. Driessen, M. S. da Costa, and W. N. Konings. 1999. Homeostasis of the membrane proton permeability in *Bacillus subtilis* grown at different temperatures. *Biochem. Biophys. Acta* 1419:97–104.

# Section 2

## Evolution of DHA and the Membrane

For the first approximately two billion years of evolution on earth, biospheric energy flux, or power, was likely dramatically reduced relative to the modern era. This era of energy restriction seemingly had low levels of primary production and low catabolic energy yields associated with the prevalence of anaerobic metabolism. We suggest that the primary mode of membrane lipid evolution during this period reflected the overwhelming needs of cells to prevent energy loss. Membrane lipid structures suggested as representative of each period are as follows.

*The Early Anaerobic Period:* Robust, isoprenoid monolayers selected during the anaerobic period based on their impermeability to protons and other cations. These lipids permit cells such as archaea to thrive in a chronically energy-starved world.

*The Anaerobic-Aerobic Transitional Period:* Dimethyl branched phospholipids and anaerobically derived monounsaturated fatty acids increasingly populate bilayer structures. These lipids provide somewhat diminished energy conservation properties but allow for enhanced membrane lateral motion needed to facilitate more rapid respiration and increase the rate of energy generation. Such lipids provided a competitive advantage for organisms seeking large and dynamic sources of energy, such as from oxygenic photosynthesis.

We suggest that an increase in biospheric energy flux and catabolic yields following the evolution of oxygenic photosynthesis and oxygenation of the atmosphere sparked the evolution of membranes capable of further boosting rates of energy generation. This occurred at the expense of stringent energy conservation, which loses importance in an energy-rich environment. During the age of animals DHA/EPA pathways—first important for photosynthesis in plants—became an essential building block for membranes of neurosensory cells.

*The Oxygenation Period:* Oxygenation of the atmosphere led to an expansion in structures of polyunsaturated species of fatty acids allowing cells to capture more energy, a benefit counterbalanced by a less robust permeability barrier (e.g., proton leaky). DHA/EPA might have been first selected to improve light harvesting during marine photosynthesis. The age of animals led to building efficient neurosensory cells depends on specialized membranes enriched with DHA/EPA.

In this section we first consider the evolution of the membrane in the context of energy availability throughout Earth's history, followed by a consideration of membrane and lipid evolution at the biochemical and genetic levels, and lastly we consider the evolution of omega-3 fatty acids for specific cellular functions.

# 2 Darwinian Selection of the Fittest Membrane Lipids: *From Archaeal Isoprenoids to DHA-Enriched Rhodopsin Disks*

Hypothesis: Production and conservation of energy are primary driving forces behind evolution of fatty acid structure with archaeal isoprenoids and omega-3s representing extreme forms.

Membrane evolution presumably begins with the first living cells containing lipid-based membranes and continues to the present. This evolution covers over four billion years including the largely anaerobic period, itself encompassing more than half this time. As the title of this chapter suggests, we assert that changes in the biosphere throughout evolution have acted to select lipid chains providing the most beneficial membrane architecture to the cell or organism as it adapts to an ever-changing environment. Foremost among these changes was the "oxygen revolution." Molecular oxygen, $O_2$, has been called the molecule that made the world, and its impact on membrane lipid structures is no exception. As shown in Figure 2.1, the evolution of $O_2$-dependent pathways of membrane lipid biosynthesis, creating an explosion of new lipid conformations, is intimately linked to oxygenation of the biosphere. Clearly, this great atmospheric shift is a critical driver of membrane lipid evolution, but not the only one. We suggest that most anaerobic pathways of membrane lipid synthesis, including saturated, monounsaturated, methyl-branched, cyclopropane fatty acids, as well as isoprenoid chains of archaeal membranes, and cholesterol-like molecules such as hopanoids and cholesterol precursors up to squalene, all likely evolved during the anaerobic period. For simplicity, only the increase in complexity of fatty acid structures is shown in the illustration. Oxygenation of the atmosphere created conditions for evolution of fatty acid desaturases, enzymes that use molecular $O_2$ for insertion of double bonds creating many new conformations. The further processing of squalene to cholesterol essential for growth of all animal cells is also dependent on $O_2$. While the balance of total fluxes for membrane fatty acids in the extant biosphere occurring via anaerobic versus aerobic pathways has yet to be established, the main point is that the strictly anaerobic period followed by transitional

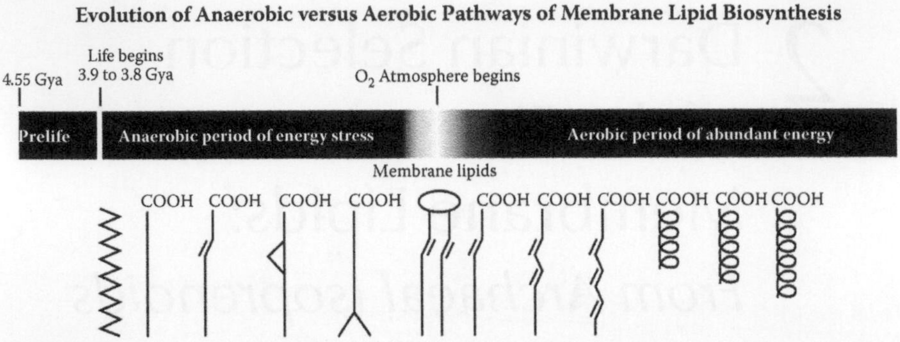


**FIGURE 2.1** Schematic geologic timeline for the development of fatty acid structures comprising membranes. The rise of oxygen ($O_2$) in the biosphere is suggested to have led to an explosion of new polyunsaturated fatty acid conformations such as DHA and EPA for building increasingly sophisticated membranes. The time scale of life on Earth is divided into three major parts: the anaerobic period beginning ~3.8 gigayears (Gya), believed to have been characterized by low energy flux and availability to the biosphere; followed by a semianaerobic, transitional period (starting around 2.2 Gya) of undetermined length in which more energy was becoming available; and finally full oxygenation when energy flux and availability increased dramatically in the biosphere. Membrane lipid structures suggested as characteristic of each period are displayed and can be interpreted in terms of the bioenergetic and specialized biochemical needs of cells/organisms living during these periods. Note that membrane lipids synthesized by anaerobic pathways tend to contribute energy conservation properties to the membrane, forming a robust permeability barrier that conserves energy. Oxygenation led to evolution of polyunsaturated chains such as DHA/EPA, which favor rapid membrane motion and were likely harnessed initially to generate more energy. Thus, the history of Earth and the evolution of membrane lipid structures are viewed as being closely linked.

semioxygenated periods likely served as an evolutionary proving ground for lipid shapes still prevalent in many cells today.

From an ecological/bioenergetic perspective, a major leap in cellular energy production might have occurred in the transitory period between anaerobic versus aerobic periods when pools of electron acceptors other than $O_2$ are believed to have become more plentiful. These include inorganic ions, such as sulfate, nitrate, and ferric iron. It has been hypothesized that evolution of electron transport chains similar to those operating in bacteria, chloroplasts, and mitochondria today might have predated full oxygenation of the biosphere. Once again, the driving force for evolution of these so-called anaerobic electron transport chains is believed to be based on the relatively large increase in energy yield delivered by these alternative electron chains. It is also clear that electron transport chains and the membranes that house these components have coevolved over a very long time frame—ample time for Darwinian selection of the bioenergetically "fittest" lipid structures needed to maximize cellular energy production. For example, we propose that a simple structural change involving the attachment of two kinked-shaped/bulky-shaped chains instead of the normative one to the glycerol head group of phospholipids might be an evolutionary milestone in boosting energy production. This change does not require $O_2$ and perhaps occurred during a transitionary period. Thus, studies on the role of lipid structures in bioenergetics seem a good starting point for understanding why cells/organisms living in different environments choose specific lipid chains such as DHA/EPA for their membranes.

Only in relatively recent times has the full extent of membrane lipid contribution to ecology come to light, and novel membrane structures primarily associated with the uncultivated majority of microbes are reported regularly. Perhaps the biggest shock came in the mid-1970s when the central dogma that all cellular membranes are based on fatty acids as seen in bacteria, plants, and animals was disproved, a key discovery in defining archaea as a distinct life form. For membrane biologists these results caused a rethinking regarding fundamental contributions of membrane lipids to bioenergetics (Figure 2.2). Indeed, isoprenoid-based membranes of archaea were discovered to lack fatty acid chains altogether. For many years, fundamental research on the nature of isoprenoid versus fatty acid–based membranes seemed to run on parallel tracks, but in the 1990s this separatism began to give way to a more unified approach. This transition in thinking is perhaps best seen in the papers of W. N. Konings and colleagues who developed a unified concept of the membranes' contributions to prokaryotic ecology based on essential cation (i.e., protons and sodium) permeability properties. These researchers proposed that the isoprenoid-based membranes of archaea are far more robust as a permeability barrier against uncoupling of electrochemical gradients than fatty acid-based bacterial membranes. This idea has been extended to help define the importance of the membrane in the ecology of archaeal cells, which are now recognized as being ubiquitous in the biosphere and capable of outcompeting bacteria in many environments. In extending this concept, we have proposed recently that the evolutionary success of archaea as a group in colonizing most niches on Earth is linked not only to their unique biochemistry, but also to their robust membranes. For example, archaea grow over a temperature range approaching –4°C to 121°C, the latter end being well beyond the range suitable for bacteria with fatty acid–based bilayers. We propose a term, *proton/sodium fidelity*, to project the positive role that membrane lipid architecture plays in cation permeability with high proton fidelity defined in terms of a membrane architecture that creates a strong permeability barrier against proton leakiness and futile cycling of cations. The general absence of ether-linked isoprenoid membranes in bacteria, plants, and animals raises the possibility that this class of membrane, while suitable for the lifestyles of Archaea, might be somehow biochemically unsuitable for other life forms. Indeed, according to our recent unified concept dealing with omega-3 enriched bilayers, the conformations and properties of these highly unsaturated fatty acid chains seem to be at opposite poles compared to archaeal isoprenoid-based membranes, thus defining extremes in membrane evolution from isoprenoids to omega-3 fatty acids.

## 2.1 BIOENERGETICS AS THE DRIVER OF EVOLUTION OF LIPID STRUCTURES

All prokaryotes face a crucial energy dilemma at the cytoplasmic membrane. Each cell must expend energy to maintain a chemiosmotic potential, which is used to drive basic cellular processes. The membrane functions as the barrier for this potential,

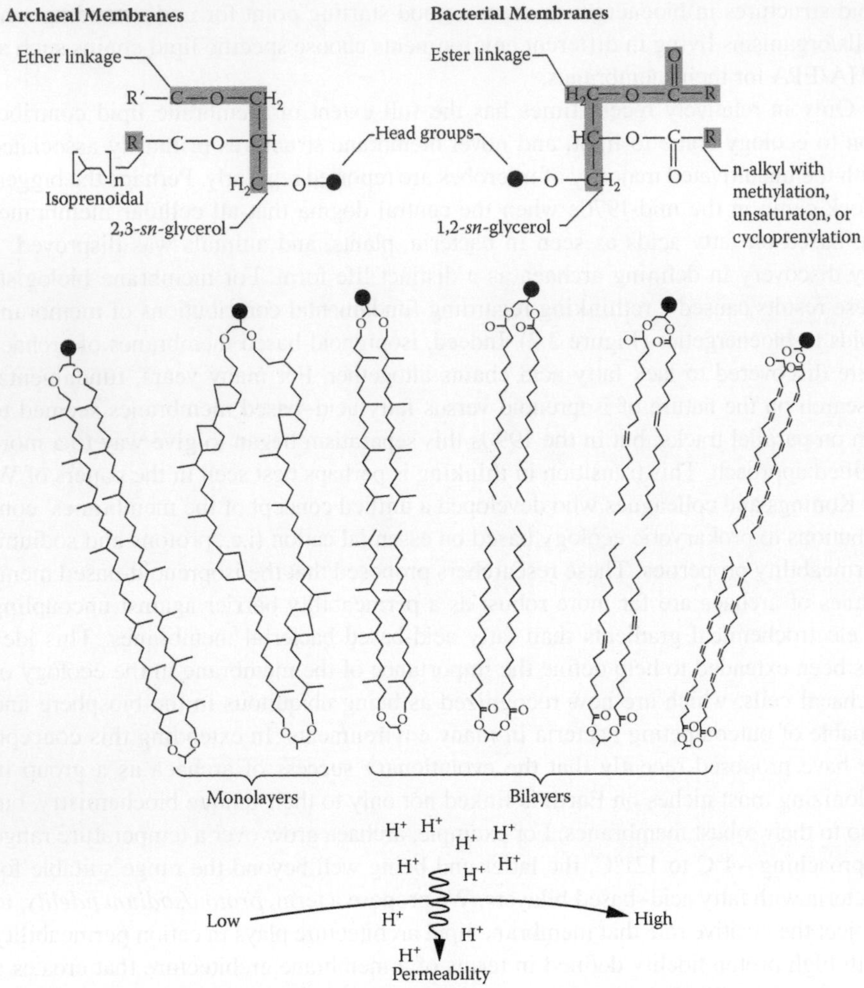


**FIGURE 2.2** Membranes of the Archaea and Bacteria. The basic chemical structures of archaeal and bacterial membrane lipids are shown to illustrate the typical chemical differences. Examples of intact membrane structures used by Archaea (left) and Bacteria (right) are shown below, including the monolayers that are produced by some archaea and the highly unsaturated membranes produced by some bacteria. The arrow (bottom) indicates a general trend of increasing permeability to ions, such as protons and sodium. Eukaryotic membranes such as those of plants and animals are similar to those of bacteria and are formed of fatty acids, including omega-3 chains. (From Valentine, D. L., *Nature Rev. Microbiol.*, 5:316–323, 2007. With permission.).

and the inadvertent passage of ions across the membrane—futile ion cycling—is a direct energy loss for the cell. Organisms require membrane fidelity to avoid futile ion cycling, and the term high proton fidelity introduced above refers to a robust membrane permeability barrier against this critical cation. Efflux of ions from the cell has been shown to account for approximately half the energy demand of resting mammalian cells and is probably a main component of energy loss for active cells in nature. This point is covered in more detail later.

One important distinction between archaea and bacteria is the chemical structure of lipids composing the cytoplasmic membrane (Figure 2.2). Bacterial lipids typically consist of fatty acids esterified to a glycerol moiety, whereas archaeal lipids typically consist of isoprenoidal alcohols that are ether-linked to glycerol. Stereochemical configurations about the glycerol moiety are also different. Archaeal membranes are less permeable to ions than bacterial membranes, and tetraether-based membranes are less permeable than diether-based membranes. These membranes reduce the amount of futile ion cycling *in vivo* and provide an energetic advantage to archaea—less energy is lost during the maintenance of chemiosmotic potential. The advantages of the archaeal membrane have been clearly shown in hyperthermophiles, halophiles, acidophiles, and lipid vesicles derived from these cells, and it seems reasonable to simply extend this principle to include other archaea.

## 2.2 DO ARCHAEA HAVE AN ACHILLES' HEEL?

Archaea lack sufficient lateral membrane motion to drive electron transport at the same rates as bacteria. The physiological or ecological significance of motion in archaeal lipids has been researched in moderate detail, with the conclusion that constraints on lateral motion exhibited by these lipids are likely beneficial to the archaeal cell in terms of proton fidelity of their membranes. For example, the membranes of thermophilic archaea are clearly capable of maintaining fidelity even at temperatures as high as 121°C. Several structural features of archaeal lipids have been proposed to account for the high levels of proton energy conservation displayed by these membranes. For example, liposomes prepared from lipids characteristic of thermophiles were found to be 6- to 120-fold less permeable to protons and other solutes than bacteriallike liposomes. It was shown that the crucial factor ensuring low permeability is cyclopentane rings in phytanyl side chains, which limit the mobility in the midplane hydrocarbon region. The substitution of ether for ester bonds appears to provide an additional barrier that specifically impedes the flux of protons. Two additional mechanisms that may involve motion include: (1) glycosylation of the glycerol and nanitol backbones; it has been proposed that hydrogen bonds between the glycosyl head groups stabilize membrane structure by reducing lateral lipid mobility; and (2) Thermophilic, acidiphilic, and mesophilic marine and soil archaea, which are known to synthesize monolayer membranes generated by covalently linking a single phytanyl chain to glycerol moieties located on opposite sides of the membrane. These tetraether-based monolayers set the standard for proton fidelity, allowing these cells to live in such extreme conditions. Studies using synthetic phytanyl vesicles suggest that the restricted motion of chain segments accounts for the lower proton permeability of branched-chained lipid membranes. Studies of liposomes from a thermoacidophilic archaeon indicate that the lateral diffusion rates of tetraether liposomes have a membrane viscosity at 30°C, which is considerably higher than that of diester phospholipids in the liquid crystalline phase. These results suggest that the membranes of archaeal liposomes are laterally immobile compared to fatty acid-based membranes. Therefore, it is clear that the molecular dynamics of archaeal lipids are radically different from those of straight-chain lipids found in other cells. The main difference is that the lateral motion at the membrane surface seems to be stalled, whereas individual chains are likely to exhibit some mobility within the membrane while being tethered to their relatively immobile head groups. These data support the concept that the restricted lateral motion of archaeal lipids is advantageous with respect to both proton and sodium fidelity and is thus critical in the ecology of these cells (Figure 2.2).

Halophilic archaea respire organic substrates, such as amino acids, as energy sources, and also employ a bilayer membrane structure, properties atypical of many archaea. These organisms have evolved electron transport chains that are remarkably similar to respiratory chains of bacteria such as *E. coli*. Horizontal transfer from bacteria to archaea is likely responsible and includes genes governing electron transport components, including lipophilic electron carriers (i.e., quinones) and electron transport enzyme complexes, such as cytochromes and terminal oxidases. Even plastocyaninlike carriers characteristic of chloroplasts that allow rapid shuttling of electrons via the aqueous phase might have been acquired by horizontal gene transfer. It can be argued that the presence of plastocyaninlike carriers, which move in the aqueous phase at rates much greater than lipophilic carriers, such as quinones, is an indication that electron shuttling continues to be a rate-limiting factor for energy production even after evolution of the bilayer structure seen in halophilic archaea. We suggest that acquisition of electron transport chains from bacteria is only part of the solution for maximizing energy production by this important group of archaea. That is, housing bacterialike electron transport chains in classic archaeal membranes lacking in lateral motion is predicted to accomplish little in terms of improved energy output required to compete against halophilic bacteria. However, the coevolution of bilayer structures (Figure 2.3) that presumably enhance lateral motion in halophilic archaea appears to enable these archaea to outcompete bacteria and thrive in hypersaline environments. Thus, the need to optimize energy conservation and generation in the presence of extreme salinity might help explain the bacteriallike membrane adaptations characteristic of halophilic archaea.

Additional similarities link the bioenergetic strategies of halophilic archaea and respiratory bacteria. Like bacteria, halophilic archaea have evolved a large battery of catabolic enzymes for uptake, degradation, and energy production coupled with oxidation of a variety of organic molecules as sources of energy, carbon, and nitrogen. Rhodopsin, which traps sunlight for energy, is characteristic of halophilic archaea, serving as an auxiliary, but not primary, source of adenosine triphosphate (ATP). That is, it is likely that halophilic archaea in the natural environment depend most heavily on organic compounds as electron donors driving their bacteriallike electron transport chains. Electrons are transferred to $O_2$ or various alternative electron acceptors, such as trimethylamine oxide. Again, the ability to use a variety of alternative, anaerobic electron acceptors in place of $O_2$ is similar to bacteria.

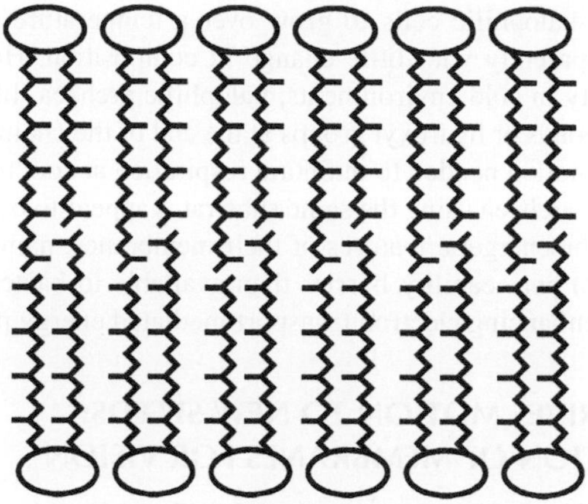


**FIGURE 2.3** Halophilic archaea produce zipper-shaped membrane bilayers needed to thrive in the saltiest environments on Earth. A zipper-shaped bilayer structure is believed to provide a robust permeability barrier against extremely high differential concentrations of $Na^+/K^+/H^+$, essentially saving energy while simultaneously providing sufficient motion for operation of a respiratory chain that resembles those from bacteria. Genomic analysis shows that many archaea have pirated electron transport genes from bacteria to populate their membranes. We suggest that convergent evolution resulted in halophilic and other Euryarchaeota replacing their monolayer structures in lockstep with their radiation to lower temperatures and alternate forms of metabolism, which set the stage for acquisition by horizontal transfer of electron transport genes from bacteria. The net effect is evolution of an intermediate class of membrane between archaeal monolayers and bacterial bilayers supporting a respiratory lifestyle in a highly saline environment.

Since halophilic archaea live in extremely saline environments where they almost always outcompete bacteria, it can be argued that their bilayer architecture is inherently more robust against futile cycling of $Na^+/H^+$. However, we believe that the bilayer structure of halophilic archaea is a double compromise both in terms of cation fidelity (i.e., lying between isoprenoid monolayers and bacterial bilayers) and efficiency of electron transport. This "intermediate" membrane structure is proposed to have selective value in terms of energy production compared to tetraether-based designs while exhibiting some weakness in terms of energy conservation compared to tetraether membranes. It is interesting to note that many halophilic archaea have evolved a "zipper" or "tongue and groove" bilayer structure (Figure 2.3) that likely increases cation fidelity important in such highly saline environments. The most common zipper structure is built using a 20-carbon isoprenoid as the short chain versus a 25-carbon longer chain. These asymmetrical chains present in opposite leaflets can be used to create the zipper structure. This membrane architecture might serve as a permeability shield against essential cations (i.e., $Na^+/H^+$) while being sufficiently antibonding because of protruding methyl groups to permit sufficient membrane lateral motion needed for enhancing electron transport. One of the most striking properties of this class of archaeal bilayers is their relatively broad thermal profile that allows halophilic cells to grow over a temperature range from about 4–5°C to 51°C, apparently with little change in composition. However, for maintaining functionality in cold environments, halophilic archaea, like bacteria, insert additional double bonds or hydroxyl groups at the end of the chain. This is proposed to increase lateral motion needed to maintain respiration at cool temperatures.

Thus, halophilic archaea using the same substrates appear to outcompete bacteria by optimizing the bioenergetic features of their membranes, namely, a more robust cation (i.e., $H^+/Na^+$) permeability barrier than available to bacteria, and sufficient lateral motion for enhancing electron transport-mediated energy production.

## 2.3 DHA DRIVES MOTION TO NEW SPEEDS: EVOLUTION OF MEMBRANES FOR VISION

The evolution of DHA-enriched membranes capable of detecting a single quantum or photon of light in the human eye occurred relatively late during the age of animals and is a striking example of Darwinian selection of the fittest membrane lipids. In this section we jump far ahead from the world of bacteria to evolution of perhaps the most vital sensory system for our survival—vision—and, more specifically, membranes designed to detect dim light. We focus on the conformations of membrane lipids that lie at the heart of the trigger mechanism for the visual cascade (Figure 2.4). Molecular evolutionists have traced the history of critical visual proteins working as receptors for detecting light signals. These workers have also laid out in broad strokes the necessity of a unique kind of stacked or complex membrane structure needed to house sufficient amounts of the light-sensing protein rhodopsin. Thus, eyes, ranging from the simplest to the most complex, detect light via a membrane antenna structure featuring bilayer after bilayer stacked one on top of the other. One view is that these multilayered membranes work simply as a physical scaffolding for inserting greater numbers of light-sensing proteins. This interpretation is likely correct but may not be the complete story. For example, this model does not explain why rod cells of the eye, whose outer segment is filled with a stack of about 1000 rhodopsin disk membranes (Figure 2.4A), are composed of a significant amount of the most highly unsaturated lipid building blocks found in nature. These phospholipids contain di-DHA, which means that two DHA chains are present per structural unit with a total of 12 double bonds. What is the biochemical explanation of the need for this highly unsaturated structure in vision?

**FIGURE 2.4 (Opposite)** (A color version of this figure follows p. 140.) Vision depends on DHA-enriched membranes that house the rapidly moving, light-sensing protein rhodopsin that triggers the visual cascade. (A) Perception of dim light occurs using strategically placed membrane disks acting as microantennae in the outer segment of rod cells in the eye. Approximately 1000 DHA-enriched membrane disks act as the light-sensing apparatus. (B) DHA-enriched membranes enable rhodopsin to rotate and move laterally at rapid speed. This facilitates collisions between rhodopsin and its G protein transducin, firing the visual cascade. (C) Rhodopsin is shown to be surrounded by DHA phospholipids in an image generated by chemical simulation. In addition to driving motion of rhodopsin, DHA phospholipids might also be involved in the biochemistry of rhodopsin. [(A) Courtesy of Frank Mueller and (C) courtesy of Scott Feller.]

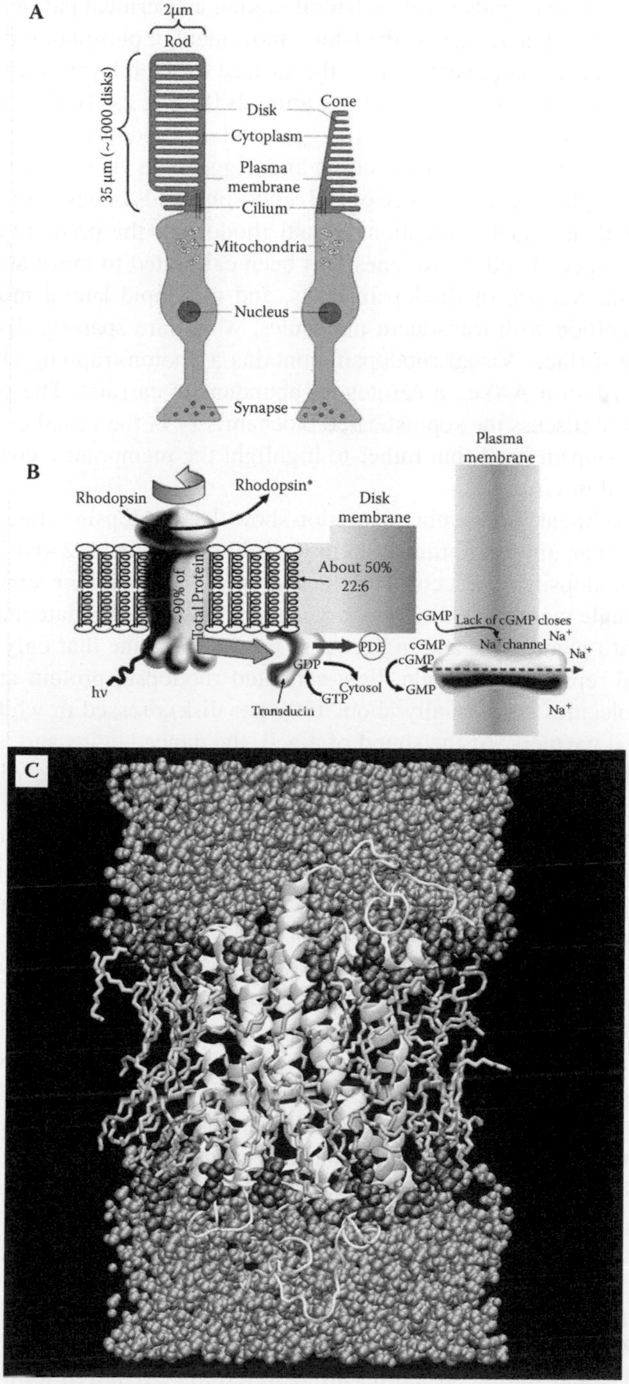
A
2μm
Rod
Cone
35 μm (~1000 disks)
Disk
Cytoplasm
Plasma
membrane
Cilium
Mitochondria
Nucleus
Synapse
B
Rhodopsin
Rhodopsin*
Disk
membrane
Plasma
membrane
~90% of
Total Protein
About 50%
22:6
cGMP
Lack of cGMP closes
Na+ channel
Na+
cGMP
cGMP
PDE
GDP
GTP
Cytosol
GMP
Transducin
hv
C

As discussed in the context of archaea above, we argue that Darwinian selection of the fittest membrane lipids involves lateral motion as a critical parameter. Lateral motion in this context is defined as the lateral movement of phospholipids, enzymes, and other membrane components across the surface of the bilayer. Importantly, lateral motion reaches its zenith in rod cells of animals (Figure 2.4B). The biochemistry of these disks, which house the visual trigger protein rhodopsin, is understood in great detail. Visual rhodopsin absorbs light and triggers the visual cascade. The artist's sketch highlights the importance of DHA phospholipids believed to drive lateral motion essential in visual perception. Visual rhodopsin, the predominant protein in these highly specialized membranes, has been calibrated to move at exceptional speed across the surface of rhodopsin disks, and this rapid lateral motion allows rhodopsin to collide with transducin molecules, which are sparsely distributed on the membrane surface. Visual rhodopsin contains a photon-trapping chromophore derived from vitamin A (i.e., a carotenoid abundant in carrots). The point of this drawing is not to discuss the sophisticated biochemistry of the visual cascade or the function of rhodopsin itself, but rather to highlight the membrane's contribution to motion essential in vision.

Classic experiments on membrane motion show that rhodopsin rotates and moves laterally faster than any membrane protein yet calibrated (Figure 2.4B). To illustrate just how fast rhodopsin moves, consider the analogy of a dance floor representing the surface of a single membrane disk and large enough to accommodate many molecular dancers with plenty of space in between. Further imagine that only one dancer dressed in red represents a single, light-activated rhodopsin protein among many transducin molecules (i.e., usually about 1000 per disk) dressed in white and waiting as potential partners. At the sound of a bell, the dance begins and is completed only when the red dancer has made contact with at least 500 of the white dancers randomly scattered over the disk surface. Compressing this entire sequence into a time frame of 1 second illustrates rhodopsin's lateral motion across the surface of the bilayer. What is more amazing is that the journey of rhodopsin requires lateral motion through the oily membrane itself. We suggest that this dance would not proceed at this rate if the DHA tails of phospholipids of these disks were replaced by more saturated tails. Obviously, these membranes have evolved the viscosity characteristics allowing rapid motion that triggers vision to occur. In the detailed biochemical picture, DHA chains are shown as molecular springs paired with a stick-shaped or linear chain (e.g., stearic acid or 18:0). In reality, a significant number of lipids in rhodopsin membranes contain the di-DHA structure. As discussed in Chapter 7, an oil composed of di-DHA is unique and retains its liquid properties down to temperatures similar to the coldest found on the surface of the Earth (i.e., –50°C). For many years, it was believed that such low viscosity oil is found only in the eye where it is needed to trigger the visual cascade, but now it is known that an analogous phospholipid, di-EPA, with a total of 10 double bonds and a freezing point perhaps lower than di-DHA is required for efficiency in neuromuscular function of the well-studied nematode *Caenorhabditis elegans*. Obviously, highly unsaturated phospholipids such as di-EPA are not confined to rod cells, since *C. elegans* has not evolved eyes. Instead, di-EPA and similar structures are believed to be required for building neurons that work efficiently at cool temperatures (e.g., 15°C). It is now recognized that phospholipid structures with properties similar to di-DHA are abundant in nature and play important roles such as harvesting light in plants (see Chapter 11).

## 2.4 SUMMARY

Darwinian selection of the fittest membrane seems to distinguish archaea from bacteria, mitochondria, and chloroplasts, with archaea specializing in immobile but highly energy-conserving lipid chains, in contrast to bacteria and eukaryotes whose bilayers are laterally mobile but inherently more proton leaky. This view is consistent with the recent proposal that archaea, in general, can be classified as adapted to a lifestyle involving energy stress. Obviously, bacteria also live in energy-limited environments but have the alternative of building membranes with greater lateral motion, which permits faster rates of energy production. This strategy, which can be considered to be a mechanism against energy stress, may allow bacteria to outcompete archaea in environments where energy substrates are transient or simply more abundant. This simple distinction is becoming blurred due to horizontal transfer of genes from bacteria to archaea and vice versa. Thus, the membrane itself seems to be a cardinal structure helping to define the thermodynamic ecology of archaea versus bacteria. In essence, the most energy-conserving membrane structures characteristic of several groups of archaea lack lateral motion, which limits the value of bacteriallike electron transport–mediated energy production in these cells. In one sense, archaeal membranes without lateral motion can be visualized as being overly specialized in terms of energy conservation. The opposite appears to be true of bacteria, mitochondria, and chloroplasts whose membranes seem best suited to drive rapid rates of energy production often at the expense of energy conservation. DHA/EPA bilayers can also be viewed as being overly specialized for energy production at the expense of energy conservation.

Halophilic archaea, which often live in environments where energy substrates commonly used by bacteria are abundant, seem to be striving to acquire lateral motion characteristics of bacterial membranes. The lines separating archaeal versus bacterial membranes are further blurred by the finding that many bacteria have evolved structural lipids, such as cholesterol-like hopanoids. These molecules are thought to stabilize membrane structure and might be considered as remnants of the isoprenoid pathway used by archaea. Energy-transducing membranes of chloroplasts and mitochondria often are enriched with polyunsaturated chains that we believe help maximize energy output and are bacterialike with respect to bioenergetics, consistent with the bacterial origins of these organelles. An increasing supply of available energy driven by oxygenation of the biosphere can account for selecting cells that grow fast but tend to waste more energy, in contrast to archaea, which grow slowly but are highly energy conserving. Between these two prokaryotic life forms, it is difficult to pick a winner because both types of cells appear to be imminently successful in the biosphere today.

Thus, Darwinian selection of the fittest membrane lipid chains seems intimately tied to the broader ecological history of cellular life in the biosphere, where the supply of available energy was dramatically increased as the result of oxygenation of the atmosphere. This major event permitted cells with leaky membranes to dominate energy-rich environments in contrast to many archaea, which still are dominant under energy-starved conditions. In short, oxygenation of the biosphere is proposed to have led to a new form of bioenergetics in which a certain amount of waste is tolerated or compensated for by more robust electron transport chains. We contend that membrane fatty acid conformations that enhance rates of collisions of electron transport components played a key role in driving the evolution of bacteria, plants, and animals. Indeed, it has been proposed that the rise of humans is intimately linked to energy stress forged by dramatic fluctuations in $O_2$ levels in the atmosphere that have occurred periodically during the age of animals. Finally, the story of evolution of membrane lipid shapes does not end with bioenergetics and has led to specialized bilayers, for example, those enriched in DHA that are essential for sensory systems such as vision and development of the brain and nervous system. However, at the biochemical level, we assert that the molecular roles of DHA/EPA are universal among living cells ranging from deep-sea bacteria to neurons.

## SELECTED BIBLIOGRAPHY


Anderson, R. E. 1970. Lipids of ocular tissues. IV. A comparison of the phospholipids from the retina of six mammalian species. *Exp. Eye Res.* 10:339–344.

Chabre, M. 1985. Trigger and amplification mechanisms in visual phototransduction. *Annu. Rev. Biophys. Biophys. Chem.* 14:331–360.

Damsté, J. S. S., M. Strous, W. I. C. Rijpstra, et al. 2002. Linearly concatenated cyclobutane lipids form a dense bacterial membrane. *Nature* 419:708–712.

Daniel, R. M., and D. A. Cowan. 2000. Biomolecular stability and life at high temperatures. *Cell. Mol. Life Sci.* 57:250–264.

Deamer, D., J. P. Dworkin, S. A. Sandford, M. P. Bernstein, and L. J. Allamandola. 2002. The first cell membranes. *Astrobiology* 2:371–381.

Deppenmeier, U., A. Johann, T. Hartsch, et al. 2002. The genome of *Methanosarcina mazei*: Evidence for lateral gene transfer between Bacteria and Archaea. *J. Mol. Microbiol. Biotechnol.* 4:453–461.

de Rosa, M., and A. Gambacorta. 1988. The lipids of Archaebacteria. *Prog. Lipid Res.* 27:153–175.

Gliozzi, A., A. Relini, and P. L.-G. Chong. 2002. Structures and permeability properties of biomimetic membranes of bolaform archaeal tetraether lipids. *J. Memb. Sci.* 206:131–147.

Kashefi, K., and D. Lovely. 2003. Extending the upper temperature limit of life. *Science* 301:934.

Miljanich, G. P., L. A. Sklar, D. L. White, and E. A. Dratz. 1979. Disaturated and dipolyunsaturated phospholipids in the bovine retinal rod outer segment disk membrane. *Biochim. Biophys. Acta* 552:294–306.

Poo, M., and R. A. Cone. 1974. Lateral diffusion of rhodopsin in the photoreceptor membrane. *Nature* 247:438–441.

Valentine, D. L. 2007. Adaptations to energy stress dictate the ecology and evolution of the Archaea. *Nature Rev. Microbiol.* 5:316–323.

Valentine, R. C., and D. L. Valentine. 2004. Omega-3 fatty acids in cellular membranes: A unified concept. *Prog. Lipid Res.* 43:383–402.

van de Vossenberg, J. L. C. M., A. J. M. Driessen, W. D. Grant, and W. N. Konings. 1999. Lipid membranes from halophilic and alkali-halophilic Archaea have a low $H^+$ and $Na^+$ permeability at high salt concentration. *Extremophiles* 3:253–257.

van de Vossenberg, J. L. C. M., A. J. M Driessen, and W. N. Konings. 1998. The essence of being extremophilic: The role of the unique archaeal membrane lipids. *Extremophiles* 2:163–170.

van de Vossenberg, J. L. C. M., T. Ubbink-Kok, M. G. L. Elferink, A. J. M. Driessen, and W. N. Konings. 1995. Ion permeability of the cytoplasmic membrane limits the maximum growth temperature of bacteria and archaea. *Mol. Microbiol.* 18:925–932.

# 3 Coevolution of DHA Membranes and Their Proteins

Hypothesis: Sensory perception in humans would not be possible without DHA and other polyunsaturated chains that drive rapid collisions among or between membrane proteins and enzymes.

Membranes and their proteins have coexisted for billions of years, ample time for evolution of sophisticated partnerships. These include the evolution of membrane lateral motion introduced in Chapter 2. For humans, the partnership between DHA-enriched membranes and their receptors is at the heart of sensory perception. Indeed, recent studies show that the largest superfamily of membrane proteins in humans depends on a collision-type mechanism driven by lateral membrane motion. An important member of this superfamily, rhodopsin, has already been discussed and is known to move at exceptional speed across the membrane surface. Whereas modulation of membrane lateral motion at the level of fatty acid structure represents an important mechanism for regulating activity of some essential membrane proteins and processes, many membrane-bound enzymes work independently of lipid motion.

## 3.1 DID MOTION OR LACK OF MOTION PREVAIL IN MEMBRANES OF PROTOCELLS?

The ubiquity of lipoidal membranes among life forms points to an early evolutionary emergence, a view supported by the geochemical record of organic molecules. While it is not known when or how membranes evolved, one interpretation of the earliest stages of cellular life posits separate or independent evolution of spherical membrane vesicles or ghosts forming spontaneously from primitive lipoidal chemicals, such as fatty acids present in the "primordial soup." The building blocks needed to create these vesicles may have formed on the icy coating of comets travelling through space with energy for lipid synthesis being powered by cosmic radiation, or perhaps by abiotic polymerization of methane in the Earth's atmosphere or at the planet's surface, or from carbon dioxide in hydrothermal systems. Whether from these or other sources, the preformed building blocks for assembly of primitive lipoidal vesicles resembling cellular ghosts likely appeared early in the Earth's history. Thus, the distinction between laterally mobile and immobile membranes may have emerged during the prelife or early life period, perhaps even setting the stage for an early divergence of Archaea and Bacteria, though little evidence exists to support or refute this assertion.

Membrane vesicles form spontaneously and like cellular membranes chemically distinguish the inside milieu from the external environment. In essence, the membrane affords protection for physical and chemical events occurring in the interior of the lipid vesicle, which may be less likely or unsuitable to the outside environment. Some membrane evolutionists also suggest that as primitive forms of chemical catalysts evolved those with lipophilic character might have been entrapped or fastened themselves to the inside of the membrane not only for protection, but also as a matrix for an early form of organized chemical catalysis. It also seems reasonable that membrane vesicles having evolved some internal specialization were subject to environmental stresses, such as temperature, which had the power to destroy the integrity of the vesicle, essentially defeating the "inside" advantage. Thus, the proto-cellular era of early life may have already been an important proving ground for selection of the fittest membrane lipids. However, the presumed lack of selectivity in lipid acquisition, the lack of complex lipid biosynthetic pathways, and the requirement for compatible lipids in proto-membranes likely limited the complexity and specialization of protomembranes.

## 3.2 WHICH CAME FIRST—PROTEINS OR MEMBRANES?

Among prokaryotic evolutionists, the current thinking is that Archaea and Bacteria shared a common ancestor, the "last universal common ancestor (LUCA)." A variety of provocative proposals have been published regarding the nature of LUCA, including the idea that this was not an ordinary cell since it lacked lipid membranes. These studies raise interesting questions concerning the evolution of the cellular membrane and its proteins. The basis of this quandary is that for a membrane to function in a cell, it must be endowed with at least a minimal battery of transport systems, but it is unclear how such systems could evolve in the absence of a membrane. Previous researchers have proposed the concept of the minimal cell as a reductionist strategy for charting the pathways of cellular evolution and gauging the necessity of cellular machinery. This seems to be a fruitful approach and has led to a number of interesting concepts on the evolution of membranes. For example, the minimal cell alive today has surprisingly few transporters for its essential amino acids needed as building blocks for protein synthesis. These cells lack all of the genes for amino acid synthesis and must import them through the cell membrane from the medium. Indeed, the minimal membrane seems to be only sparsely populated with transporters, suggesting that if necessary there are primitive or nonselective mechanisms to transport important chemicals across the membrane.

There are other data that support the view of a "minimal membrane" in early cells. One example involves a cell formerly called *Streptococcus faecalis* whose membrane is simplified because it lacks a functional respiratory chain and depends on fermentation for ATP production. Numerous transporters are present and are active, some energized by proton electrochemical gradients and others by ATP. However, the battery of proton-fueled metabolite pumps can be eliminated by destroying or uncoupling the protonic potential of these cells using the powerful ionophore gramicidin, discussed in detail later. This strong energy decoupler short-circuits the proton gradient, thus creating a ripple effect deenergizing many transporters, disrupting pH homeostasis and blocking essential systems of cellular osmoregulation, deactivating the cell's defenses against $Na^+$ toxicity, and rendering flagella useless. Deactivating so many critical membrane proteins would reasonably be expected to kill these cells, but this is not the case. These cells display active growth provided the growth medium is adjusted to meet their new requirements: a high concentration of external $K^+$, the absence of excessive $Na^+$, a pH above 7.5, and a sufficiently high concentration of amino acids and vitamins. These cells are clearly crippled with respect to their ability to maintain the normal cytoplasmic pH and ionic composition, to accumulate metabolites, and to cope with a variety of stresses. However, the fact that these cells grow and divide normally, producing progeny morphologically indistinguishable from controls, shows that a bacterial cell can be constructed without functioning proton energized transporters/systems.

The impact of gramicidin on *Streptococcus faecalis* was reported by Harold in 1986, and more recent work has revealed how gramicidin itself acts as a primitive membrane pore and enables survival of these cells. The current interpretation of these data centers on cation transport via the molecular thread of water now known to be held in the core of gramicidin functioning as a membrane-spanning nanotube. The prediction that protons move rapidly through this nanotube was confirmed by showing that the proton gradients in these cells have collapsed. At the same time, the normal range of acidic pH tolerated by these cells is severely restricted to near neutrality or slightly above. This is explained on the basis of rapid movement of $H^+$ from the medium via gramicidin to excessively acidify the cytoplasm, quickly lowering the pH of the cytoplasm below a critical threshold level required for supporting cellular enzymatic machinery (i.e., pH homeostasis is destroyed). The requirement for low $Na^+$ confirms that gramicidin nanotubes also transport $Na^+$ into the cell, essentially equilibrating inside and outside concentrations. This case provides a clear example showing that $Na^+$ above a certain threshold level in the cytoplasm is toxic to the point of inhibiting growth and likely causing cellular death. The high, external $K^+$ requirement confirms both that $K^+$ is conducted through these membrane nanotubes and also that $K^+$, in contrast to $Na^+$, is a compatible solute/osmoregulant supporting cytoplasmic metabolism, such as protein synthesis, while balancing osmotic pressure.

The growth requirement for high levels of amino acids in the medium might be explained as a mechanism to help drive these substrates into the cell, diminishing the need for transporters and strong electrochemical gradients usually required for energizing uptake of these essential catabolites. However, the uptake of amino acids is complicated because these metabolites are transported by both $H^+$ and ATP-mediated pumps with the ATP battery of pumps remaining active after $H^+$-fueled pumps are deactivated.

## 3.3 IONOPHORES BEHAVE AS PRIMITIVE TRANSPORTERS AND SOME DEPEND ON THE PHYSICAL STATE OF THE MEMBRANE

The concept that ion-conducting antibiotics such as gramicidin might provide the simplest models for primitive biological transport deserves more attention and is expanded on next. Ionophores have been widely used as probes in studies of bioenergetics and models of ion transport, but we're not aware that this database has been used previously as a way around the membrane quandary outlined above. Ionophores have two distinct modes of action with gramicidin, as already introduced, working as a membrane-spanning nanotube in contrast to mobile, lipophilic carriers such as valinomycin (Figure 3.1). Valinomycin is highly selective for $K^+$ ions and is a small cyclic peptide forming a bracelet-like cage around $K^+$. Nonpolar residues form a hydrophobic shell that generates the membrane-soluble nature of this molecule. Valinomycin and its $K^+$ complex diffuse freely across the membrane's lipid region, essentially shuttling $K^+$ in downhill fashion from a region of high $K^+$ to a less concentrated region. Thus valinomycin serves as a $K^+$ carrier with levels of this ion on either side of the bilayer driving the direction of flow. No exogenous energy is required besides a chemiosmotic potassium gradient. However, the $K^+$ uncoupling activity of valinomycin grinds to a halt as the membrane enters the gel phase. This is an important point likely applicable to lipophilic electron carriers such as ubiquinone. We

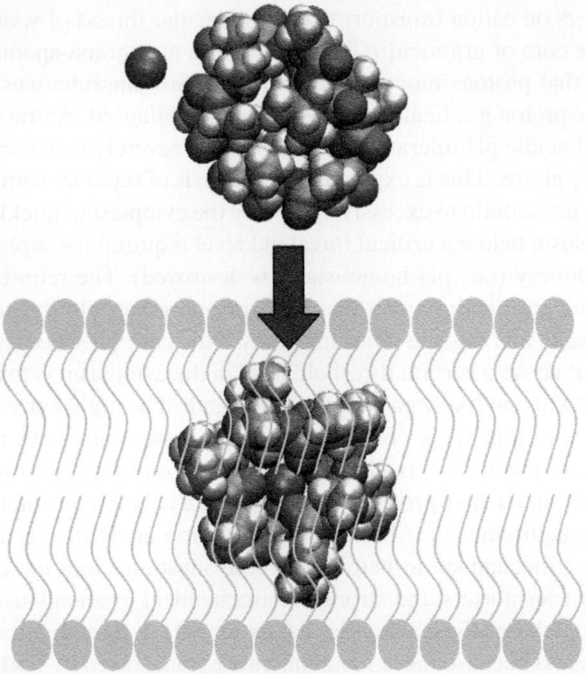

**FIGURE 3.1** (A color version of this figure follows p. 140.) Valinomycin is a membrane-acting antibiotic that kills bacteria by uncoupling essential $K^+$ gradients. This lipophilic, cyclic peptide enters the membrane of target cells, forms a molecular cage around a $K^+$ ion, and shuttles across the bilayer, releasing the $K^+$ ion on the opposite surface. This mode of action transports $K^+$ across the membrane in the direction of the chemiosmotic gradient, effectively deenergizing the $K^+$ gradient and wasting cellular energy. Of particular interest here, valinomycin depends on the fluidized state of the bilayer and loses its uncoupling capacity when membranes enter the gel phase. This is one of the simplest examples of the importance of the fluid state of the membrane. (Courtesy of Borislava Bekker and Toby W. Allen, personal communication.)

thus infer that electron transport chains in bacteria, chloroplasts, and mitochondria are blocked in the gel phase.

Gramicidin forms channels that give passage to all the small univalent cations: $H^+$, $K^+$, $Rb^+$, and $Na^+$, but not to divalent ions such as $Ca^{2+}$. The peptidic monomer chain of gramicidin consists of 15 alternating L and D amino acids all with hydrophobic side chains. Within the lipid bilayer, the chain coils into a helix, 1.2 to 1.5 nm in length with a central pore 0.4 nm in diameter holding a molecular thread of water (discussed in detail in Chapter 8). Two gramicidin molecules joined end to end make a nanotube just long enough to span a phospholipid bilayer. It is interesting to note that gramicidin channels remain open even if the surrounding lipid enters the hardened gel phase, in contrast to molecular carriers such as valinomycin.

When in evolutionary history did ionophores emerge? It is not yet possible to deduce an evolutionary history for peptidic ionophores. However, there are several lines of evidence consistent with an ancient history. Ion channels such as gramicidin are peptides in a chemical sense but are not coded by genes like other proteins and do not require ribosomes for biosynthesis. Instead, the joining of specific amino acids in peptide linkage is carried out by a large enzyme as a template. The type of biochemistry involved in making small peptides on templates is much simpler than ribosome-based systems and seems more primitive in that it is easier to envision how monomers could be chemically activated, held on a surface, and joined together to form a primitive pore/transport structure. Also, the lack of monomer specificity is evident in ionophores that include lactic acid along with D-amino acids in the final structure. There are many similarities to the general biochemistry of DHA synthesis by the polyketide route detailed in the next chapter. The point is that synthesis of ionophores is another example of convergent evolution of peptide synthesis that is far simpler than ribosome-based protein synthesis. It is interesting to speculate in this case that "simpler means earlier" and that the first primitive peptide transporters were made by a catalytic mechanism more chemical than enzymatic. There is a second aspect of ionophore structure consistent with an ancient history. This is seen in the case of gramicidin A, which has six D-amino acid isomers in contrast to L-isomers found exclusively in most proteins. Interestingly, chemical synthesis of amino acids such as occurs in the Miller–Urey experiment simulating the primordial soup results in a D-L-racemic mixture. Early cells might have been able to harness combinations of D or L isomers to form unique, bioactive structures such as lipid-soluble ionophores with novel properties still evident today.

The relative simplicity of synthesis and structure of other peptidic ionophores compared to protein pumps is further suggestive of an ancient origin for this molecular class. For example, valinomycin is actually a trimer, composed of a monomer repeated three times. Gramicidin A is more sophisticated, being 15 residues in length, but requires dimerization (noncovalent) to form the mature nanotube. Gramicidin S is a cyclical decapeptide composed of two identical repeated sequences. It seems reasonable that membranes were first populated by these simple, often-repetitive, structures as primitive transporters prior to evolution of transport proteins.

In essence, we are proposing that the first cells with lipid membranes might have been virtually "nude" of complex metabolite/ion-pumping proteins common today, instead using a more rudimentary form of ion circulation involving peptidic/nonpeptidic ionophores. Complex proteins clearly provide sufficient competitive advantage that they are ubiquitous today, an evolutionary route that may have first seen increasingly sophisticated classes of ionophores, with true proteins eventually replacing all of these structures. Thus, we envision three stages of evolution of transporters: molecular pores created by defects in the membrane lipid architecture itself → ionophore stage → protein stage. Today, the only known roles of ionophores seem to be as chemical warfare agents among microorganisms, although under specific environmental conditions the life of a cell in the laboratory can be forced to become "dependent" on these primitive transporters.

## 3.4 MANY ARCHAEAL MEMBRANE PROTEINS ARE LATERALLY IMMOBILE, BUT SOME CAN SPIN

The concept that Archaea are ancient cells is not a new idea and indeed gave rise to their original name, Archaeabacteria, from "ancient bacteria." However, the current thinking is that both Archaea and Bacteria shared a common ancestor. Regardless of which domain arose first, it is clear that archaeal membranes as a platform for housing critical enzymes are very different compared to bacteria. Genomic analysis shows that archaea depend on membrane enzymes for survival as much as bacteria, but there are dramatic differences in the dynamics of proteins embedded in archaeal membranes. For example, studies of archaeal rhodopsin, a light-trapping protein characteristic of halophilic archaea, show that this molecule forms large, ordered aggregates or patches on the membrane surface. It is difficult to imagine how individual rhodopsin molecules trapped within these large aggregates, called purple membranes, could depend on lateral motion for activity. In other words, this energy-transducing membrane protein is effectively a stationary or immobile enzyme. Interestingly, individual rhodopsin molecules when introduced into more fluid bacterial membranes retain their catalytic function, showing that this membrane-bound protein also works when diluted in membranes displaying greater lateral motion.

Another essential membrane-bound enzyme found in archaea as well as all other life forms is ATP synthase (Figure 3.2). This sophisticated enzyme complex is an example of a molecular machine that captures and transduces the energy held in proton gradients, yielding ATP. This is a reversible reaction in which ATP can also be dephosphorylated to pump excess protons from the cytoplasm for the purpose of maintaining pH homeostasis and so forth. The microrotor or miniaturized turbine in the core of ATP synthase rotates rapidly when fueled by protons as energy source. The protein rotor seems to spin in interface with membrane lipids for only a part of each rotation. During the remainder of each turn the rotor interfaces with a membrane-spanning ATP synthase subunit perhaps behaving as a molecular-bearing created by the enzyme itself. The main point is that ATP synthase does not appear to depend appreciably on the physical state, displaying activity in a variety of membranes encompassing a wide range in levels of lateral motion. This mode of action of one of the most important enzymes in bioenergetics is consistent with the emergence of ATP synthase early in evolutionary history—in the anaerobic period when membranes were likely less mobile, and long before the oxygenation of the biosphere. From an ecological

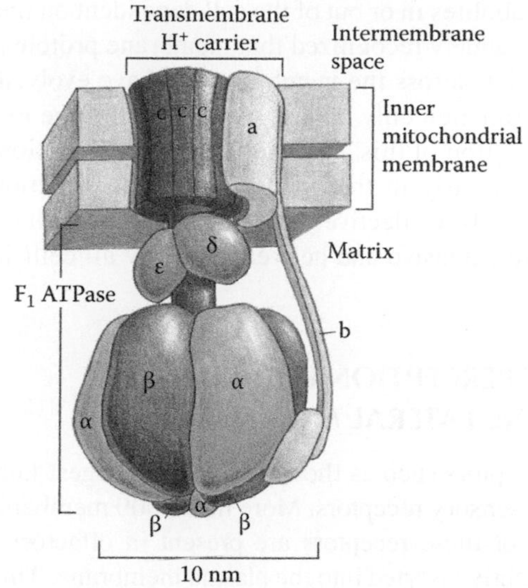


**FIGURE 3.2** ATP synthase is an example of an integral membrane rotary-protein whose mode of action is independent of the physical state of the bilayer. That is, the molecular turbine blades required for energy transduction spin in either relatively laterally immobile membranes (e.g., archaea) or laterally mobile membranes (e.g., DHA-enriched mitochondria or chloroplasts). This mode of action is consistent with early evolution of this universal enzyme during the anaerobic period when membranes are suspected to have been largely laterally immobile. (Copyright Alberts et al., *Molecular Biology of the Cell,* 2008. With permission of Garland Science/Taylor & Francis LLC.)

perspective, the evolution of high fidelity membrane lipids along with proteins that work in these relatively immobile membranes permits cells such as archaea to live in chronically energy-stressed environments, a lifestyle dependent on preventing energy loss at the membrane level. Even though many archaea grow slowly, growth rates alone are not a measure of the ecological significance of this life form. Indeed, organisms that grow slowly but have an active metabolism may over long time periods become the dominant players in driving cycling of matter in the biosphere.

We suggest that coevolution of the matrix function of membrane lipids with proteins is driven by Darwinian selection of the fittest partnership. We further speculate that the earliest forms of membrane proteins arose during the anaerobic period and are survived by the immobile class seen in archaea today—proteins that have evolved to work in a laterally immobile membrane matrix. This membrane state is compatible with the evolution of essential rotary enzymes, such as ATP synthase, whose motion is little affected by the physical state of the membrane. Also, it is important to recall that the stationary matrix function characteristic of many archaea and perhaps considered a more primitive kind of membrane still contributes multiple, fundamental biochemical advantages to the cell. These include the energetic benefits already discussed, physical protection from proteolysis, as well as "sidedness," which allows proteins such as transporters (e.g., ATP synthase in reverse) to move metabolites in or out of the cell dependent on the orientation of the pump itself. It is now widely recognized that membrane proteins orient themselves in a particular direction across the membrane and have evolved specific domains that are located and function only on one specific side of the membrane. Again, it seems likely that evolution of this asymmetry of mode of action of the membrane protein itself occurred early in the history of life, as directionality would have been critical to early cells. Collectively, the benefits to the cell of even the simplest membrane matrix are extensive and help explain why all cells have evolved lipid-based membranes.

## 3.5 SENSORY PERCEPTION REQUIRES MEMBRANE LATERAL MOTION

G protein–linked receptors such as rhodopsin are the largest family of cell-surface or membrane-bound sensory receptors. More than 1000 members have been defined in mammals. Many of these receptors are present in olfactory cilia (Figure 3.3). These proteins are tightly inserted into the plasma membrane. This class of protein is referred to as integral membrane proteins, and movement of membrane lipids drives their lateral motion. Thus, lateral motion of these receptors depends not on their own structures, but rather on the lateral mobility contributed by various structures of fatty acid chains. DHA is proposed to represent the extreme form of membrane lateral motion.

G protein–linked receptors initiate the cellular responses to an enormous diversity of signaling molecules, including hormones, neurotransmitters, and odorants.

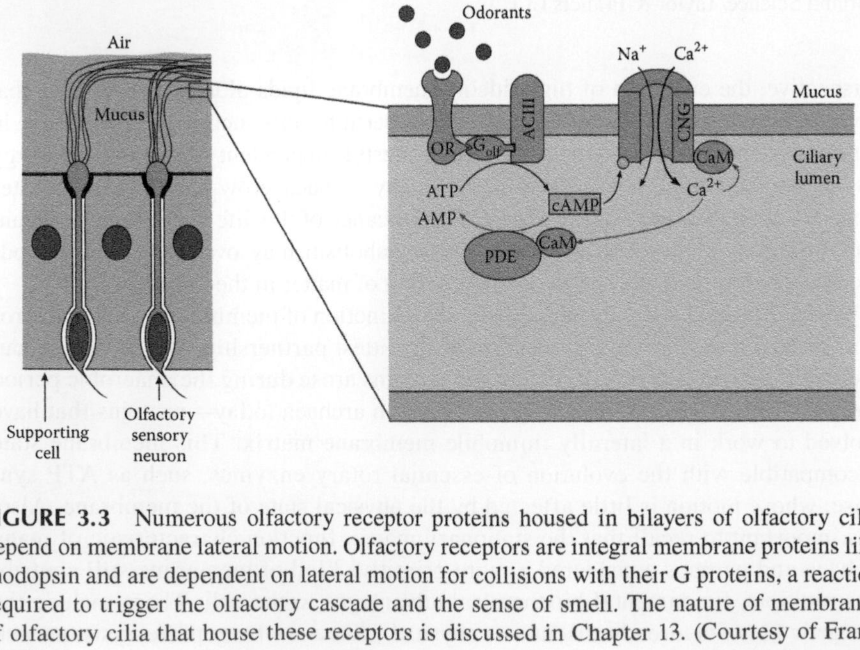


**FIGURE 3.3** Numerous olfactory receptor proteins housed in bilayers of olfactory cilia depend on membrane lateral motion. Olfactory receptors are integral membrane proteins like rhodopsin and are dependent on lateral motion for collisions with their G proteins, a reaction required to trigger the olfactory cascade and the sense of smell. The nature of membranes of olfactory cilia that house these receptors is discussed in Chapter 13. (Courtesy of Frank Mueller.)

For example, about 350 different odorant receptors are required for smell. Despite the chemical diversity of the signaling molecules that bind to them, all G protein–linked receptors whose amino acid sequences are known have a similar structure and are closely related evolutionarily. They consist of single polypeptide chains that thread back and forth across the bilayer seven times.

A detailed analysis of the cellular signaling cascades triggered by these receptors suggests the importance of the lipid matrix that houses these receptors (see Figure 3.3). The key point is that receptors transmit their information via G proteins, which are also embedded in the bilayer. This interaction between membrane receptor and its G protein partner defines what is sometimes referred to as the trigger reaction for the cascade in question. Once again properties of the bilayer itself are proposed to modulate the rates of trigger reactions. It is important to note that G proteins, which are composed of three different polypeptide chains, α, β, and γ, are anchored to the protein at least partly by a lipid chain that is covalently attached to the γ subunit. This anchoring is believed to be strong enough to subject the G protein complex to the same rates of lateral motion as its receptor. Thus, we can envision that both proteins are in motion for the purpose of effective collisions between the two components.

## 3.6 DOES RHODOPSIN MOVE FASTER IN A DHA (22:6) VERSUS DPA (22:5) BILAYER?

Rigorous exclusion of dietary precursors of DHA in rats causes a significant shift in composition of fatty acids in the brain. This shift involves replacement of DHA with docosapentaenoic acid (DPA; 22:5), which has one fewer double bond on the methyl end of the chain. Studies of dynamics of rhodopsin in DPA-enriched membranes show significant biochemical differences, including a reduction in rates of motion of rhodopsin. These results open "Pandora's box" regarding the molecular roles played by DHA/DPA and indeed unsaturation in general in driving membrane motion.

Some background information is needed to set the stage for this discussion. Historically the motional state of membranes was described on the basis of fluidity, the inverse of viscosity. As discussed previously and in Chapter 7, the term fluidity is controversial for several reasons and was essentially redefined by van de Vossenberg (see van de Vossenberg references in Chapters 1 and 2) as being a combination of permeability and a viscosity term referred to here simply as motion. However, interpreting studies with rhodopsin moving in a DPA membrane seems to require yet a further refinement of the definition of fluidity, or at least an acceptance that similar lipid chains may exhibit dramatic differences in their motional state. A closer look at the conformations of DHA versus DPA chains revealed by chemical simulation analysis shows a possible molecular explanation to account for biochemical differences between DHA and DPA membranes. This scenario involves the greater opportunity for lipophilic bonding exhibited by the comparatively longer stretch of saturated hydrocarbon chain at the methyl end of DPA (i.e., six carbons in length) versus a three-carbon chain at the methyl end of DHA. What this means is that DPA is expected to engage in a greater amount of weak lipid-lipid bonding with neighboring chains. We hypothesize that even a relatively short stretch of saturated chain at the end of DPA chains is sufficient to increase chain-chain interaction and quantitatively slow lipid and thus protein motion (i.e., rhodopsin–G protein collision frequency is reduced).

## 3.7 DHA PHOSPHOLIPIDS LIBERATE MEMBRANE ENZYMES/SUBSTRATES TRAPPED IN LIPID RAFTS

A growing list of enzymes from both exothermic and endothermic organisms can be classified as being dependent on antifreeze/antiviscosity properties of DHA phospholipids to prevent enzymes/components from being trapped in gel-phase lipids. Only a brief description is given here since Stillwell and others have reviewed this field recently (Stillwell 2006; Stillwell et al. 2005). Note that either the membrane-bound enzyme or its substrate or both can be "inactivated" once motion is stalled. For example, human membrane-bound lipases or other enzymes targeting phospholipids can be separated from their phospholipid substrates that become trapped in gel-phase lipids. Indeed, the cell's ability to modulate the mosaic nature of its membranes represents a distinct mechanism of regulating enzyme action. The class of enzymes functional only in the liquid or conversely only in lipid rafts can be said to "sense" changes in membrane motion, which is amplifiable via changes in enzymatic activity.

The phosphoinositol sensory cascade serves as an example of the relationship between lipase and its substrate. This cascade features two distinct collisional events dependent on membrane motion. The first, similar to the trigger reaction for vision, involves the collision on the inner membrane surface of a classic seven-span receptor with its G protein. A second collisional event between membrane-associated phospholipase and its substrate, a specific class of membrane phospholipid, is required to sustain the cascade. It is important to note that the membrane phospholipid substrate in this reaction is both membrane bound and highly unsaturated. The latter property favors the localization of phospholipase in a fluid region of the bilayer. The presence of the membrane-bound form of phospholipase in the liquid region would allow the lipase access to its substrate. Reaction rates are known to become dependent on DHA phospholipids, apparently to free enzymes trapped in lipid rafts. Note that the opposite case has also been reported in which membrane enzymes are specific for substrates trapped in lipid rafts.

## 3.8 HAVE SOME MEMBRANE-BOUND ENZYMES EVOLVED DEPENDENCE ON A FLUID LIPID ENVIRONMENT FOR BIOCATALYSIS?

The case for lateral motion as a precondition for rapid electron transport or signaling between distinct membrane components is reasonably well established, as discussed above. The question of whether other classes of enzymes depend on lateral motion is an interesting "gray zone," specifically for membrane proteins that work as self-contained molecular machines. The proteins in question include some of the most abundant and important enzymes in the cell, such as ATP/adenosine diphosphate (ADP) exchange protein of mitochondria, $Na^+/K^+$ ATPase, and $K^+$ gates of excitatory cells. Crystal structures have been solved for a related cation transporter, $Ca^{2+}$ ATPase, and show that this enzyme is indeed a molecular machine with large and cyclical conformational changes taking place deep in the lipid region of the membrane. We pose the following question regarding such structures: Would such large, rapid, and repetitive conformational changes involving lipid-bound domains of this protein still occur if the enzyme were trapped in a lipid raft? The same scenario applies to other proteins whose modes of action involve lipid-bound protein domains that are both dynamic and work against the matrix lipids surrounding the enzyme (e.g., opening and closing as envisioned for ADP/ATP exchange protein).

## 3.9 PHOSPHOLIPID-DEPENDENT ENZYMES

Convincing biochemical-genetic data that negatively charged headgroups of membrane phospholipids are important in membrane biochemistry in bacteria such as *E. coli* has now been shown in dramatic fashion to apply to animals. This research from the laboratory of Nobel Laureate R. Mackinnon involves voltage-gated channels (i.e., gates) in neurons that have coevolved with DHA-enriched membranes to regulate ion conductance essential in electrophysiology (Schmidt, Jiang, and Mackinnon, 2006). Such voltage-dependent processes rely on the action of protein domains known as voltage sensors that are embedded on/inside the cell membrane and contain an excess of positively charged amino acids that react to an electric field (Figure 3.4). How does the membrane create an environment suitable for voltage-sensing domains? Recently, researchers have found that a voltage-dependent $K^+$ channel is dependent on the negatively charged phosphodiester of phospholipid molecules. These data led to the hypothesis that membrane phospholipids provide stabilizing ionic interactions between positively charged voltage-sensor arginine residues and negatively charged phosphodiester groups. According to this model, the membrane works in partnership with the $K^+$ gate by providing an appropriate environment for the stability and operation of the paddle-like voltage-sensing machinery. Whereas this elegant research targets voltage-gated channels, it might be possible to generalize this sort of electrostatic mechanism to include other environmental stimuli, as well as a wide range of membrane proteins/channels found in many life forms.

## 3.10 DHA AS A SPACE-FILLING SEALANT AROUND MEMBRANE PROTEINS

We once observed an experiment in a Norwegian laboratory in which membrane vesicles derived from a halophilic archaea were preloaded with radioactive amino acids, stored at 2°C, and sampled for leakage over a time course of several months. The result is that even after about 6 months little amino acid leakage, high inside to low outside, was observed. This is a testament to the robust permeability barrier formed by archaeal lipids but goes further than this. These vesicles, formed from whole cells, still contained numerous membrane-bound proteins. Obviously, these membranes formed a tight seal of impermeability around these sometimes large enzyme complexes. This experiment highlights an important aspect of membranes

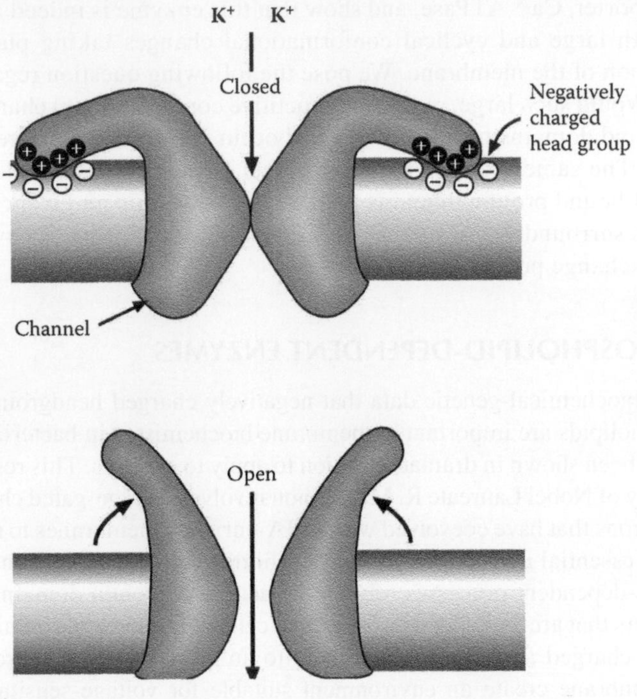


**FIGURE 3.4** Lipid headgroups are involved in opening and closing of a voltage-gated $K^+$ channel through ionic interactions. A simplified view of how the positively charged voltage sensor "paddle" of a $K^+$ channel is modulated by headgroup negative charge is displayed. In this model, in which the artist duplicates the paddle to highlight this effect, an electrical charge severs the ionic bonds between positive charges on the peptide forming the paddle (i.e., arginine residues) and negatively charged headgroups, causing the $K^+$ channel to open.

that has received relatively little attention, specifically, what creates the permeability seal at the interface between membrane proteins and their surrounding lipids. The Norwegian experiment conducted by Kjell Andersen (personal communication) shows that in the case of immobile proteins an extremely tight permeability barrier (i.e. high $H^+/Na^+/K^+$ fidelity) is possible. It is tempting to extrapolate this finding to all membrane proteins, but caution is warranted, especially for proteins that move rapidly over the membrane surface.

In the following paragraphs we address the issue of sealing around membrane proteins by developing the concept that highly mobile proteins might carry their seals with them. Furthermore, we hypothesize that polyunsaturated chains, especially DHA, often make the best sealants against cation leakage, especially when maintenance of rapid motion is required.

One line of evidence for a tight association of lipids with proteins comes from the study of annular lipids—those that are copurified with their proteins. The tight binding of annular lipids to their proteins has been observed in crystal structures of membrane proteins using X-ray crystallography. That is, phospholipid-protein binding is tight enough to persist throughout rigorous biochemical procedures used to obtain enzyme crystals.

Data consistent with these concepts come from several sources. For example, biochemists have found that numerous membrane-bound enzymes when "delipidated" lose biochemical activity. In many cases, activity is restored by reconstitution with native or chemically defined phospholipids. Often there is a preference for polyunsaturated chains. For example, lipids surrounding $Na^+/K^+$-ATPase extracted from neurons of a cold-adapted crab showed a dramatically increased level of EPA compared to a preparation from beef brain: 21.2% EPA in crab compared to only traces in beef brain. The enrichment of DHA in annular lipids of bovine $Na^+/K^+$-ATPase were around 13% DHA. The levels of saturated chains in crab axons were significantly lower than beef brain. One interpretation of this data is that EPA helps maintain molecules of $Na^+/K^+$-ATPase in a "fluidized" or motional state in the cold-adapted crab axon. $Na^+/K^+$-ATPase of beef axons also likely requires a fluidized membrane environment, maintained by relatively high enrichment with DHA.

Studies of the roles of DHA phospholipids as annular lipids are most advanced in the case of rhodopsin. In this case chemical modeling shows a tight tongue and groove structure formed by DHA chains filling grooves on the surface of rhodopsin. This further demonstrates the space-filling nature of DHA. There are at least three mechanisms to explain the tight fit between DHA phospholipids and their proteins. In the case of rhodopsin, DHA phospholipids as annular lipids might directly modulate catalytic properties of the protein itself or alternately work indirectly to help propel the protein as it searches for its G protein partner on the membrane surface. However, DHA enrichment of energy transducing membranes such as bacteria, chloroplasts, and mitochondria raises another issue. Recall that electron transport complexes of mitochondria have been calibrated to move rapidly, albeit somewhat slower than rhodopsin. Also recall that the surface landscape of mitochondria is highly cluttered with large numbers of proteins that continually collide with one another. The point is that such rapidly moving proteins or their collisions might generate defects in the proton permeability barrier, defects that might be instantaneously filled by DHA phospholipids. In other words the third idea involves a self-sealing mechanism in which DHA chains form a barrier against cation permeability defects caused by motion and collisions among proteins. We recognize that there are few data to support these ideas and offer them to generate discussion in this area.

## 3.11 SUMMARY

By drawing on and synthesizing membrane research covering a wide range of life forms, a new perspective of the complex biochemical partnership between membrane enzymes and their matrix phospholipids is emerging. The current picture is that many membrane enzymes have coevolved with and become dependent on phospholipids, both lipid chains and head groups, to achieve maximum rates of membrane biocatalysis. This provides the cell with an important mechanism to control biochemical activity. Perhaps it is time to give serious consideration to the idea of upgrading the status of membrane phospholipids to the level of cocatalysts. This proposal is based on the conventional biochemical rules concerning biocatalysis where cocatalysts are defined on the basis of increased rates of enzymatic reactions. Recall that previous concepts of the role of cocatalysts are based largely on enzymatic reactions occurring in the aqueous phase. The distinction here is that membrane biochemistry occurs in a completely different solvent, the lipid phase. Thus, the solvent becomes a key player in reaction rates. We are not making any claims that membrane lipids themselves are catalysts in a classic sense. However, there are some clues in the literature that a single DHA phospholipid working like a membrane ice-breaker may "catalyze" the breakup of many neighboring phospholipids, preventing formation of aggregates of gel-phase lipids needed to enhance enzyme reactions (see Chapter 7). Finally, research probing the partnership between membranes and their proteins has revealed that DHA-phospholipids might bind to specific sites of enzymes and directly modify catalytic powers of the protein.

## SELECTED BIBLIOGRAPHY


Alberts, B., D. Barry, J. Lewis, M. Raff, K. Roberts, and J. D. Watson, 1994. *Molecular Biology of the Cell,* 3rd ed. New York: Garland Publishing, Inc.

Castresana, J., and M. Saraste. 1995. Evolution of energetic metabolism: The respiration-early hypothesis. *Trends Biochem. Sci.* 20:443–448.

Diaz, O., A. Berquand, M. Dubois, et al. 2002. The mechanism of docosahexaenoic acid-induced phospholipase D activation in human lymphocytes involves exclusion of the enzyme from lipid rafts. *J. Biol. Chem.* 277:39368–39378.

Eldho, N. V., S. E. Feller, S. Tristram-Nagle, I. V. Polozov, and K. Gawrisch. 2003. Polyunsaturated docosahexaenoic vs. docosapentaenoic acid: Differences in lipid matrix properties from the loss of one double bond. *J. Am. Chem. Soc.* 125:6409–6421.

Feller, S. E., K. Gawrisch, and T. B. Woolf. 2003. Rhodopsin exhibits a preference for solvation by polyunsaturated docosahexaenoic acid. *J. Am. Chem. Soc.* 125:4434–4435.

Fyfe, P. K., K. E. McAuley, A. W. Roszak, N. W. Isaacs, R. J. Cogdell, and M. R. Jones. 2001. Probing the interface between membrane proteins and membrane lipids by X-ray crystallography. *Trends Biochem. Sci.* 26:106–112.

Gibson, N. J., and M. F. Brown. 1993. Lipid headgroup and acyl chain composition modulate the MI-MII equilibrium of rhodopsin in recombinant membranes. *Biochemistry* 32:2438–2454.

Grossfield, A., S. E. Feller, and M. C. Pitman. 2006. A role for direct interactions in the modulation of rhodopsin by ω-3 polyunsaturated lipids. *Proc. Natl. Acad. Sci. U.S.A.* 103:4888–4893.

Harold, F. M. 1986. *The vital force: A study of bioenergetics*. New York: W. H. Freeman and Company.

Huber, T., A.V. Botelho, K. Beyer and M. F. Brown. 2004. Membrane model for the G protein-coupled receptor rhodopsin: Hydrophobic interface and dynamical structure. *Biophys. J.* 86: 2078–2100.

Hutchison, C. A., III, S. N. Petersen, and S. R. Gill et al. 1999. Global transposon mutagenesis and a minimal Mycoplasma genome. *Science* 286:2165–2169.

Jeffrey, B.G., H.S. Weisinger, M. Neuringer and D. C. Mitchell. 2001. The role of docosahexaenoic acid in retinal function. *Lipids* 36:859–871.

Litman, B. J., S.-L. Niu, A. Polozova and D. C. Mitchell. 2001. The role of docosahexaenoic acid containing phospholipids in modulating G protein-coupled signaling pathways: Visual transduction. *J. Mol. Neurosci.* 16:237–242; discussion 279–284.

Mulkidjanian, A. Y., K. S. Makarova, M. Y. Galperin, and E. V. Koonin. 2007. Inventing the dynamo machine: The evolution of the F-type and V-type ATPases. *Nature Rev. Microbiol.* 5:892–899.

Mushegian, A. R., and E. V. Koonin. 1996. A minimal gene set for cellular life derived by comparison of complete bacterial genomes. *Proc. Natl. Acad. Sci. U.S.A.* 93:10268–10273.

Niu, S.-L., D. C. Mitchell, S.-Y. Lim, et al. 2004. Reduced G protein-coupled signaling efficiency in retinal rod outer segments in response to n-3 fatty acid deficiency. *J. Biol. Chem.* 279:31098–31104.

Oesterhelt, D., and W. Stoeckenius. 1973. Functions of a new photoreceptor membrane. *Proc. Natl. Acad. Sci. U.S.A.* 70:2853–2857.

Schmidt, D., Q.-X. Jiang, and R. MacKinnon. 2006. Phospholipids and the origin of cationic gating charges in voltage sensors. *Nature* 444:775–779.

Schmidt, D., and R. MacKinnon. 2008. Voltage-dependent $K^+$ channel gating and voltage sensor toxin sensitivity depend on the mechanical state of the lipid membrane. *Proc. Natl. Acad. Sci. U.S.A.* 105:19276–19281.

Stillwell, W. 2006. The role of polyunsaturated lipids in membrane raft function. *Scand. J. Food Nutr.* 50:107–113.

Stillwell, W., S. R. Shaikh, M. Zerouga, R. Siddiqui, and S. R. Wassall. 2005. Docosahexaenoic acid affects cell signaling by altering lipid rafts. *Reprod. Nutr. Dev.* 45:559–579.

Toyoshima, C., and H. Nomura. 2002. Structural changes in the calcium pump accompanying the dissociation of calcium. *Nature* 418:605–611.

Wang, H., and G. Oster. 1998. Energy transduction in the F1 motor of ATP synthase. *Nature* 396:279–282.

Yasuda, R., H. Noji, K. Kinosita, Jr., and M. Yoshida. 1998. F1-ATPase is a highly efficient molecular motor that rotates with discreet 120 degree steps. *Cell* 93:1117–1124.

# 4 Convergent Evolution of DHA/EPA Biosynthetic Pathways

Hypothesis: The convergent evolution of aerobic and anaerobic pathways for DHA/EPA biosynthesis suggests substantial benefits from these chains for bacteria, plants and animals.

For many years it was believed that plants and animals were the only organisms that benefitted from the incorporation of DHA/EPA into their membrane phospholipids. The desaturase pathway is well established and features molecular oxygen ($O_2$) as substrate for introduction of multiple double bonds along the chains. The discovery of DHA/EPA in membranes of bacteria along with a new anaerobic pathway (Figure 4.1) necessitate a rethinking of the evolutionary history of DHA/EPA synthesis.

These discoveries represent efforts among several laboratories, with the first hints of a novel biosynthetic mechanism reported by deep-sea microbiologists (Delong and Yayanos, 1986). Biochemical and genomic analyses show that EPA/DHA synthetases are very similar, being large, apparent membrane-associated, enzyme complexes requiring reducing power in the form of reduced pyridine nucleotides but with no requirement for molecular $O_2$ as substrate for double bond formation. DHA/EPA synthetases seem to catalyze anaerobic insertion of multiple double bonds forming either DHA or EPA, but not both. This anaerobic pathway permits DHA/EPA to be produced without the consideration of $O_2$ supply. Thus, DHA/EPA may benefit cells living in both aerobic and anaerobic environments, suggesting a more fundamental role for these chains than previously suspected. A striking feature of DHA/EPA synthetase enzymes involves an array of multiple acyl carrier protein (ACP)-like groups present on the large open reading frame (ORF) 6 subunit of the complex, believed to serve as docking sites for growing chains attached to the enzyme. Among bacterial proteins, two of the subunits of DHA/EPA synthetases are among the largest at over 280 Da. These large proteins are composed of multiple domains essential for enzymatic catalysis. ACP groups are covalently attached, but their topological arrangement on the enzyme surface is unknown. Immature chains are envisioned to advance along an ACP-mediated assembly line requiring multiple steps, including chain elongation, generation of double bonds, and ultimately termination of mature chains as discussed below. Thus, it is clear that the polyketide mechanism for synthesis of DHA/EPA by bacteria differs fundamentally from the aerobic pathway used by plants and animals and is a classic example of convergent evolution (i.e., two distinct biochemical pathways leading to identical end products).

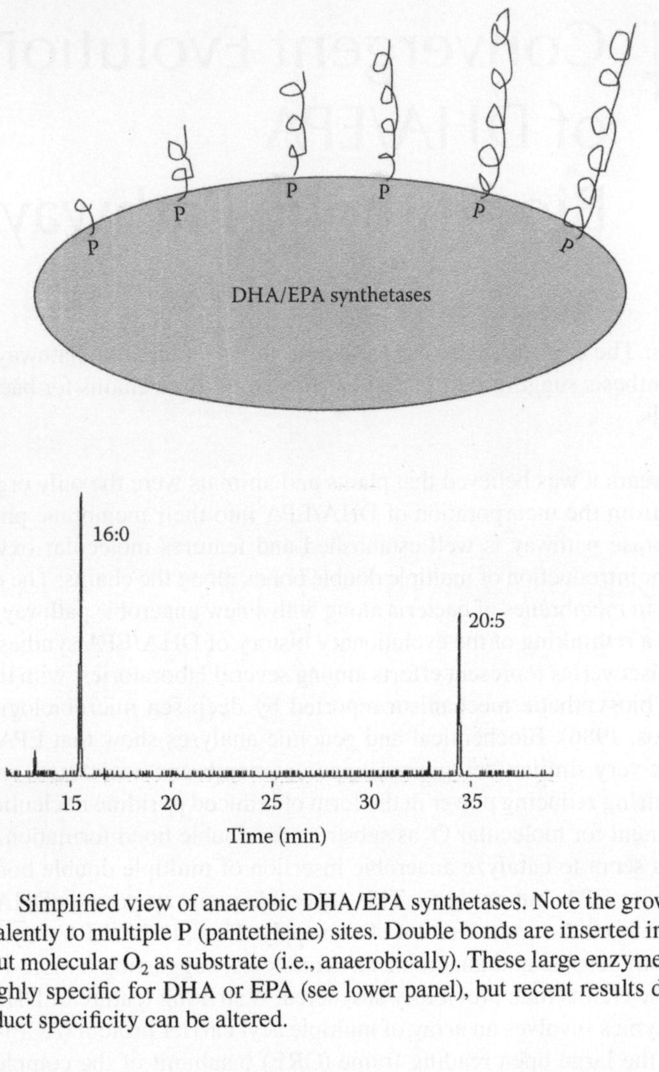


**FIGURE 4.1** Simplified view of anaerobic DHA/EPA synthetases. Note the growing chains attached covalently to multiple P (pantetheine) sites. Double bonds are inserted into growing chains without molecular $O_2$ as substrate (i.e., anaerobically). These large enzyme complexes tend to be highly specific for DHA or EPA (see lower panel), but recent results demonstrate that end-product specificity can be altered.

## 4.1 DOMAIN ANALYSIS OF DHA/EPA GENE CLUSTERS

Genomic analysis (Figure 4.2) shows that marine bacteria have organized DHA/EPA genes into gene clusters, a typical arrangement seen with other biosynthetic enzymes in numerous bacterial genomes. At the next level of genetic fine structure, domain analysis of the proteins encoded by the ORFs essential for DHA/EPA production reveals the presence of domain structures characteristic of polyketide synthetase (PKS)-like systems used widely for biosynthesis of a variety of natural products, such as antibiotics. Figure 4.2 shows a summary of some of the domains, motifs, and also key homologies detected by searching for matches in gene data banks. Because EPA is different from many of the other substances known to be produced by PKS-like pathways, that is, it contains five *cis*-double bonds, spaced at three carbon intervals along the molecule, a PKS-like system for synthesis of EPA/DHA is not expected. Further, searches of data banks using the domains present in the *Shewanella* EPA ORFs reveal that several are related to proteins encoded by a PKS-like gene cluster found in the cyanobacterium *Anabeana*. The *Anabeana* PKS-like genes have been linked to the synthesis of a long-chain ($C_{26}$), hydroxyl-fatty acid found in a glycolipid layer of heterocysts, specialized cells important in $N_2$ fixation.

ORF 6 of *Shewanella*'s EPA gene cluster contains domains or motifs, many common to keto acyl synthetases (KAS), proteins important in fatty acids biosynthesis in bacteria. ORF 6 contains a malonyl-CoA:ACP acyl transferase (AT) domain. Sequences near the active site motif suggest it transfers malonate rather than methylmalonate, that is, it resembles the acetate-like ATs. Following a linker region, there is a cluster of five repeating domains, each ~100 amino acids in length, which are homologous to PKS-like ACP sequences. Each contains a pantetheine binding site motif. The presence of five such ACP domains has not been observed previously in fatty acid synthases (FAS) or PKS-like systems. Near the end of the protein is a region that shows homology to β-keto-ACP reductases (KR). It contains a pyridine nucleotide binding site motif.

The *Shewanella* ORF 8, another large protein, begins with a KAS domain, including active site and ending motifs. The best match in the data banks is with the *Anabeana* HglD. There is also a domain that has sequence homology to the

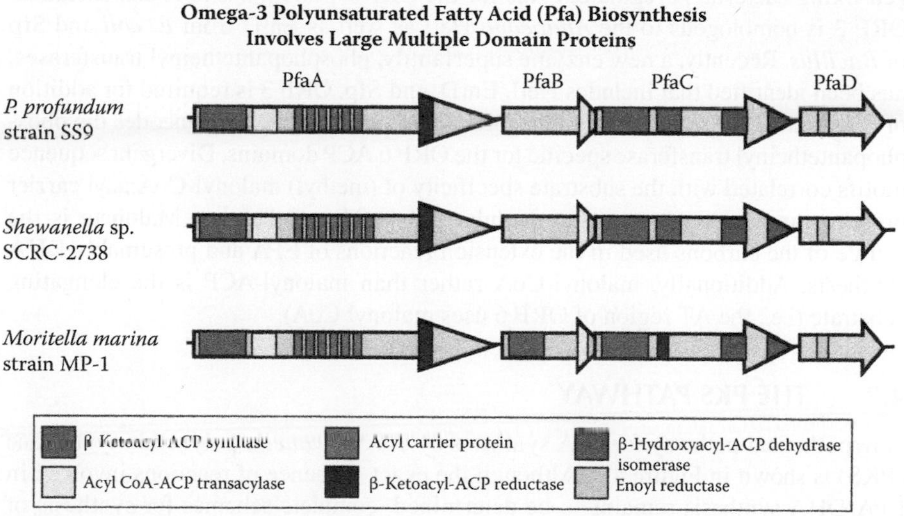


**FIGURE 4.2** Anaerobic, DHA/EPA genes from several bacteria. Domain analysis shows a striking similarity between gene clusters coding for DHA (*Moritella marina*) or EPA (*Photobacterium profundum/Shewanella* sp.) synthetase enzyme complexes. Various biochemical motifs are described in the text. The *Shewanella* EPA cluster has been transferred and expressed in a recombinant of *E. coli*, as have DHA genes from *Moritella*, and efforts continue to express these genes in plants to create "fish oil crops," described in a later chapter. (Reproduced from Valentine, R. C. and Valentine, D. L., *Prog. Lipid Res.*, 43:383–402, 2004. With permission from Elsevier.)

N-terminal half of the *Anabeana* HglC. This region also shows weak homology to KAS proteins, although it lacks the active site and ending motifs. It has the characteristics of the so-called chain length factors (CLF) of type II PKS-like systems. ORF 8 also has two domains with homology to β-hydroxyacyl-ACP dehydrases, which catalyze removal of water forming a *cis*-double bond in its place. The best match for both domains is with *E. coli* FabA, a bifunctional enzyme critical for anaerobic, monounsaturated fatty acid synthesis, which carries out both the dehydrase reaction and an isomerization (*trans* to *cis*) of the resulting double bond. The first dehydrase domain contains both the active site histidine and an adjacent cysteine implicated in catalysis mediated by FabA, again a critical reaction for insertion of double bonds in fatty acid chains under anaerobic conditions. The second dehydrase domain has the active site histidine but lacks the adjacent cysteine. Searches with the second dehydrase domain also show matches to FabZ, a second *E. coli* dehydrase, which does not possess isomerase activity.

The best match of the C-terminal half of ORF 7 is with a C-terminal portion for the *Anabeana* HglC, but the N-terminal half of ORF 7 has no significant matches in the data banks. The C-terminal domain contains an acyl-transferase (AT) motif. These data suggest that ORF 7 may function as a thioesterase, perhaps for cleaving or transferring mature EPA chains from their final ACP binding site.

ORF 9 is homologous to an ORF of unknown function in the *Anabeana* Hgl cluster. It also exhibits a very weak homology to NifA, a regulatory protein in nitrogen-fixing bacteria. A regulatory role for the ORF 9 protein has not been excluded. ORF 3 is homologous to the *Anabeana* HetI as well as EntD from *E. coli* and Sfp of *Bacillus*. Recently, a new enzyme superfamily, phosphopantetheinyl transferases, has been identified that includes HetI, EntD, and Sfp. ORF 3 is required for addition of β-alanine (i.e., pantetheine) to the ORF 6 protein. Thus, ORF 3 encodes the phosphopantetheinyl transferase specific for the ORF 6 ACP domains. Divergent sequence motifs correlated with the substrate specificity of (methyl) malonyl-CoA:acyl carrier protein transacylase domains in modular polyketide synthetases. Malonate is the source of the carbons used in the extension reactions of EPA and presumably DHA synthesis. Additionally, malonyl-CoA rather than malonyl-ACP is the elongating substrate (i.e., the AT region of ORF 6 uses malonyl CoA).

## 4.2 THE PKS PATHWAY

A hypothetical pathway for EPA synthesis by the *Shewanella* polyketide synthetase (PKS) is shown in Figure 4.3. Although the exact sequence of reactions involved in EPA/DHA synthesis remains to be determined, complete schemes for synthesis of EPA by the *Shewanella* PKS can be proposed, and one such scheme is provided. The identification of protein domains homologous to the *E. coli* FabA protein, and the observation that bacterial EPA synthesis occurs anaerobically, provide evidence for a mechanism where insertion of *cis*-double bonds occurs through the action of a bifunctional dehydratase/2-*trans*, 3-*cis* isomerase. In *E. coli*, condensation of the 3-*cis* acyl intermediate with malonyl-ACP requires a particular ketoacyl-ACP synthase and this may provide a rationale for the presence of two PKS in the *Shewanella* gene cluster. However, the PKS cycle extends the chain in two-carbon increments while the double bonds in the EPA product occur at every third carbon. This disjunction can be solved if the double bonds at Δ14 and Δ8 of EPA are generated by 2-*trans*, 2-*cis* isomerization followed by incorporation of the *cis* double bond into the elongating fatty acid chain. The enzymatic conversion of a *trans* double bond to the *cis* configuration without bond migration is known to occur, for example, in the synthesis of 11-*cis*-retinal in the retinoid cycle. Although such an enzyme function has not yet been identified in the *Shewanella* PKS, it may reside in one of the

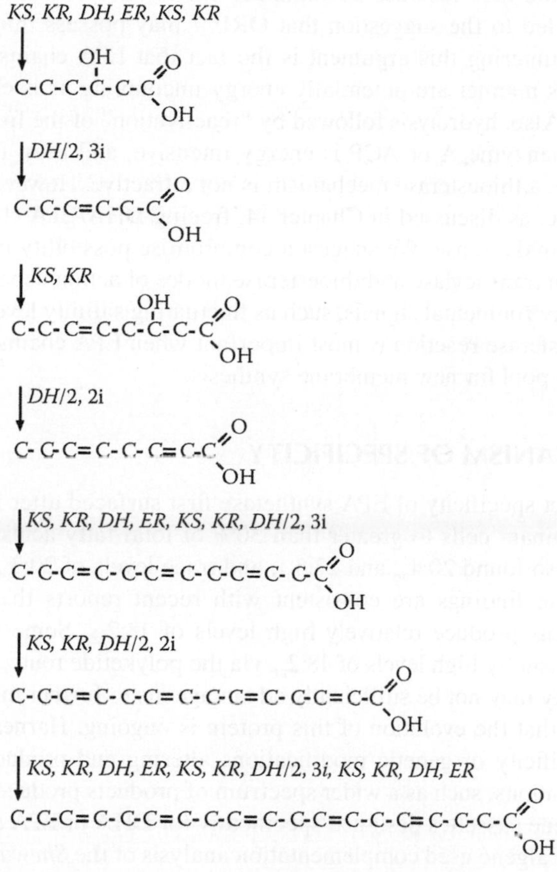


**FIGURE 4.3** Polyketide mechanism of DHA/EPA synthesis. DHA/EPA join a growing list of important secondary metabolites, such as certain antibiotics, that are synthesized via a "polyketide" process. For synthesis of DHA/EPA, this involves cycles of chain elongation followed by multiple steps leading to the next cycle of elongation. The covalent binding of growing chains to a template, the surface of a large enzyme complex, is a common mechanism among polyketide synthetases. In the case of DHA, six double bonds must be created at exact locations along the chain while maintaining isomeric specificity (i.e., *cis* configuration) across each double bond. As yet there are no chemical synthesis routes that match this same level of precision, which explains why DHA and EPA are not available as bulk chemicals. (Courtesy of Daniel Facciotti and Jim Metz.)

unassigned protein domains. Simple modification of the reaction scheme allows for DHA synthesis in the eukaryotic fungus *Schizochytrium* and the DHA-producing bacterium *Moritella marina*. Possible modification of the end-product specificities of DHA/EPA synthetases, usually quite specific, is discussed below.

According to the proposed pathway of EPA synthesis outlined in Figure 4.3, double bonds are inserted in essentially a sequential fashion as the EPA chain matures, but there are unanswered questions regarding how mature chains leave the complex. As mentioned above, ORF 7 represents an abbreviated acyl transferase domain missing two amino acid residues essential for exclusion of water from the active site. These data led to the suggestion that ORF 7 may possess thioesterase activity. However, countering this argument is the fact that EPA chains freed into the cytoplasm in this manner are potentially energy uncoupling as is characteristic of free fatty acids. Also, hydrolysis followed by "reactivation" of the free fatty acid for attachment to coenzyme A or ACP is energy intensive, and thus, from a bioenergetic perspective, a thioesterase mechanism is not attractive. However, from a symbiotic perspective, as discussed in Chapter 14, freeing DHA/EPA chains for export to the host does make sense. We suggest a compromise possibility in which ORF 7 switches between transacylase and thioesterase modes of action depending on as yet undetermined environmental signals, such as fluctuating salinity levels. We propose that the acyltransferase reaction is most important when EPA chains are shunted to the phospholipid pool for new membrane synthesis.

## 4.3 MECHANISM OF SPECIFICITY

A lack of product specificity of EPA synthetase first surfaced after increasing EPA levels in recombinant cells to greater than 30% of total fatty acids. In addition to EPA, we have also found $20{:}4_{n6}$ and $22{:}5_{n3}$, and trace levels of $20{:}3_{n6}$, $20{:}1_{n9}$, $18{:}3_{n6}$, and $18{:}2_{n6}$. These findings are consistent with recent reports that certain EPA-producing mutants produce relatively high levels of $18{:}2_{n6}$. Some natural isolates also produce unusually high levels of $18{:}2_{n6}$ via the polyketide route, suggesting that greater specificity may not be sufficiently advantageous to further modify this complex protein, or that the evolution of this protein is ongoing. Harnessing flexibility of enzyme specificity or genetic modifications altering end products has obvious industrial applications, such as a wider spectrum of products produced from a single oil crop using gene cassettes designed specifically for DHA or EPA and so forth.

Scientists at Calgene used complementation analysis of the *Shewanella* EPA gene cluster to help determine the specificity of individual subunits (Figure 4.4). In this experiment, plasmid vectors representing various deletions of the *Shewanella* gene cluster are cotransformed with plasmids carrying individual DHA genes. Figure 4.4 is a depiction of various complementation experiments resulting in production of DHA versus EPA. On the right is the largest chain produced (i.e., EPA or DHA) in an *E. coli* strain containing the *Moritella* or *Shewanella* genes shown on the left. The hollow boxes indicate ORFs from *Shewanella*. The solid boxes indicate ORFs from *Moritella*. Note that only *Moritella* ORF 8 shifts synthesis toward DHA, although specificity is not high since recombinant cells also produce similar levels of EPA. The conclusion we draw is that ORF 8, the second largest "subunit" compared to

ORF 6, is an important determinant of DHA specificity. However, since a fine-structure mutational map of the DHA gene cluster is not yet available, it is too early to tell whether other genes might also be important in specificity.

## 4.4 SUMMARY

It is now clear that DHA/EPA-producing bacteria, as well as certain eukaryotic fungi such as *Schizochytrium* (a candidate for industrial production of DHA), have evolved a radically different biochemical pathway for synthesis of DHA/EPA involving a polyketide mechanism, an anaerobic process. The finding of two distinct and complex pathways for biosynthesis of DHA/EPA in nature is consistent with important roles of these chains in membranes. Attention now shifts from "how" DHA/EPA is made to "what" properties contributed by these molecules are important enough to warrant the evolution of convergent pathways. Because DHA/EPA are produced by single-cell organisms that obviously lack specialized DHA-essential membranes such as those found in humans, a fundamental role of these chains in membrane biochemistry is proposed.

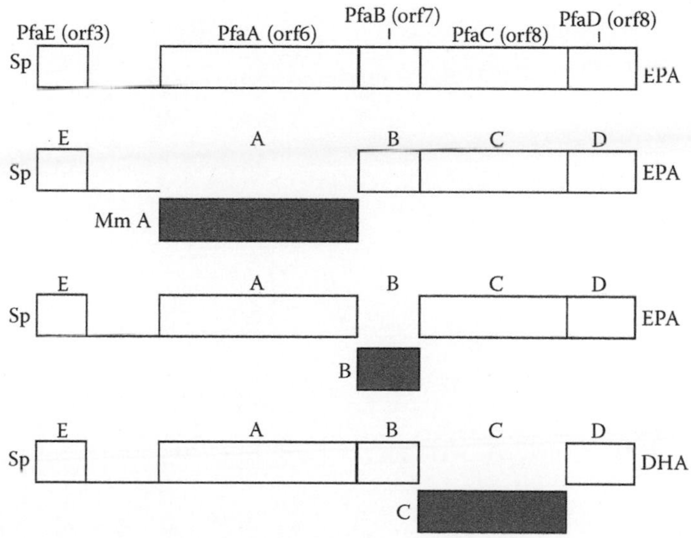


**FIGURE 4.4** Specificity of DHA/EPA biosynthesis. Scientists at Calgene used genetic complementation analysis to help define which genes are most important in product specificity. Recombinants of *E. coli* carrying plasmids with various deletions of the EPA gene cluster of *Shewanella putrefaciens* (Sp, unfilled rectangles) are complemented by a second plasmid harboring various DHA genes of *Moritella marina* (Mm, gray rectangles). Note that the products produced by the various gene combinations are shown on the right. Also note that this is the first demonstration of DHA biosynthesis using recombinants of *E. coli*. (Courtesy of Daniel Facciotti, Jim Metz, and Vic Knauf.)

## SELECTED BIBLIOGRAPHY

Allen, E. E., and D. H. Bartlett. 2002. Structure and regulation of the omega-3 polyunsaturated fatty acid synthase genes from the deep-sea bacterium *Photobacterium profundum* strain *SS9*. *Microbiology* 148:1903–1913.

De Long, E. F., and A. A. Yayanos. 1986. Biochemical function and ecological significance of novel bacterial lipids in deep-sea prokaryotes. *Appl. Environ. Microbiol.* 51:730–737.

Facciotti, D., J. G. Metz, and M. Lassner. 1998. Production of polyunsaturated fatty acids by expression of polyketide-like synthesis genes in plants. United States Patent 6,140,468.

Johns, R. B., and G. J. Perry. 1977. Lipids of the marine bacterium *Flexibacter polymorphus*. *Arch. Microbiol.* 114:267-271.

Okuyama, H., Y. Orikasa, T. Nishida, K. Watanabe, and N. Morita. 2007. Bacterial genes responsible for the biosynthesis of eicosapentaenoic and docosahexaenoic acids and their heterologous expression. *Appl. Environ. Microbiol.* 73:665–670.

Valentine, R. C., and D. L. Valentine. 2004. Omega-3 fatty acids in cellular membranes: A unified concept. *Prog. Lipid Res.* 43:383–402.

Yazawa, K. 1996. Production of eicosapentaenoic acid from marine bacteria. *Lipids* 31:297–300.

# 5 Membrane Evolution in a Marine Bacterium: *Capitalizing on DHA for Energy Conservation in Seawater*

Hypothesis: Deep-sea bacteria have evolved DHA chains in part as a suitable permeability barrier against $Na^+$.

The marine world is one in which energy sources for bacteria are often highly limited and environmental conditions, such as extreme cold, create additional energy stress. The DHA-producing bacterium *Moritella*, described here, is found in the deepest parts of the ocean (Figure 5.1) and has evolved to live in a world dominated by multiple stresses that limit energy production and hinder energy conservation: extreme cold, high salinity, and extreme hydrostatic pressure. Indeed, deep-sea bacteria have become so specialized that they have developed a dependency on the chemical and physical environment of the deep sea. They often die when conditions are changed even slightly and can only be cultivated under stringent laboratory conditions characteristic of their deep marine world. These bacteria are classified as psychrophiles, or cold extremophiles. *Moritella* has evolved both conventional and novel mechanisms for capturing energy, and the purpose of this chapter is to demonstrate how DHA-enriched membranes allow these cells to capitalize on the 0.5 M $Na^+$ in seawater, for energy conservation. This mechanism of $Na^+$ bioenergetics requires an intimate partnership with its DHA-enriched membrane.

## 5.1 ENERGY LIMITATION PLAGUES THE LIFE OF A DEEP-SEA BACTERIUM

Genomic and physiological analyses show that deep-sea bacteria depend on electron transport chains for energy production and conservation even at temperatures below 0°C. At first glance it is difficult to imagine how electron transport is possible under such extreme conditions. The picture becomes even more puzzling when considering that certain DHA-producing bacteria have maximum growth temperatures around 10°C and quickly die when exposed to temperatures even a few degrees warmer. Indeed, it was found almost 50 years ago that one of the most studied of DHA-producing isolates, normally displaying a temperature range from about 0°C to 21°C,

**FIGURE 5.1** (A color version of this figure follows p. 140.) Discovery of DHA-producing bacteria is intimately linked to the search for life in the deep ocean. (A) The deep-diving remotely operated vehicle, *Kaiko*, from Japan, was designed to search for life in the deepest part of the ocean, such as the Mariana Trench (>30,000 feet in depth). (B) Sediment samples taken by *Kaiko's* robotic arm contained DHA-producing bacteria, confirming the discovery in 1986 by DeLong and Yayanos (see reference in Chapter 5), who first described the isolation of DHA-containing bacteria from water samples taken from the deep ocean. (Photos courtesy of JAMSTEC.)

stops growing above 10°C when the $Na^+$ content of the culture medium is diluted. The relationship between salinity and DHA is strengthened by the finding discussed below that the percentage of DHA chains in the bilayer of these cells is upregulated by increasing salinity in growth media. Earlier physiological data begin to make sense when considered from the perspective of genomics. It seems clear that DHA-producing cells specialize in a respiratory lifestyle of bioenergetics characterized by electron transport to oxygen as well as a variety of alternate terminal electron acceptors, such as trimethylamine oxide (TMAO). A glycolytic pathway is also likely to be operational in these cells but seems less important than respiration. Genomic analysis shows that DHA-producing bacteria run electron transport reactions in the cold that are surprisingly similar to those of *E. coli* or mitochondria. However, an evolutionary change has occurred in which $Na^+$ ions play a more important role in energy production. An extreme case of $Na^+$ bioenergetics is discussed in Chapter 15, in which the entire electron transport chain including ATP synthase has evolved to accept $Na^+$ as primary bioenergetic cation. Note that this total makeover has not yet been observed in DHA- or EPA-producing bacteria, which have modified some, but not all, of their proton pumps to accommodate $Na^+$.

The era of bioenergetics as a trendsetter in biological thinking is long past, but major advances in bioenergetics are still being made. A case in point is the concept that uncoupling of proton and sodium electrochemical gradients is a universal property of membrane lipids themselves. These studies show that fatty acid structure is a major player not only in energy production but in energy conservation as well. Our attention was drawn to the hypothesis made by several investigators that energy conservation at the membrane level primarily concerns leakage of critical cations such as $H^+$ and $Na^+$. For example, one of the leading bioenergeticists of the twentieth century, E. Racker (reviewed by Haines; see Chapter 8 references), calculated that red blood cells spend a majority of their total energy budget on maintaining $Na^+$ homeostasis. We suggest that $Na^+$ leakage across membranes of a DHA-producing bacterium living in full seawater, along with the high energy cost of effluxing these toxic molecules, constitutes a significant fraction of the cellular energy budget in these bacteria. We further suggest that these cells have evolved membrane-based mechanisms for conserving $Na^+/H^+$ gradients. The historical background leading to the concept that membrane lipids are crucial in energy conservation was developed in Chapter 2, and further detail is provided in Chapter 8. Here we consider DHA-enriched membranes adapted to support respiration under deep-sea conditions. The membranes' contributions to bioenergetics in deep-sea bacteria are divided into two parts, energy production and energy conservation. Description of antifreeze properties of DHA chains, leading to enhanced electron transport, is mostly deferred to Chapter 7, with conservation highlighted here.

## 5.2 *MORITELLA* HAS EVOLVED POWERFUL $Na^+$ EFFLUX PUMPS

There is interesting correlative evidence emerging from genomic analysis that DHA plays an important role(s) in $Na^+$ circulation in marine bacteria. As background, the marine bacteria *Moritella* (DHA) and *Shewanella* (EPA) are closely related and are sometimes referred to as "*E. coli* of the ocean" because of their frequent isolation from seawater. As discussed in the previous chapter, there is sufficient compatibility between these marine isolates and *E. coli* to allow expression and function of DHA/EPA genes in recombinants of *E. coli*. The emphasis here shifts from DHA genes themselves to a battery of supporting genes, identified by genomic analysis, which we suggest are necessary for *Moritella* to harness the benefits of DHA *in vivo*. However, even with the coevolution of supporting genes, DHA-producing *Moritella* are still dependent on a stringent set of environmental conditions for growth. The point is that these conditions are believed to strongly influence membrane architecture. Previous researchers have defined *Moritella* as a classic example of a psychrophilic cell. As discussed below, this is only part of the story, since these cells also require unusually high levels of salinity and thus can also be considered as moderate halophiles or "osmophiles." These cells can also be classified as "barophiles" based on their deep-sea origin and the finding that environmental conditions can be manipulated such that growth becomes dependent on hydrostatic pressure. However, in spite of their extremophilic nature, *Moritella* are remarkably similar to *E. coli* at the genome level. Clues about the selective advantages and evolution of DHA membranes are emerging from analysis of the genome of *Moritella* as described next.

Genomic analysis indicates that *Moritella* has evolved $Na^+$ bioenergetics, in which chemiosmotic gradients of sodium ions (high outside/low inside) play a central role in energizing cellular functions (Figure 5.2). Normally protons are the cation of choice for cellular energy production and use as discussed already, but an increasing number of bacteria have been found to couple electron transport to efflux of $Na^+$ ions. *Moritella* has evolved mechanisms that allow these cells to capitalize on the chemiosmotic energy of surrounding seawater, specifically the energy held in $Na^+$ gradients (high outside/low inside). Recall that $Na^+$ is toxic when present

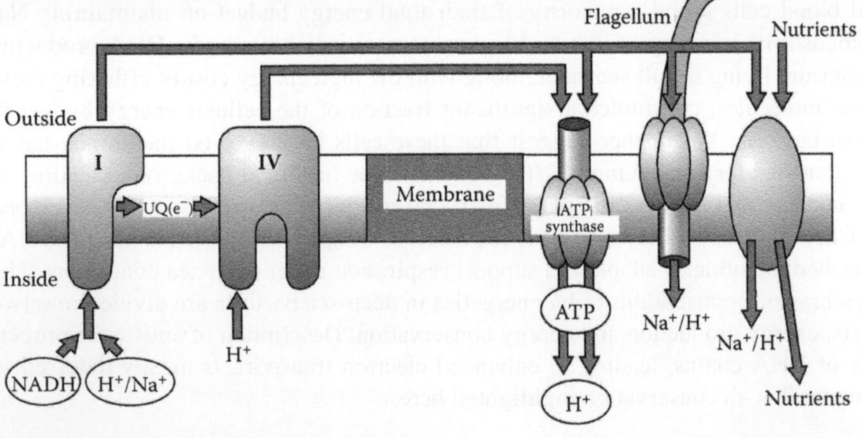


**FIGURE 5.2** Genomic analysis shows that *Moritella* has evolved a DHA membrane capable of performing specific bioenergetic roles. That is, the molecular architecture of DHA-enriched membranes is proposed to be more effective in supporting $Na^+$ bioenergetics compared to proton bioenergetics. Powerful antifreeze properties of DHA phospholipids are also believed to be harnessed for enhancing energy production in *Moritella*, living in a deep-sea environment (see Chapter 7).

above threshold levels in the cytoplasm and must be expelled/effluxed/detoxified at considerable energy cost to the cell. Thus, *Moritella* live in a toxic, saline world, but they have evolved to capitalize on the most abundant toxic cation. Genomic analysis shows that numerous kinds of marine bacteria are able to harness seawater $Na^+$ to drive numerous reactions, which, broadly speaking, can be subdivided into four types of critical $Na^+$-dependent membrane processes as follows:

- Energizing nutrient uptake enzymes, such as amino acid transporters
- Energizing efflux of toxic substances, such as antibiotics
- Fueling flagellar motors
- Driving the molecular turbine of the rotary enzyme ATP synthase for production of ATP

Note that in all four examples, the flow of $Na^+$ is inward or downhill with respect to the high levels of $Na^+$ in seawater. Since all membranes also leak $Na^+$, this means that $Na^+$ will soon build up to toxic levels unless effluxed from the cell. As already mentioned, $Na^+$ efflux, especially against a strong gradient, requires a significant energy input. It is now clear that marine bacteria have evolved $Na^+$ bioenergetics, which involves powerful $Na^+$ efflux pumps, some of which are coupled directly to catabolism. A key biochemical adaptation involves switching former $H^+$ pumps to use $Na^+$ as follows:

- Membrane-bound decarboxylases, which double as $Na^+$ efflux pumps
- $Na^+$-translocating NADH-quinone dehydrogenases (Complex I–like)
- $Na^+$-translocating cytochrome C/quinol terminal cytochrome oxidases (Complex IV–like)

*Moritella* has evolved the first two of these mechanisms, allowing these cells to harness the elevated $Na^+$ present in seawater to energize key membrane processes and then replenish these $Na^+$ gradients by using their strong $Na^+$ efflux pumps. Proton chemiosmotic gradients also remain important for energizing many critical membrane processes. Interestingly, some nutrient transport enzymes are fueled by $H^+$ specifically or $Na^+$ specifically, while others accept either cation.

However, none of the above reactions would be possible without a membrane architecture that functions as a permeability barrier for conserving both $H^+$ and $Na^+$ gradients. As already introduced in Chapter 2, it is clear that membrane lipids play a critical role as a permeability barrier in conserving cellular energy. This is where DHA comes into the picture and we hypothesize that the molecular architecture of DHA membranes under conditions of the deep sea are effective as a permeability barrier especially for conserving energy as $Na^+$ gradients. In other words, we propose that DHA membranes and $Na^+$ bioenergetics are linked in a beneficial arrangement. As discussed in detail in Chapter 8, $Na^+$ is three to four orders of magnitude less permeable than $H^+$ in spontaneously crossing the bilayer. This dramatic difference in permeability is explained by the larger size and greater mass of $Na^+$ relative to $H^+$. The larger bulk of $Na^+$ slows migration rates across the membrane, and the low mass of $H^+$ allows the bypass of select energetic barriers over short distances, through quantum tunneling. This large difference in permeability is countered by the dramatic differences in concentration, with seawater $Na^+$ at 0.5 M and $H^+$ at $\leq 10^{-7}$ M. For *Moritella* and other marine bacteria the effective shielding of $Na^+$ influx seems to serve as a selective force in evolving DHA membranes. The forging of DHA into a suitable membrane architecture as a permeability barrier against both $Na^+$ and $H^+$ might require further evolutionary changes involving specialized phospholipid structures, as discussed below and in Chapter 12. A regulatory linkage between $Na^+$ concentration and DHA levels in membranes is considered next.

## 5.3 DHA/EPA SYNTHESIS IS OSMOREGULATED

There is increasing evidence that polyunsaturated phospholipids enhance salinity tolerance in bacteria, plants, and animals. For example, increasing the levels of linoleic acid (18:2) over oleic acid (18:1) in membranes of a cyanobacterium improved salt stress tolerance. The general concept that emerges from these important studies is that polyunsaturated fatty acids might work by providing a critical lipid environment as a defense against salinity stress. These observations led us to the idea that highly unsaturated chains, such as DHA, also might be important in salinity/energy stress tolerance in *Moritella*. Note that salinity stress shares certain earmarks with energy stress. This concept is explored in this section using regulatory and genomic analysis with the goal of understanding the timing and mechanisms for modulation of DHA synthesis. Recall that regulatory analysis has been successful in revealing the purpose or need for various metabolites with unknown function. The basis for this analysis is our observation that increasing osmotic pressure, especially at near maximum growth temperature, can strongly upregulate levels of DHA/EPA incorporated into bacterial membranes (Figure 5.3).

As background, it is widely known that many marine bacteria require seawater or a dilution of seawater for growth, but the mechanism is unknown and there is little information linking this requirement to membrane lipid structure. During the course of studies of an EPA recombinant of *E. coli* we noticed that salt (i.e., NaCl) levels roughly equivalent to seawater are required both for EPA synthesis and growth. We also observed the absolute growth requirement for NaCl in an EPA recombinant of *E. coli* grown at 22°C. We noted a lack of growth in the plate of Luria broth (LB) medium in which NaCl is omitted from the recipe. We also noted that EPA recombinant cells tolerate high salt. When grown at 24°C EPA recombinant cells require significantly more NaCl. The absolute requirement for NaCl for growth is gradually relieved as temperatures drop from 24°C toward 18°C, becoming stimulatory but not essential at 18°C. This shows an interesting linkage between osmoregulation and temperature, but this point is not discussed further here. The NaCl requirement in EPA recombinant cells is a generalized osmotic effect based on the finding that sucrose and KCl replace NaCl. Later experiments showed that seawater equivalent to about 0.44 M NaCl also stimulates EPA synthesis in a marine *Shewanella* sp. (Figure 5.3B), but is essential for growth only in cold-sensitive mutants described in Chapter 7. Seawater also stimulates DHA production in the marine isolate *Moritella marina* MP-1 (Figure 5.3C). Previous workers showed that synthesis of DHA is strongly downregulated as temperatures approach 20°C. However, it was not expected that increasing the

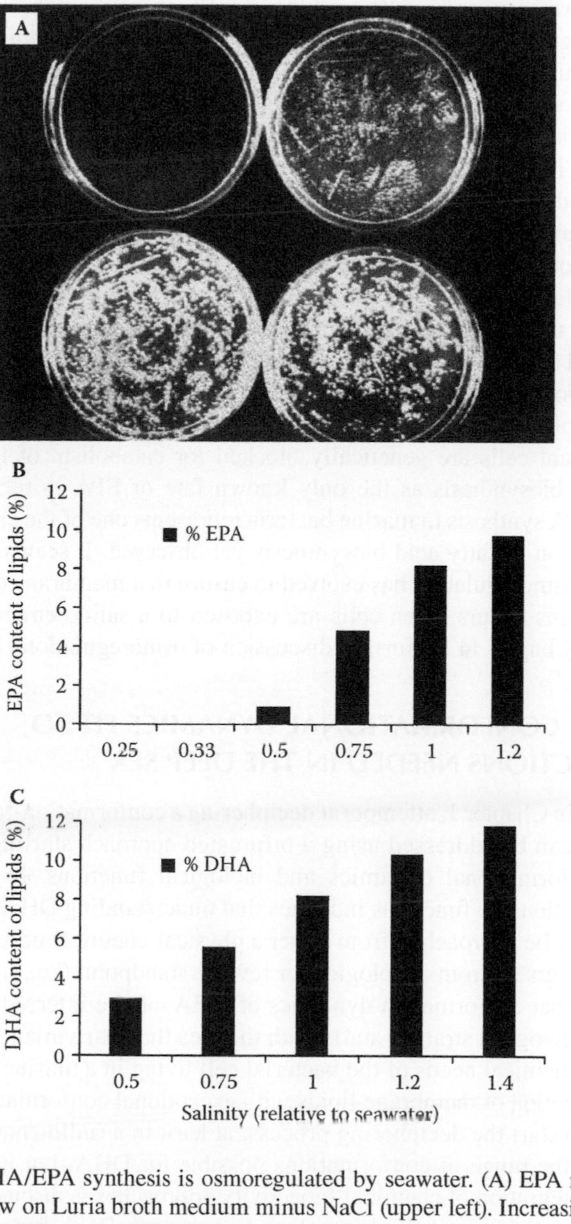


**FIGURE 5.3** DHA/EPA synthesis is osmoregulated by seawater. (A) EPA recombinant of *E. coli* fails to grow on Luria broth medium minus NaCl (upper left). Increasing NaCl levels from 0.1 M (upper right), 0.15 M (lower left), and 0.2 M (lower right) restores growth at 22°C for 4 days. This finding led us to test positive regulation of DHA/EPA in marine bacteria by seawater. (B) *Shewanella pneumatophori* (formerly *S. putrefaciens SCRC-2738*) grown on marine agar minus sea salts and supplemented with increasing levels of sea salts. Incubation at 31°C for 48 hours. (C) *Moritella marina* MP1 grown as for *Shewanella* except incubation is at 20°C for 4 days. Note that osmotic upregulation is most pronounced in cells cultured near their maximum growth temperatures. This datum is one of the clearest examples yet of a tight linkage between DHA/EPA levels and salinity.

salinity of seawater given to cells growing at 20°C would somehow overpower the strong downregulatory signals generated by increasing temperatures, and instead strongly upregulate DHA synthesis.

The current working model based on genomic analysis is that $K^+$, the major cationic constituent in microbes, plants, and animals, behaves as a "second intracellular messenger" in modulating DHA/EPA synthesis in marine bacteria. $K^+$ levels are tightly regulated in *E. coli* because of the crucial roles played by $K^+$ in osmoregulation. Genomic analysis shows that a similar system of $K^+$ osmoregulation operates in DHA/EPA-producing bacteria. We have observed that osmotic strength equivalent to seawater modulates EPA/DHA synthesis in each of three bacteria investigated so far. We previously showed that EPA synthetase supplies essentially all the unsaturated chains required for growth of an EPA recombinant of *E. coli*. This explains why NaCl is needed for growth of EPA recombinant cells with this genetic background, because in the absence of EPA alternative fluidizing chains are not available. Also, recall that EPA recombinant cells are genetically blocked for catabolism of fatty acids leaving membrane biosynthesis as the only known fate of EPA in recombinant cells. Thus, DHA/EPA synthesis in marine bacteria represents one of the clearest examples of osmoregulation of fatty acid biosynthesis yet observed. It seems reasonable that this system of osmoregulation has evolved to ensure that membrane enrichment with DHA/EPA chains occurs when cells are exposed to a saline environment such as seawater. (See Chapter 14 for further discussion of osmoregulation.)

## 5.4 DHA CONFORMATIONAL DYNAMICS FIT TO FUNCTIONS NEEDED IN THE DEEP SEA

As introduced in Chapter 1, attempts at deciphering a conformation-function relationship for DHA can be addressed using a bifurcated approach starting independently with both conformational dynamics and biological functions. In our shorthand, DHA conformations ⇔ functions indicates that understanding DHA conformational dynamics might be approached from either a physical-chemical perspective (i.e., left to right) or conversely from a biological or reverse standpoint (i.e., right to left). That is, we suggest that conformation dynamics of DHA can be inferred from biological activity. This biological strategy starts with the idea that Darwinian selection, driven by critical biochemical needs of the bacterial cell living in a marine environment, is behind the evolution of membrane lipids with exceptional conformational dynamics. It is possible to start the deciphering process, at least in a rudimentary way, without fully defining the range of conformations possible for DHA, but it helps to have a picture. The integration of chemical-bioactivity approaches is discussed next.

Using physical-chemical thinking alone to approach DHA function in deep-sea bacteria has great potential, but has been controversial. This is shown by contradictory conclusions regarding DHA conformations reported by different researchers over the past 50 years, in chronological order:

Helical Shape → Molecular Springs → Rigid Conformation → Highly Flexible, Space-Filling Conformations → → → Conformations most useful in deep-sea bacteria.

We accept that the flexible, space-filling conformations of DHA phospholipids currently emerging in physical-chemical studies are the best fit for explaining bioactivity of DHA chains in bacteria and serve as a starting point for discussion here. For example, Scott Feller, a physical chemist at Wabash College, has generated videos from chemical dynamic simulations showing the dynamic conformations of DHA phospholipids. We highly recommend that readers view Feller's video (which can be accessed at http://www.persweb.wabash.edu/facstaff/fellers/), since it better demonstrates DHA conformational dynamics than do our words here (see references in Chapter 1). In the phospholipid structure illustrated by the video a DHA chain is covalently attached to its headgroup and partnered with a saturated, straight chain of 18:0 (stearic acid), which also serves as an important frame of reference. This is a naturally occurring structure found, for example, in rhodopsin disk membranes. Studies by membrane biophysicists such as William Stillwell who specialize in the biophysics of DHA membranes have described unique properties of DHA bilayers that are consistent with chemical models (Stillwell and Wassall 2003; see also Chapter 3 references). Several aspects of these models caught our attention with an eye toward understanding the evolution of DHA membranes in deep-sea bacteria such as *Moritella*:

1. Short length—DHA displays a short overall length with respect to 18:0.
2. Compact shape—DHA shows greater compaction compared to 18:0.
3. Curvature—DHA shows marked curvature at the methyl end.
4. Flexibility—DHA displays great flexibility compared to 18:0.
5. Partner interaction—Apparent interactions with its partner 18:0 are strong enough to alter the conformation of these normally relatively straight chains.
6. Flip-flop—DHA is suspected to undergo relatively rapid flip-flop in bilayers (reported by Stillwell and colleagues).

A biological interpretation for these six aspects of DHA conformational dynamics (see Chapter 2 references) is integrated into the lifestyle of *Moritella* as follows:

1. Short length—This aspect of DHA suggests thin membranes in *Moritella*. Note that 18:0 is virtually absent in *Moritella* membranes being replaced by chains significantly shorter in length (i.e., chains ranging from 13:0 to 16:0 in length). This raises the possibility that DHA chains partnering with shorter chains in bacterial phospholipids are even more compacted than shown in the DHA-stearic acid model.
2. Compact shape—This aspect of DHA lipids implicates bulkiness as a defense against $Na^+$ leakiness. (See 3 below.)
3. Curvature—This aspect of DHA lipids acts to concentrate mass at the methyl end and adds bulkiness where most needed for preventing ion transport, and is antibonding, important in maintaining motion. This observation leads us to predict that DHA is more effective than EPA at extreme depths.
4. Flexibility—This tendency of DHA suggests many favorable as well as unfavorable conformations possibly modulated by heat, cold, pressure, and

ionic environment. Helps illuminate the need for a rigid set of environmental conditions to maintain physiological activity of DHA membranes.

5. Partner interaction—DHA interacts with other lipid molecules in such a way as to destabilize weak hydrophobic bonding interactions and thus break apart lipid rafts caused by cold and high pressure. Prediction: Small amounts of DHA phospholipids might work "catalytically" to break up gel-phase lipids in *Moritella.* Note that DHA/EPA are often present in relatively small levels in bacterial membranes, consistent with the idea that a few molecules go a long way.
6. Flip-flop—This property of DHA adds to the growing list of clues that DHA membranes are leaky for protons and must be carefully managed in proton energy transducing membranes.

Point 6 deserves further discussion because of implications for better understanding of why *Moritella* has switched to a $Na^+$ economy. Excess lipid flip-flop in a membrane is a toxic event since it serves to transport protons across the membrane decoupled from energy conservation—a form of futile cycling. Reducing the rate of flip-flop might help to explain why *Moritella* has evolved a regulatory system in which DHA phospholipid levels in membranes rise in lock step with NaCl levels. Some background on DHA flip-flop is first required.

First of all, rates of flip-flop of di-DHA phospholipids were found to be about 116 times greater than control membranes composed of phospholipids, similar to species found in bacteria such as *E. coli.* Many di-acyl phospholipids in *E. coli* contain a straight saturated chain such as 16:0 or 18:0 paired with a monounsaturated chain such as 16:1 or 18:1. As discussed in Chapter 9, *E. coli* does not produce DHA and might even be inhibited by DHA. We have estimated, based on reported rates of di-DHA flip-flop and a total number of 6 million phospholipid molecules forming the outer leaflet of the cytoplasmic membrane of *E. coli*, that about 6000 protons/second would cross the bilayer of a DHA recombinant of *E. coli* through a flip-flop mechanism. This compares to about 60 protons/second in normal cells. It seems unlikely that this level of energy uncoupling would fully deenergize an *E. coli* cell, at least under standard laboratory culture conditions. However, a flip-flop mechanism might affect *Moritella* or other cells under conditions of energy stress. There is no evidence that *Moritella* synthesizes di-DHA, but based on gross fatty acid compositional data DHA likely pairs with chains as much as four C-C bonds shorter than used in the DHA membrane model described above. This means that flip-flop rates might be relatively fast in such thin membranes. Alternatively, the finding of cardiolipin (CL) synthetase in the genome along with some other clues from the literature raises the possibility that these cells might produce DHA-cardiolipin. For example, significant amounts of EPA have been found to be incorporated into CL of three different genera of EPA-producing bacteria. Interestingly, CL species with up to three out of four unsaturated chains including EPA were detected. Also EPA is found to pair with 16:1 or 18:1 in diacylglycerides, yielding some of the most highly unsaturated phospholipids yet detected in bacteria. As described in Chapter 12, DHA-cardiolipin with four DHA tails has been reported in marine animals. Recall that cardiolipin is formed by joining two di-acyl phospholipids together yielding a single joined headgroup with a total of four acyl chains. In animals DHA chains enter DHA-cardiolipin via a remodeling pathway. The presence of a similar pathway in *Moritella* is an interesting possibility with an as-yet undetermined impact on flip-flop rates and other membrane properties.

The main point arising from this consideration of DHA flip-flop is that conventional DHA phospholipids may be inherently leaky for protons and exhibit high flip-flop rates. This might represent a weak link for a bioenergetic system in *Moritella* based solely on proton circulation and would serve as a selective pressure toward $Na^+$ bioenergetics. The last and perhaps most interesting comment deals with a possible mechanism of energy conservation in which $Na^+$ levels and DHA function are directly linked. According to this model, high levels of $Na^+$ ions in seawater compete with protons for negative charges on DHA phospholipids, with their bulkier groups hindering flip-flop. Alternatively, greater osmolality could force tighter electrostatic interactions between phosphate headgroups and the adjacent aqueous phase, thus increasing the energetic barrier for flip-flop. In either case, this model correctly predicts that *Moritella* might closely coordinate synthesis of DHA phospholipids and $Na^+$ concentration—a direct relationship as observed.

## 5.5 SUMMARY

Approximately 70% of total fatty acid chains in membranes of *Moritella* are unsaturated, including about 60% monounsaturates and 10 to 17% DHA. Monounsaturated chains include significant amounts of 14:1 expected to increase membrane fluidity. Overall, the average number of double bonds per fatty acid chain in the bilayer of *Moritella* is about 1.2, a number more than double that of *E. coli*. *Moritella* membranes are also unusual in having only about 1% *cis*-vaccenate ($18{:}1_{n7}$) which eliminates the possibility of forming di-$18{:}1_{n7}$ phospholipids such as seen in cold-adapted cells of *E. coli*. Also, *Moritella* does not have genes for forming cholesterol-like molecules commonly found in bacteria that produce membranes with comparable unsaturation indexes. About 28% of the total chains in *Moritella* membranes are straight, saturated, and significantly shorter (i.e., 14:0 and 16:0) compared to 16:0/18:0, which partners with DHA in phospholipids of rhodopsin disk membranes. Cyclopropane fatty acids are not present, nor are branched-chain fatty acids. Note that branched fatty acids are found in high levels in membranes of EPA-producing *Shewanella*. Based on the above fatty acid composition, a number of properties concerning the molecular architecture of *Moritella* membranes are illuminated. For example, from bulk composition alone it is clear that the strong bias in unsaturated chains increases the probability that a significant fraction of the membrane surface is composed of di-unsaturated phospholipids (e.g., a monounsaturated chain paired with DHA). Such structures are considered (in Chapter 7) as being powerful membrane antifreeze because of their low phase transition temperatures. Note that pairing a DHA chain with a monounsaturated partner raises the maximum number of double bonds per phospholipid to as high as seven in *Moritella* compared to one to two seen in *E. coli*. Thus, *Moritella* is predicted to synthesize DHA and other phospholipids as powerful membrane antifreeze. Experiments described in Chapter 7 confirm this idea.

Finally, genomic analysis involving several hundred different bacteria living in a variety of niches shows that DHA-enriched membranes are both rare and highly specialized, being largely confined to the marine genus *Moritella*. DHA-enriched membranes of *Moritella* have evolved to function at temperatures approaching 0°C in seawater and often under extremes of hydrostatic pressure. A close linkage has now been established between seawater and DHA levels. A mechanism of osmoregulation of DHA synthesis is proposed. *Moritella* has evolved a system of $Na^+$ bioenergetics focusing attention on DHA membranes as a key player in $Na^+$ circulation (i.e., generation and conservation of chemiosmotic gradients of $Na^+$). We also hypothesize that the molecular architecture of DHA membranes of *Moritella* enhances electron transport in the extreme cold. Lessons from physical-chemical studies of DHA-containing phospholipids provide interesting clues linking DHA conformational dynamics to functions important in bacteria. The present chemical model is expected to apply to DHA-cardiolipin if and when it is found in bacteria. This information provides a springboard for interpreting bioactivity of DHA across all life forms such as neurons, as discussed in the next chapter.

## SELECTED BIBLIOGRAPHY

Allakhverdiev, S. I., Y. Nishiyama, I. Suzuki, Y. Tasaka, and N. Murata. 1999. Genetic engineering of the unsaturation of fatty acids in membrane lipids alters tolerance of *Synechocystis* to salt stress. *Proc. Natl. Acad. Sci. U.S.A.* 96:5862–5867.

Epstein, W. 1986. Osmoregulation by potassium transport in *Escherichia coli. FEMS Microbiol. Rev.* 39:73–78.

Freese, E., H. Rütters, J. Köster, J. Rullkötter, and H. Sass. 2009. Gammaproteobacteria as a possible source of eicosapentaenoic acid in anoxic intertidal sediments. *Microb. Ecol.* 57:444–454.

Kogure, K. 1998. Bioenergetics of marine bacteria. *Curr. Opin. Biotechnol.* 9:278–282.

Methé, B.A., K. E. Nelson, J. W. Deming, et al. 2005. The psychrophilic lifestyle as revealed by the genome sequence of *Colwellia psychrerythraea* 34H through genomic and proteomic analyses. *Proc. Natl. Acad. Sci. U.S A.* 102:10913–10918.

Stillwell, W., and S. K. Wassall. 2003. Docosahexaenoic acid: Membrane properties of a unique fatty acid. *Chem. Phys. Lipids* 126:1–27.

Takami, H., A. Inoue, F. Fuji, and K. Horikoshi. 1997. Microbial flora in the deepest sea mud of the Mariana Trench. *FEMS Microbiol. Lett.* 152:279–285.

Tokuda, H., and K. Kogure. 1989. Generalized distribution and common properties of $Na^+$-dependent NADH: Quinone oxidoreductases in Gram-negative marine bacteria. *J. Gen. Microbiol.* 135:703–709.

Vezzi, A., S. Campanaro, M. D'Angelo, et al. 2005. Life at depth: *Photobacterium profundum* genome sequence and expression analysis. *Science* 307:1459–1461.

Wood, J. M. 1999. Osmosensing by bacteria: Signals and membrane-based sensors. *Microbiol. Mol. Biol. Rev.* 63:230–262.

Yayanos, A. A. 1995. Microbology to 10,500 meters in the deep sea. *Annu. Rev. Microbiol.* 49:777–805.

Yayanos, A. A., A. S. Deitz, and R.V. Boxtel. 1982. Dependence of reproduction rate on pressure as a hallmark of deep-sea bacteria. *Appl. Environ. Microbiol.* 44:1356–1361.

Yayanos, A. A., S. Ferriera, J. Johnson, S. Kravitz, A. Halpern, K. Remington, K. Beeson, B. Than, Y.-H. Rogers, R. Freidman, and J. C. Venter. *Moritella* sp. PE36. Submitted June 2007 to the EMBL/GenBank/DDBJ databases.

Zerouga, M., L. J. Jenski, and W. Stillwell. 1995. Comparison of phosphatidylcholines containing one or two docosahexaenoic acyl chains on properties of phospholipid monolayers and bilayers. *Biochim. Biophys. Acta* 1236: 266–272.

# 6 Evolution of DHA Membranes in Human Neurons

Hypothesis: The dynamic conformations of DHA are harnessed in membranes of neurons to conserve energy and disrupt lipid rafts, in essence maximizing the efficiency of excitatory biochemistry.

From an evolutionary perspective it is clear that neurons are a specialized class of essential cells that evolved to support animal life in contrast to plant life. This traces the evolution of neurons back to the dawn of animals and long after $O_2$-evolving plants had evolved. Nevertheless, gauging from their membrane lipids, neurons tend to be more plantlike in composition. That is, neurons of most animals, including mammals, birds, reptiles, and even a lowly worm such as *C. elegans,* are often highly enriched with DHA/EPA. However, there are important classes of animals such as insects and even flatworm relatives of *C. elegans* that lack DHA/EPA altogether. This leads us to conclude that neuronal membranes are subject to Darwinian selection of the fittest membrane composition relevant to the neurosensory needs of different groups of animals. It is also concluded that DHA/EPA are not fundamental for neurosensory function across all animals.

Having made the case that DHA/EPA are not absolutely required for neurosensory function, it is interesting to look more broadly at environmental or ecological parameters that might have led to lipid compositions characteristic of human neurons. We begin with a brief discussion of neurons restricted to function over the temperature range –2°C to 2°C. Recall that neurons have evolved to function over an ~50°C range of temperature (i.e. –2°C to ~50°C) in different animals with ultra-cold-adapted neurons being represented by one of the most abundant populations in the biosphere—marine krill. These shrimplike animals live their entire lives at temperatures around 0°C and depend on cold-adapted neurons for food gathering, egg laying, reproduction, motion, and other life functions. Krill membranes in general contain among the highest levels of DHA/EPA found in nature, which is not surprising given their cold environment and diet of DHA/EPA-rich phytoplankton. Little is known about the composition of neuronal membranes in krill. However, researchers have succeeded in dissecting and analyzing the large neuron of lobster tail muscle, which is enriched with EPA and DHA. Thus, we are faced with the quandary of explaining why neurons separated by up to a 40°C difference in temperature (i.e., -2°C to 37°C) require DHA/EPA.

In contrast to lobsters, membranes of mammalian neurons are enriched with DHA over EPA. Indeed the bulk of the DHA in humans is found in the plasma membranes and synaptosomes of neurons forming the neurosensory system. In addition to having DHA phospholipids as structural lipids, mammalian neurons are distinguished in terms of their often dramatically elongated and multibranched morphology, and also their unique biochemistry. The biochemistry of neurons features extensive energy expenditure to drive specialized electrophysiological processes. That is, neurons are among the most energy intensive cells in the body based on the following information (Figure 6.1):

- The brain as a 3-pound machine consumes about 20% of total energy supply.

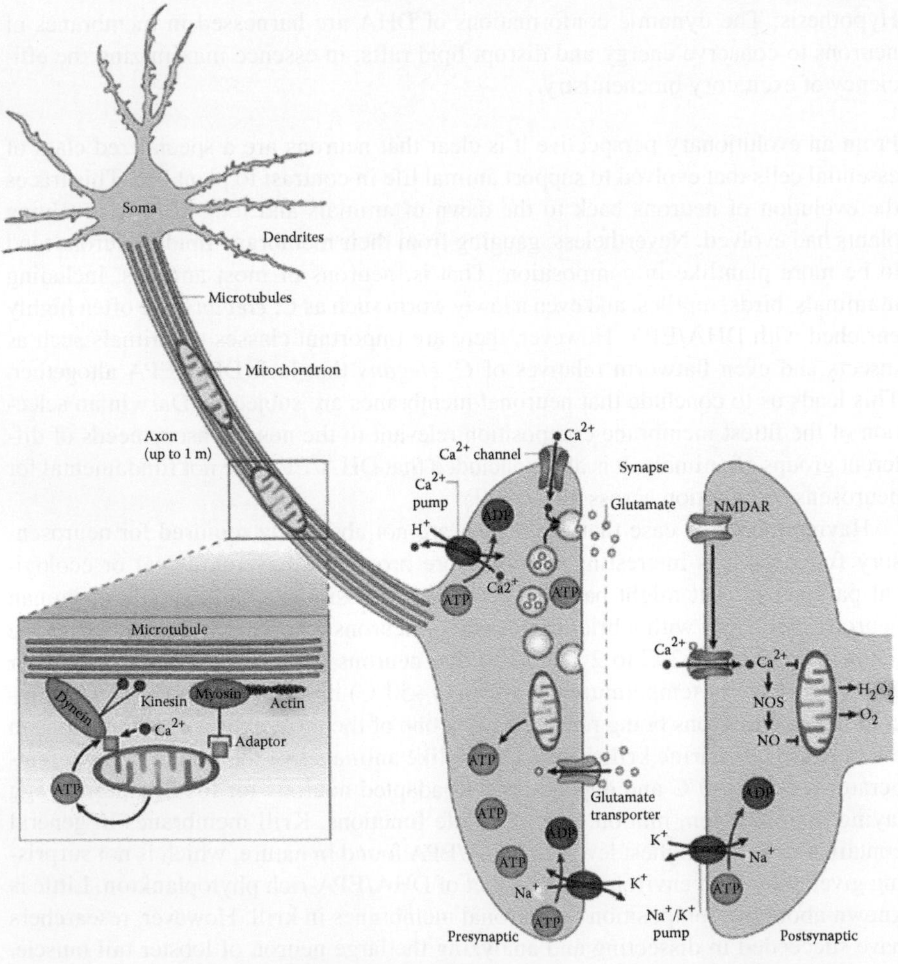


**FIGURE 6.1** Neurons require a high and consistent input of energy to function. Several well-known ATP-requiring steps important in functions of neurons are shown. Other important energy-dependent processes not shown include repairing DHA membranes, synthesis of new mitochondria, protein turnover, etc. Thus neurons are defined as being among the most energy-consuming cells in the body, a point discussed in Chapters 19 and 20. (Reproduced from Knott, A. B. et al., *Nature Rev. Neurosci.*, 9:505–518, 2008. With permission.)

- A significant proportion of the total energy budget of neurons is likely used to energize cation pumps and maintain $Na^+/K^+$ homeostasis.
- Numerous processes essential for excitatory biochemistry including membrane repair and long distance transport of metabolites are energy intensive.
- Neurons in specialized regions of the brain seem to be operating near their maximum energy-producing capacity (see Chapters 19 and 20).

Thus, evolution of DHA-based mammalian neurons might be defined at least in part based on a bioenergetic system functioning near full capacity, essentially pushed to maximally energize excitatory biochemistry. In essence, neurons are forced to run at a state of chronic energy stress. Preventing the loss of energy is equivalent to making more energy, and we suggest that DHA membranes are harnessed for this role. However, energy conservation is likely only part of the story. For example, DHA membranes also function as a physical matrix for housing numerous critical excitatory proteins and enzymes, many of whose mode of action is thought to depend on motion. As already introduced in Chapter 2, motion and cation fidelity are proposed to be inversely related, suggesting these membranes have evolved a composition to both maximize motion and manage strong $K^+/Na^+$ gradients. According to this scenario, if membrane motion is also a measure of neuronal efficiency, as we suggest, then the need for motion comes with additional energy cost. Once again energy seems to serve as a selective pressure in neuronal membranes, which is another way of saying that neuronal efficiency is intimately linked to cellular energetics, which in turn is linked to evolution of DHA-based membranes.

## 6.1 DHA MAY REDUCE $Na^+$ LEAKAGE INTO NEURONS

The high abundance of DHA in phospholipids of neuronal membranes raises questions of what constitutes an acceptable level of cation permeability, and how much interplay exists between $H^+/K^+/Na^+$. The contribution of the plasma membrane to the permeability barrier against high $Na^+$ outside and high $K^+$ inside is shown in Figure 6.2. A porous membrane would permit uncoupling of these gradients and dissipate cellular energy through futile cycling. As with marine bacteria, we suggest that a selective pressure for minimizing cellular energy loss effectively increases cellular energy, thereby providing a competitive advantage. This concept of DHA in bioenergetics, introduced in the previous chapter, is a recurring theme that seemingly applies to neurons.

The specialized biochemistry carried out by neurons is energy intensive, and we argue that a reduction in the rates of $K^+/Na^+$ transmembrane leakage provides enormous benefit for these energy-stressed cells. Several energy-intensive processes plague these cells, including neurotransmitter pumping, synaptic biochemistry in general, repair processes from growth of new mitochondria, membrane recycling, and energized transport of cellular components traveling the long distances common in neurons. Excessive protein misfolding/damage can also create a major energy sink, and energy wasted through ATP-dependent processing of misfolded proteins can further tap the cellular energy balance. We argue that energy loss from futile ion cycling is an equal or greater sink, but one that can be modulated by membrane lipid

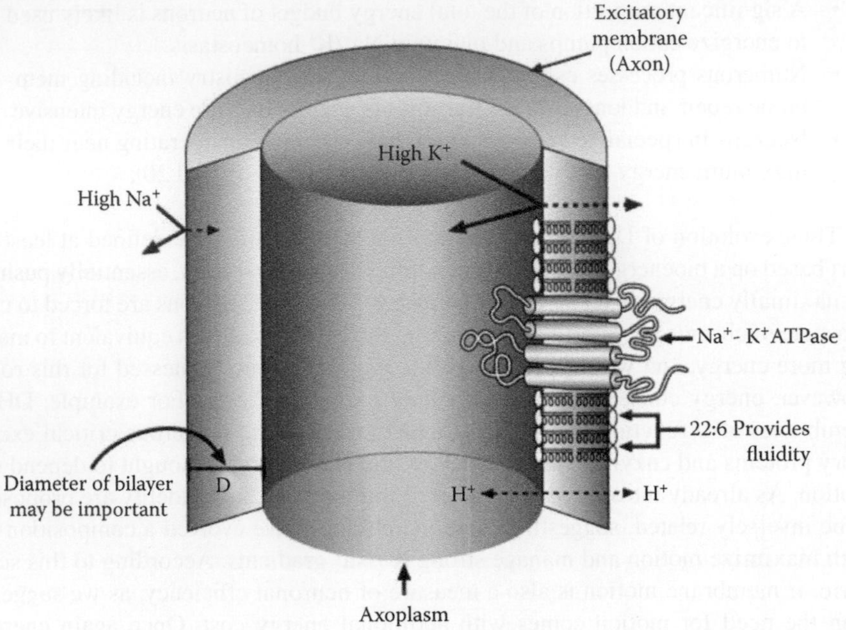


**FIGURE 6.2** Artist's sketch of a DHA-enriched membrane of a sensory axon. DHA is proposed to contribute to energy conservation in neurons by acting as a satisfactory permeability barrier against futile cycling of excitatory cations (i.e., $Na^+/K^+$) while being a mediocre barrier for $H^+$ (see Chapter 8). DHA phospholipids are also proposed to work by disrupting lipid rafts, thus maintaining membrane lateral motion believed to be important in neurosensory perception (see Chapter 7).

composition. The root of salinity stress in neurons comes from human blood serum, which carries the "sea within," or ~0.15 M $Na^+$. That is, cells are bathed by high $Na^+$ outside, balanced by high $K^+$ inside. Neurons must continuously expend energy to maintain this "cation homeostasis." Studies with membrane vesicles have shown that leak rates of protons ($H^+$) are up to $10^4$-fold faster than $Na^+$, but a simple calculation shows that high outside $Na^+$ levels, such as found in serum (0.15 M $Na^+$ vs. $10^{-7.4}$ M $H^+$), are predicted to create a perpetual state of $Na^+$ stress/energy stress across neuronal membranes. Energy-dependent transporters are essential for detoxifying the cytoplasm, generating $Na^+$ gradients for nutrient uptake, and maintaining water balance. Thus, futile cycling of $Na^+$ across the expansive surface area of neural axonal membranes is predicted to consume an unexpectedly large amount of the cellular energy budget. The high energy cost for driving neuronal function may be behind the need for nurse cells (i.e., glial cells) responsible for mediating and aiding neuronal functions such as helping provide a continuous stream of energy-rich substrates to maximize mitochondrial energy output. Once again we argue that the evolution of DHA membranes is tied to salinity and a balance of lateral mobility versus energy conservation.

Neurons synthesizing DHA-enriched bilayers face a delicate balancing act between rapid lateral motion and ion permeability. There appears to be no perfect solution and some energy loss through futile cation cycling is likely to always occur (see Chapter 8). Chemical oxidation of DHA also seems inevitable in aerobic cells at warm temperatures (Chapter 9), and oxidative stability must be further considered in this balancing act. According to this tripartite concept, neurons face the need to optimize membrane motion to suit biochemical needs, minimize cation permeability to prevent excessive energy loss, and prevent oxidative damage of their highly reactive membranes. We again draw the analogy to deep-sea bacteria that face the same problems (Chapter 5) but that have come to depend on stringent environmental conditions found only in the deep sea to maintain the efficacy of their DHA membranes. Clearly there are differences too, though we argue the parallels are worth considering. For example, the function of DHA-enriched membranes of neurons might have also evolved to be dependent on the immediate external as well as internal environments of these cells. For example, it is predicted that plasma membranes of neurites (i.e., axons and dendrites) will turn out to be leaky for protons. It is further predicted that this weakness is overcome in part because of the action of generalized pH homeostasis maintained in brain fluid, which essentially sets the internal/external pH gradient of neurons to near zero. The idea is not that pH homeostasis in the cytoplasm of neurons is not important, but rather that this need is met in part by a preexisting generalized mechanism of pH modulation at the whole organism level—effectively communalizing this energetic cost and partially removing this energy-intensive function from these specialized cells. According to this idea the availability of an overarching mechanism for maintaining cytoplasmic pH helped set the stage for evolution of DHA-enriched bilayers as permeability barriers needed primarily to protect circulation of $Na^+/K^+$ without the selective pressure of simultaneously maintaining proton gradients. The inverse might be true of mitochondria of neurons whose membranes must instead protect proton gradients. This idea leads to the prediction that DHA will be targeted away from inner membranes of mitochondria where protons, in contrast to $Na^+/K^+$, set the standard of permeability. These scenarios provide a biochemical framework for understanding the evolution of specialized DHA-enriched membranes in neurons.

## 6.2 DO DHA PLASMALOGENS SHIELD CATIONS?

Cells have evolved many mechanisms for enhancing cation fidelity and overall efficiency of their DHA membranes. These include modifying the linkages between acyl chains with their headgroups. For example, DHA plasmalogens are present in high levels in neuronal membranes and often consist of a DHA chain in conventional ester linkage paired with a saturated chain in ether linkage (i.e., about 50% of ethanolamine glycophospholipids in an adult human brain). It has been proposed that plasmalogens contribute important biophysical/biochemical properties to neural bilayers. In particular, the perpendicular orientation of the *sn*-2 acyl chain often occupied by DHA, arachidonic acid (20:4), or linoleic acid (18:2) at the neural membrane surface and the lack of a carbonyl group at the *sn*-1 position is proposed to affect the hydrophilicity of the headgroup. This results in the stronger intermolecular hydrogen bonding between headgroups, in essence drawing neighbors closer together. Thus, the ether linkage may affect lipid packing and consequently reduce permeability against $Na^+/K^+$. Plasmalogens have previously been assigned a variety of possible functions in neural membranes, ranging from antioxidants to precursors for eicosanoid hormones, but we are not aware that a fundamental role in energy conservation has been considered. The main reason for thinking about plasmalogens in an energy conservation role is based on studies that show that conventional (i.e., ester-linked) DHA phospholipids might be a mediocre permeability barrier against cations, as described in Chapter 8. Thus, DHA plasmalogen structure is proposed to serve as a second line of defense against futile dissipation of $Na^+/K^+$ gradients, again an energy-conserving mechanism. Similar roles for plasmalogens in plasma membranes of cardiac and sperm cells are discussed further in Chapter 13.

## 6.3 IMPORTANCE OF MOTION

The concept that DHA functions as an important fluidizing molecule in membranes of neurons is unsettled but gaining momentum. Attention here shifts to the roles of DHA chains as powerful antirafting molecules modulating the physical state of neuronal membranes, though admittedly the critical functions benefitting most from this action are not identified. Experimental evidence supports the view that the DHA membranes confer extraordinary lateral motion to excitatory proteins such as $Na^+/K^+$ ATPase. We suggest a simple relationship by which the combination of thermal energy and DHA membranes yields extraordinary membrane motion, which in turn acts to disrupt lipid raft formation. Notice that proteins are left out of this simplified relationship.

We envision that rapid lateral motion is needed for neuronal function in that membrane biochemistry in neurons is dependent on it; we further suggest that the extent of motion may be proportional to the signaling capability of the neuron. Neuronal systems that potentially benefit from rapid lateral motion include collision-dependent receptors or pumps, enzymes whose mode of action is inhibited in lipid rafts, synaptosome cycling, and so forth. This concept stems from classical experiments in which rhodopsin is replaced with other proteins of different sizes and shapes. Results demonstrate that it is the lipid matrix, not the protein, that is responsible for motion. It is important to note that this concept is not in conflict with the idea that DHA might play other more specific biochemical roles in modulating the mode of action of proteins working in neuronal membranes (Figure 6.3). Thus, the main hypothesis featured here is that the extraordinary mobility of DHA lipids, conferred by their structure and conformational dynamics, helps maximize lateral motion of critical neuronal proteins. Many of these proteins are intimately bound to membranes and thus depend on the properties of DHA membranes for motion. The importance of lateral motion for several hundred different olfactory receptors in olfactory neurons is discussed in Chapter 13.

## 6.4 CASE HISTORIES

Understanding the molecular biology of omega-3 fatty acids in neurons is most advanced in the case of *C. elegans* (Figure 6.4B). The basic assay used to measure neuronal efficiency in *C. elegans* involves a motility test in which rates of body bends created by the major muscle group are measured (Figure 6.4A). The rates of body bends are significantly slowed in EPA minus mutants from about three per

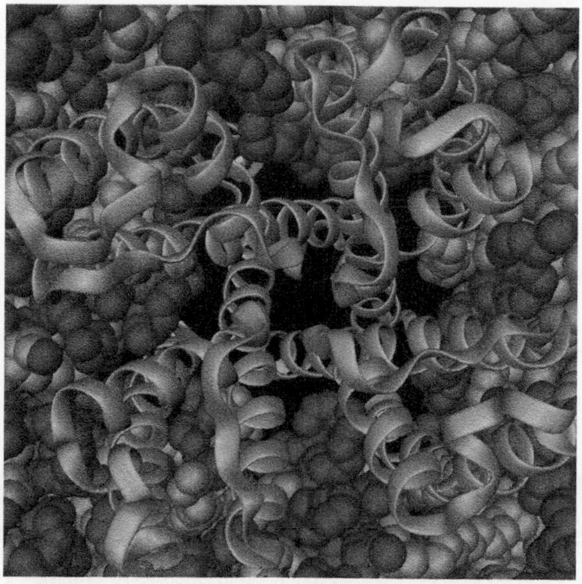

**FIGURE 6.3** (A color version of this figure follows p. 140.) Model of a $K^+$ channel (KcsA) in yellow shows a close association with DHA phospholipids (red). Antirafting properties of DHA might help maintain neuronal membrane proteins in motion, such as already described above for the G-protein–mediated receptors. Alternatively, a more direct role in enzyme catalysis can be envisioned. (Courtesy of Igor Vorobyov, Scott Feller, and Toby W. Allen, unpublished data.)

second in wild type worms to about one per second in EPA mutants. Both EPA and DHA provided exogenously restore function to these "fast muscles," but recall that this worm neither produces DHA nor feeds on DHA-producing bacteria in nature. The defect caused by EPA deficiency has been traced to the level of the synaptosomal cycle where a shortage of synaptosomes seems to slow synaptic transmissions to this rapid-fire muscle group. The biochemical roles of EPA/DHA in modulating the synaptosomal cycle remain an interesting topic for future research. Here we consider the possible biochemical roles of EPA in nematodes.

As discussed in Chapter 7, *C. elegans* synthesizes di-EPA, which has a melting point around negative 60°C, perhaps the lowest in nature. As already discussed, the precise biochemical role of EPA in synaptosomal cycling remains unknown, but reasoning by analogy leads us to suspect that rapid lateral motion might be harnessed for enhancing collision rates among sets of synaptosomal proteins—perhaps rate-limiting in recycling/function of synaptosomes. Also, the likely compact conformations of EPA chains, perhaps in the form of plasmalogens, might play an additional role as a permeability barrier or defense against $H^+$ leakage from synaptosomes. Recall that proton gradients are essential to energizing uptake of neurotransmitters in synaptosomes and that preloading with protons occurs by reversal of the ATP synthase reaction. Thus, benefits of DHA/EPA inclusion in synaptosomal membranes might be explained by mechanisms similar to bacteria where these dynamic chains are

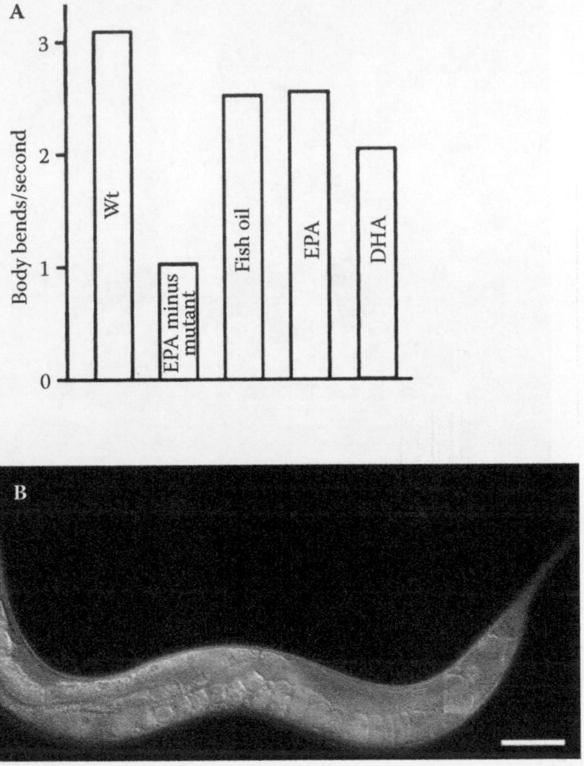


**FIGURE 6.4** Both EPA and DHA function in membranes of *C. elegans,* but only EPA is available from de novo biosynthesis. (A) Mutational analysis shows that fish oil containing both DHA and EPA stimulates neuromuscular function of EPA-minus mutants of *C. elegans* from about one body bend per second to about three. Purified samples of EPA and DHA have similar effects. Researchers previously showed that synaptosomal cycling is defective in EPA minus mutants. (B) *C. elegans* is a voracious feeder using powerful body-bending muscles for propulsion. The scale bar in the photo is 100 microns. [(A) Courtesy of Luke Hillyard and Bruce German. (B) Photo courtesy of Dr. Ian Chin-Sang, Queen's University.]

proposed to play dual roles relating to permeability and motion. Recall that certain species of nematodes, including parasitic forms living at mammalian body temperatures, do not produce either DHA or EPA and obviously do not depend on or require these chains for their neuronal membranes. This is consistent with mutant analysis, which shows that the EPA chains in *C. elegans* membranes improve efficiency of neurons but are not essential for growth, reproduction, and survival (under laboratory conditions). Again DHA/EPA provide direct benefits, improving the efficacy of synaptosomal membranes in *C. elegans.*

Thus, there are convincing clues based on data scattered in the literature that evolution of "efficiency" in neurons is closely linked to unsaturation levels of membranes surrounding these essential cells. According to these data, neurons can be divided into two broad thermal classes, ectothermic (i.e., cold-adapted) animals and endothermic animals, including humans. For example, cold-adapted membranes of

*C. elegans* (i.e., grown at 15°C) are composed of greater than 30% EPA as total fatty acids of which di-EPA (i.e., ten double bonds) makes up a significant fraction of phospholipids. Indeed, phospholipids representing essentially the complete spectrum of molecular species ranging from one to ten double bonds are present. The presence of these diverse classes with a maximum of ten double bonds has not been directly determined by analysis of neural membranes but can be inferred from bulk composition data. Recall that *C. elegans* mutationally deprived of EPA still retains the necessary biochemical pathway to form di-18:3 species of phospholipid with a maximum of six double bonds per phospholipid. This compares with a maximum of ten double bonds in di-EPA found in wild-type *C. elegans* grown under cool conditions. The levels of di-EPA species is downregulated while worms adapt to warmer temperature where EPA paired with saturated chains in phospholipids may predominate. Parasitic nematodes without EPA have the capacity of synthesizing di-18:3 with six double bonds. Insects, such as bees and fruit flies, also lack DHA and EPA for building neuronal membranes and might depend on a similar mechanism to generate low melting point phospholipids. Neurons functioning in animals living at temperatures as low as 0°C are predicted to require phospholipids of extreme fluidity, such as 18:1/DHA, discussed next.

Hungarian scientists (Buda et al., 1994; Farkas et al., 2000) have found that the DHA content of brain cells *in vivo* and *in vitro* of fishes is similar to that of human brains. DHA and arachidonic acid are predominant over EPA, which is present at much lower levels. These researchers have made the important observation that during adaptation to low temperature (i.e., 25°C to 5°C) brain cells of freshwater fish (e.g., carp) as well as whole animals synthesize about 12% more of the di-unsaturated molecular species, bound to phosphatidylethanolamine. Levels of 18:1/22:6, 18:1/20:4 and 18:1/18:1 rose two- to threefold at 5°C, whereas a major drop (i.e., almost 17%) in levels of 18:0/22:6 was shown. These data show that the gross amounts of DHA in brain cells/brains of cold-adapted freshwater fish does not change appreciably during temperature shifts, but rather that modulation of di-unsaturated molecular species such as 18:1/22:6 lies at the heart of how brain cells restructure their membranes in the cold. From the perspective of bulk fluidity it seems difficult to envision how such relatively modest changes in the level of say 18:1/22:6 versus 18:0/22:6 molecular species of phospholipids might impact physical properties of membranes of neurons. However, if it is accepted that molecular species such as 18:1/22:6 are far more powerful membrane icebreakers compared to 18:0/22:6, then a new picture emerges regarding the mechanism of disruption of lipid rafts or islands of gel-phase lipids, as described in Chapter 7.

A third case history involves synaptic vesicles of mammals in which DHA is enriched in synaptosomal membranes (Figure 6.5). These tiny organelles, which carry neurotransmitters essential for biochemistry for synaptic transmission, have evolved membranes that are literally packed with about 50 different integral membrane proteins, of which about 27 are major components. It has been calculated that these proteins are embedded in a lipid bilayer comprised of ~7000 phospholipid molecules (i.e., 14,000 acyl chains), of which ~2000 are estimated to be DHA chains. This membrane is also relatively highly enriched with cholesterol (i.e., about 5700 molecules). Uptake of neurotransmitters such as glutamate into synaptosomes is

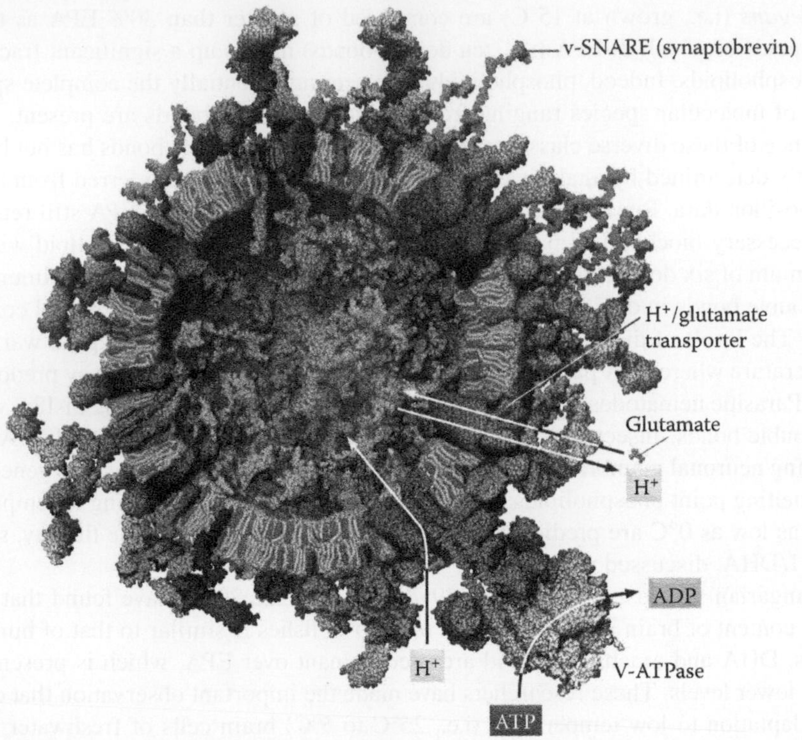


**FIGURE 6.5** Molecular anatomy of a critical DHA-enriched neurosensory membrane: the synaptosome or synaptic vesicle. In this now classic study, experts in this field pooled their expertise to create the clearest picture yet of the inner workings of a neurosensory membrane. In contemplating this model it is easy to imagine that at least some of these densely packed proteins depend on DHA for motion. Alternatively, tethering in lipid rafts might be advantageous or required by other synaptosomal proteins for purposes of docking, etc. Also, DHA might favor lipid structures most advantageous for facilitating rapid rates of synaptosomal cycling. (Reproduced from S. Takamori et al., *Cell,* 127:831–846, 2006. With permission.)

known to be dependent on energy supplied by proton electrochemical gradients, high inside and low outside. Proton gradients are generated by ATP-dependent reversal of the ATP synthase reaction. The structure of this membrane poses apparent contradictions to the DHA principle. DHA is highly fluidizing and is proposed to disrupt lipid rafts, while also being leaky for protons. In contrast, cholesterol is thought to enhance the permeability barrier against protons, but to also favor lipid raft formation. Why both components are present at such elevated levels is not clear, but it is interesting to speculate that an important role of DHA is to maintain the motion of at least some classes of synaptosomal proteins, either through a general impact on the membrane or as annular lipids specifically associated with membrane proteins. However, an equally convincing case can be made for the need to immobilize at least some synaptosomal proteins in lipid rafts, present in part due to the elevated cholesterol content of the membrane. Might a membrane mosaic model apply to these tiny vesicles? As already discussed in the case of *C. elegans* there is also the possibility that DHA phospholipids play an important but unknown structural role in facilitating rapid rates of synaptosomal cycling, in essence overcoming a potential rate-limiting step in synaptic biochemistry.

## 6.5 SUMMARY

Neurons evolved during the age of animals and for the purpose of forming a rapid communications link for coordinating different regions of increasingly complex body forms. By this time all of the modern structures of fatty acid chains for constructing membranes of neurons had presumably evolved. Comparative studies of animals show that Darwinian selection of the fittest membrane architecture applies to neurons, with DHA being far more prevalent than EPA in the nervous system of mammals (i.e., vertebrates in general) and presumed to confer improved efficiency. Historically, nutritional studies with mammals provided the first evidence of a functional hierarchy of unsaturated fatty acids in membranes of neurons. Studies of *C. elegans* support this view and provide a quantitative rather than qualitative perspective as to the biochemical roles of DHA versus EPA in neurons. It is now clear that DHA chains represent the most sophisticated level of conformational dynamics for building neurons in humans, and it is proposed that DHA contributes multiple important biochemical functions. The current picture is that the conformational dynamics of DHA are harnessed for both membrane lateral motion (e.g., disrupting lipid rafts) and as a permeability barrier especially against uncoupling of $Na^+/K^+$ gradients essential in electrophysiology, tempered by proton permeability (Chapter 8) and oxidative instability (Chapter 9). Finally, these studies are consistent with the concept that DHA enhances the efficiency of neurons, which is beneficial because of the energy intensive nature of these essential cells.

## SELECTED BIBLIOGRAPHY


Buda, C., I. Dey, N. Balogh, L. I. Horvath, K. Maderspach, M. Juhasz, Y. K. Yeo, and T. Farkas. 1994. Structural order of membranes and composition of phospholipids in fish brain-cells during thermal acclimatization. *Proc. Natl. Acad. Sci. U.S.A.* 91:8234–8238.

Chacko, G. K., F. V. Barnola, and R. Villegas. 1977. Phospholipid and fatty acid compositions of axon and periaxonal cell plasma membranes of lobster leg nerve. *J. Neurochem.* 28:445–447.

Crawford, M. A., C. Leigh Broadhurst, C. Galli, et al. 2008. The role of docosahexaenoic and arachidonic acids as determinants of evolution and hominid brain development. In *Fisheries for Global Welfare and Environment, 5th World Fisheries Congress,* ed. K. Tsukamoto, T. Kawamura, T. Takeuchi, T. D. Beard, Jr., and M. J. Kaiser, 57–76. Tokyo: TERRAPUB.

Farkas, T., K. Kitajka, E. Fodor, I. Csengeri, E. Lahdes, Y. K. Yeo, Z. Krasznai, and J. E. Halver. 2000. Docosahexaenoic acid-containing phospholipid molecular species in brains of vertebrates. *Proc. Natl. Acad. Sci. U.S.A.* 97:6362–6366.

Farooqui, A. A., and L. A. Horrocks. 2001. Plasmalogens: Workhorse lipids of membranes in normal and injured neurons and glia. *Neuroscientist* 7:232–245.

Hendricks, T. H., A. A. Klompmakers, F. S. M. Daemen, and S. L. Bonting. 1976. Biochemical aspects of the visual process. XXII. Movement of sodium ions through bilayers composed of retinal and rod outer segment lipids. *Biochim. Biophys. Acta* 433:271–281.

Knott, A. B., G. A. Perkins, R. Schwarzenbacher, and E. Bossy-Wetzel. 2008. Mitochondrial fragmentation in neurodegeneration. *Nature Rev. Neurosci.* 9:505–518.

Lesa, G. M., M. Palfreyman, D. H. Hall, et al. 2003. Long chain polyunsaturated fatty acids are required for efficient neurotransmission in *C. elegans. J. Cell Sci.* 116:4965–4975.

Takamori, S., M. Holt, K. Stenius, et al. 2006. Molecular anatomy of the trafficking organelle. *Cell* 127:831–846.

Tanaka, T., S. Izuwa, K. Tanaka, et al. 1999. Biosynthesis of 1,2-dieicosapentaenoyl-*sn*-glycero-3-phosphocholine in *Caenorhabditis elegans. Eur. J. Biochem.* 263:189–194.

Watts, J. L., and J. Browse. 2002. Genetic dissection of polyunsaturated fatty acid synthesis in *Caenorhabditis elegans. Proc. Natl. Acad. Sci. U.S.A.* 99:5854–5859.

# Section 3

# General Properties of Omega-3s and Other Membrane Lipids

DHA and EPA contribute at least three essential biochemical properties to membranes: permeability, motion, and chemical stability. The standard for permeability is set by barrier properties against spontaneous leakage or futile cycling of protons and other cations. Recall that proton gradients are the primary energy currency of the cell, with membranes being a critical gatekeeper. Thus, a more robust proton permeability barrier can be considered as a mechanism to prevent loss of cellular energy. Here we consider motion as shorthand for membrane lateral motion, which is seemingly a property of all core biological membranes. Among membranes, those with abundant DHA or EPA are considered to set the standard for rapid lateral motion, which is essential for respiration, photosynthesis, and neurosensory processes. Chemical stability refers to oxidative stability and is inversely related to membrane unsaturation. In any cell in which DHA/EPA are in contact with molecular oxygen ($O_2$), the chemical process referred to as lipoxidation occurs to such an extent that protective mechanisms become essential. Without this protection DHA/EPA can contribute to the pool of reactive oxygen species, creating a chain reaction potent enough to kill the cell.

In this section we consider motion, permeability, and chemical stability properties of membranes with DHA or EPA in the context of other cellular membranes, in order to draw out both benefits and risks associated with these omega-3s.

# 7 DHA/EPA Chains as Powerful Membrane Antifreeze

Hypothesis: The extraordinary conformational dynamics of DHA/EPA chains act to maintain and optimize motion and to disrupt gel-phase rafts in membranes of both ectothermic and endothermic organisms.

Energy-transducing membranes of mitochondria and chloroplasts, operating at a narrow range of temperatures around 0°C, lie at the heart of one of the most fertile regions of the world ocean—the Antarctic marine ecosystem. It is no coincidence that a significant fraction of total global omega-3 production occurs in this region. Evidence is summarized here that DHA/EPA are needed in such massive amounts as powerful membrane antifreeze in organisms living their entire lives at temperatures from about –2°C to 3°C. In this section we are concerned with how the dynamic conformations of DHA, EPA, and phospholipid structures with similar properties contribute to the motional state of membrane proteins, as well as to the physical state of membranes. As background, membrane surface structure is commonly described on the basis of two physical states, a liquid/fluid stage called the liquid-crystalline state, versus a hardened state called the gel phase. The term phase transition temperature refers to the temperature at which the fluid state is converted to the gel phase. One of the most popular theories in membrane biology today states that the surface of membranes is actually a lipid mosaic structure consisting of gel-phase islands or hardened patches of lipids surrounded by fluid regions. A number of factors, including temperature, protein content, abundance of cholesterol or similar molecules, and molecular structures of fatty acid chains, determine the proportion of the membrane surface that is liquid or hardened. The major focus here is on how the rapid and relatively wild conformational changes of DHA/EPA chains contribute antifreeze properties to the bilayer (Figure 7.1).

The motion of membrane components trapped in gel-phase lipids is dramatically slowed compared to the liquid phase. Thus, the motional state of essential membrane components, such as proteins and electron carriers of the electron transport chain, is a property determined by the motion status of the membrane itself. We argue that the conformational dynamics of membrane lipids such as DHA/EPA, which are antibonding with respect to neighboring chains, provide an enhanced level of motion to their membranes, more so than virtually any other membrane lipid. The underlying principles have been known and applied for many years. Indeed, the kinked or bent conformations generated by introduction of one double bond into a straight-chain saturated lipid are cited as a classic example of structure impacting function of lipid

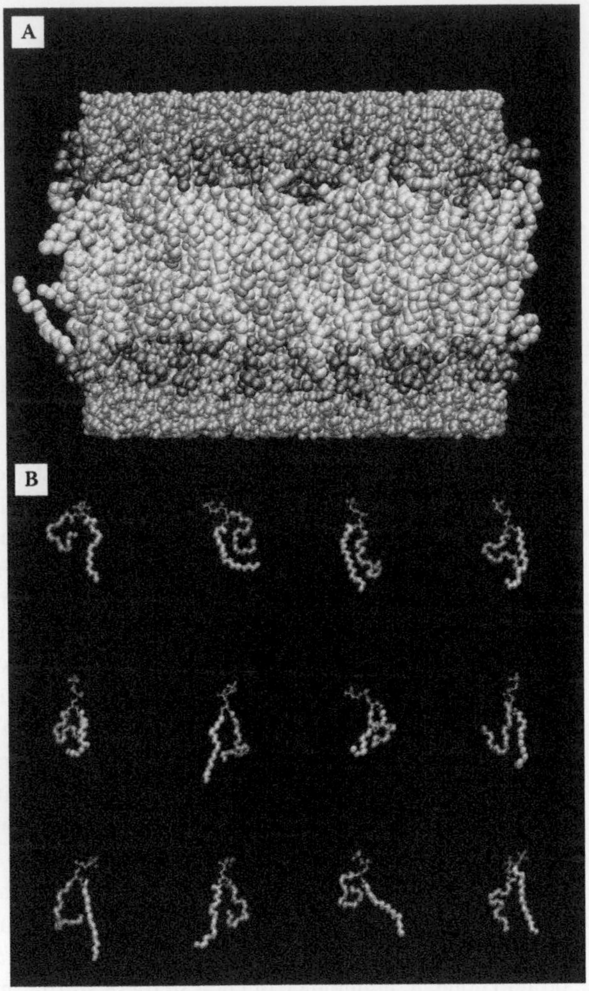


**FIGURE 7.1** (A color version of this figure follows p. 140.) DHA chains display powerful membrane antibonding properties. (A) DHA-enriched membrane bilayer generated using chemical simulation techniques (gold chains are DHA and greenish chains are 18:0). (B) Extraordinary conformations of DHA seen in this series of images are consistent with the powerful membrane antifreeze properties attributed to these highly unsaturated chains. DHA chains are shown as golden, 18:0 as greenish, and the head group as red. (Images courtesy of Scott Feller and generated for this book by Matthew B. Roark, both of Wabash College.)

chains in membranes. The term "fluidizing chains" has been applied to lipid structures that drive membranes toward the liquid state. The term "membrane fluidity" has fallen out of favor because of difficulties in defining "fluidity" in biochemical terms. It is clear that membrane fluidity is not as simple as originally thought and is actually a combination of at least two essential biochemical properties, including membrane architecture modulating permeability, discussed in the next chapter, and motional properties, covered below.

## 7.1 SURVEY OF PHOSPHOLIPIDS BASED ON PHASE TRANSITION TEMPERATURES

LIPIDAT is an Internet accessible, computerized relational database centered at Ohio State University and providing access to the wealth of information scattered throughout the literature concerning physical-chemical properties of phospholipids, including omega-3 chains. The subset of data concerning phosphatidylcholine, a class of phospholipids common in many membranes, as reviewed in 1998 covers a 43-year period and consists of 12,208 records obtained from 1573 articles in 106 different journals. The data show the impact of lipid chain length, desaturation (number, isomeric type, and position of double bonds), methyl branching, and positioning of distinct chains attached to the glycerol backbone on phospholipid properties. Attention here is focused on naturally occurring membrane phospholipids such as di-DHA, mentioned in previous chapters, as contributing extraordinary motion to rhodopsin in membrane disks. Several critical parameters governing membrane motion can be inferred from these data. For example, it seems clear from the database that attachment of unsaturated/methyl branched chains to both positions of the glycerol backbone results in low melting point phospholipids predicted to contribute low viscosity to the bilayer. Perhaps the most common mechanism to lower membrane viscosity involves increasing desaturation of the acyl chain occupying the *sn*-2 position, from one double bond to two or more. As previously mentioned, a dramatic decrease in the viscosity of membranes occurs by doubling the number of unsaturated fatty acids attached to a glycerol backbone from typically one to two. In the latter case, melting points are lowered substantially, as summarized in Table 7.1. The main point is that chemical and biological evidence is consistent and supports the

**TABLE 7.1**
**Hierarchy of Unsaturated Phospholipids Found in Natural Membranes and Relationship to Phase Transition Temperature**

| Organism/Membrane | Structure | Double Bonds | Phase Transition Temperature (°C)[a] |
|---|---|---|---|
| Rhodopsin disks | di-DHA | 12 | −67.4 |
| *C. elegans* | di-EPA | 10 | <−65 |
| | di-20:4n6 | 8 | −69.5 |
| Neurons/deep-sea bacteria | 18:0-DHA | 6 | −3.8 |
| | 18:0-EPA | 5 | −10.4 |
| Bacteria/plants/animals | 18:0-18:2n6 | 2 | −14.4 |
| | 18:0-18:3n3 | 3 | −12.3 |
| | di-18:1n9 | 2 | −18.3 |
| | di-18:1n7 | 2 | −20.5 |
| Bacteria | 18:0-18:1n9 | 1 | 6.9 |
| | di-12:0 | 0 | −1.9 |
| | di-14:0 | 0 | 23.6 |

[a] *Predicted phospholipid order to disorder transition temperature (not actually measured).*
*Source:* Modified from R. Koynova and M. Caffrey. *Biochim. Biophys. Acta*, 1376:91–145, 1998.

evolution of a hierarchy of phospholipids whose phase transition temperatures are modulated by the fatty acid chains themselves. Furthermore, this database highlights what we consider as an important missing link in relating lipid structure and function in membrane biology—mainly the evolution of a class of phospholipids that work as powerful "membrane antifreeze" in both ectothermic and endothermic organisms.

## 7.2 ECOLOGICAL DISTRIBUTION OF PHOSPHOLIPIDS WITH ULTRALOW PHASE TRANSITION TEMPERATURES

Phospholipids, such as di-DHA, that retain their liquid state at extremely low temperatures are far more abundant in nature than currently appreciated. Plants synthesize the majority of DHA/EPA as well as the largest amounts of polyunsaturated phospholipids found in nature; this topic is covered in a separate chapter on photosynthetic membranes (Chapter 11). It is interesting to note, based on total double bonds per phospholipid molecule, that land plants seldom surpass six, in contrast to marine plants, which incorporate up to twelve. As already mentioned, the incorporation of two unsaturated fatty acid chains in place of one creates a phospholipid structure with a markedly lower phase transition point. We suggest important biochemical functions for these lipids found in bacteria, plants, and animals. However, unsaturation is not the only structural modification leading to reduced phase transition temperatures, and evolution has provided a large menu of lipids with which membranes can be constructed. Some case histories are discussed next.

Low melting point phospholipids are widely distributed among bacteria and include economically important organisms as well as many pathogens. For example, symbiotic $N_2$ fixing bacteria forming root nodules on soybeans, certain photosynthetic bacteria, ubiquitous methane oxidizers, and even *E. coli* produce high amounts of phospholipids in which both *sn*-1 and *sn*-2 positions of the glycerol backbone are occupied by the 18-carbon monounsaturated fatty acid *cis*-vaccenate ($18{:}1_{n7}$). In some cases virtually all of the phospholipids of the cytoplasmic membrane are composed of di-$18{:}1_{n7}$. There is also one report of an isolate from Antarctica that synthesizes primarily di-$18{:}1_{n9}$, but the biochemistry of this pathway has not been studied. Interestingly, the levels of di-$18{:}1_{n9}$ in certain cold-adapted crop plants, such as oats and rye, also increase during cold acclimation.

Many bacteria synthesize membrane phospholipids in which methyl-branched fatty acids occupy both the *sn*-1 and *sn*-2 position of phospholipids instead of unsaturated chains; some specific examples are discussed in Chapter 10. The main point here is that di-methyl–branched phospholipids also have low melting points. As discussed in a later chapter, *anteiso*-methyl branching in combination with chain shortening generates the lowest melting point phospholipids of this class.

Among animals, cold adaptation in *C. elegans* results in a substantial increase in the amounts of 1,2 polyunsaturated fatty acid–containing phosphatidylcholine, such as di-EPA, which reaches levels of around 10% total fatty acids at 15°C. Indeed, a hierarchy of di-polyunsaturated phospholipids expected to behave as strong antifreeze molecules was detected, including structures with ten, nine, eight, seven, six, and five double bonds. Cold adaptation (i.e., 25°C to 15°C) causes the proportions of

EPA to increase from 23.6% to 32.5% in phosphatidylcholine, from 7.4% to 10.8% in phosphatidylethanolamine and from 12.9% to 19.9% in total lipids. Note that the genes for synthesis of DHA are absent in *C. elegans*, and this fatty acid is detected in membrane phospholipids only when incorporated into the diet. A significant proportion of exogenous DHA is retroconverted to EPA, which suggests that there may be a specific advantage to EPA that DHA cannot fulfill. The possible role of DHA/EPA in the evolution of a more efficient nervous system has already been described in Chapter 6.

Polar seas contain animals whose membranes are highly enriched with DHA/EPA presumably as an adaptation to life in extreme cold. For example, the combined krill populations of both Arctic and Antarctic polar regions are considered the largest source of animal biomass on Earth. An estimated 550 to 825 million tons of krill inhabit Antarctic seas. The krill population contains large numbers of mitochondria needed to support this level of biomass. At the membrane level krill contain by far the largest amounts of DHA/EPA in the animal world—about 20% to 30% total fatty acids. Mitochondria of Antarctic krill are responsible for conducting most of the respiration that occurs in the extreme cold. Indeed this massive pool of "specialized" mitochondria catalyze whole-animal rates of respiration at –2°C that are roughly one half the rate at 4°C, the latter being the upper limit for growth of this extreme cold-dependent animal. Mitochondria of northern krill have evolved to respire over the broader temperature range of this animal, which is about 6°C to 16°C. In the next section we calibrate the antifreeze properties of EPA phospholipids using an EPA recombinant of *E. coli*.

## 7.3 CALIBRATING THE FLUIDIZING POWER OF FATTY ACIDS

Given that a significant fraction of all mitochondria in the biosphere operate in the temperature range of –2°C to about 2°C, it seems clear that membranes have evolved to work in the extreme cold. In this section the potent membrane antifreeze properties of DHA/EPA are validated using bacteria as surrogates for mitochondria. Recall that components of the electron transport chain are housed in membranes whose surface is a mosaic of fluid zones versus hardened regions, and electron transport would be shut down at cold temperatures if not for the incorporation of unsaturated lipids maintaining fluid patches. The main point is that electron transport components are believed to belong to a growing list of membrane components that experience a dramatic loss of activity when trapped in gel-phase lipids. Thus, collisions that link together components of the electron transport chain are expected to stall when too much of the surface is in the gel phase, with major negative impact on energy production. This concept extends also to warmer temperatures. For example, consider the energy-transducing membranes of endothermic animals where motion of components of the mitochondrial electron transport chain is driven at least in part by relatively high temperatures. The presence of high levels of desaturation in these bilayers suggests that the purpose of these chains is to increase the proportion of the inner mitochondrial membrane surface suitable for electron transport beyond that contributed by the relatively high temperatures. In this case it may not be gel-phase lipids that impede motion, but rather the tight packing of many large electron transport complexes into the membrane to the point that these large molecules needed for energy production cover much of the surface. This phenomenon clutters the landscape, essentially creating an obstacle course for lipophilic electron carriers that must navigate around these relatively large bodies in order to shuttle electrons between complexes. A case history in which DHA phospholipids play an important role in endothermic mitochondria is covered in Chapter 12.

Components of the electron transport chain housed in the inner mitochondrial membrane have been shown to move with rapid lateral motion at 37°C. These rates are somewhat lower when compared to rhodopsin protein at this temperature, which is the fastest yet calibrated. How do ultra-cold–adapted mitochondria or deep-sea bacteria living at temperatures near 0°C maintain motion of their electron transport components against powerful forces that drive membrane structure toward the gel phase? Based on physiological studies carried out almost 40 years ago, it is clear that the electron transport chains of marine psychrophiles (i.e., cold-loving organisms), such as *Moritella marina*, are active down to temperatures approaching 0°C, a temperature that completely inhibits mammalian mitochondrial respiration. For example, cells of *M. marina* showed active rates of $O_2$ uptake down to 4°C, with research at lower temperatures being prohibited by design limitations of the $O_2$ electrode used for measuring respiration. Researchers noted that respiration rates at 4°C were about 50% the rate at the maximum growth temperature of these cells, 15°C. These data were collected years prior to the discovery of relatively high DHA levels in membranes of *M. marina*. These data can now be interpreted on the basis of powerful antifreeze properties of DHA enabling collisions among electron transport components even as temperatures approach 0°C.

While the chemical reasoning for unsaturated fatty acids enabling movement of membrane components is sound, there is a disconnect with biological systems on account of a lack of models available for comparative analysis. Important questions such as the relative impact *in vivo* of different lipid chains remain largely unanswered. In our work we have addressed this issue by analyzing growth and respiration in EPA recombinant cells of *E. coli*. In one experiment we found that cells with EPA-enriched membranes support $O_2$ uptake rates at 4°C of approximately 40% the rate at 15°C. It was also found that the rate of oxygen uptake over the temperature range of 4°C to 15°C by wild-type *E. coli* cells grown at 18°C was comparable to the rate of oxygen consumption by the EPA recombinant grown at 18°C, with the wild type synthesizing and enriching its bilayer with di-18:$1_{n7}$ phospholipids. Results from this assay suggest that di-18:$1_{n7}$ behaves as a highly fluidizing chain, roughly equivalent to EPA-containing phospholipids (i.e., 16:0/20:5). Control cells grown at 37°C, which do not produce di-18:$1_{n7}$, showed oxygen consumption rates at 4°C that were 83% lower than those of the same strain grown at 18°C. Fatty acid analysis showed that the EPA-recombinant cells grown at 37°C downregulated EPA synthesis to undetectable levels, but synthesized about 25% to 28% monounsaturated fatty acids as fluidizing chains, with the remainder of membrane lipid chains composed of saturated fatty acids. We have found that monounsaturated chains at this level are insufficient to support growth below about 30°C, in contrast to EPA-enriched membranes that are functional at much lower temperatures, as discussed above. Thus, compared on a molecule to molecule basis, EPA appears to contribute more powerful fluidizing properties compared to monounsaturated chains.

An alternative approach for calibrating antifreeze properties of fatty acids involves feeding various unsaturated chains to mutants of *E. coli* dependent for growth on acquisition of these essential chains from the medium. For example, previous researchers used unsaturated fatty acid–requiring mutants of *E. coli* to calibrate the relative fluidizing powers of the 20-carbon unsaturated fatty acids, starting with 20:1 and including 20:2 and 20:3 (Esfahani, Barnes, and Wakil, 1969). This growth assay is based on the general rule of thumb that for each drop in temperature of 1°C about 1% more monounsaturated fatty acids are required for growth of *E. coli* (Figure 7.2A). This classic method is readily modified to measure the fluidizing powers of EPA using recombinant cells, with the result that EPA is calibrated to be twice as effective as 20:1 (Table 7.2, Figure 7.2B). According to previous data, 20:2 and 20:3 are 1.2 and 1.5 times more fluidizing, respectively, than 20:1. The finding that 20:5 appears to be twice as fluidizing as 20:1 is consistent with these results. Based on these findings it is concluded that EPA chains not only supply bulk fluidity essential for growth of an EPA recombinant of *E. coli*, but each EPA chain works as a more powerful membrane fluidizer/antifreeze than other fatty acids common in membranes of bacteria. Note that the *in vivo* test for calibrating EPA is conducted at 24°C and that about 75% of total membrane fatty acids in EPA recombinant cells are saturated chains, mainly 16:0. Thus, EPA likely pairs with 16:0 in phospholipids of these cells, although direct chemical analysis has not been conducted.

Another approach for calibrating the relative fluidizing powers of EPA chains involves studies of a class of cold-sensitive mutants derived from marine isolates of *Shewanella*. One interesting set of mutants is dependent on supplementation with

**TABLE 7.2**
**Fatty Acid Composition of EPA Recombinant Cells Grown in Medium Supplemented with 0.2 M NaCl at 24°C**

| Fatty Acid | % Total Fatty Acids ($n$ = 4) | ±1 σ |
|---|---|---|
| 12:0 | 3.0 | 0.1 |
| 14:0 | 11.1 | 0.5 |
| 16:0 | 64.7 | 0.5 |
| 16:1 | 1.2 | 0.1 |
| 17.0 CFA[a] | 0.4 | 0.1 |
| 18:0 | 0.6 | 0.1 |
| $18:1_{11}$ | 0.5 | 0.1 |
| 19:0 CFA | 0.0 | 0.0 |
| 20:4 | 1.0 | 0.1 |
| 20:5 | 11.9 | 1.0 |
| 22:5 | 0.6 | 0.1 |

*Inoculant of EPA recombinant cells was spread on LB plates containing a total of 0.2M NaCl and 100 mg/mL of ampicillin. After 4 days of incubation, fresh medium was added to free cells from the plate surface and cells were collected for fatty acid analysis.*

[a] *CFA, cyclopropane fatty acids.*

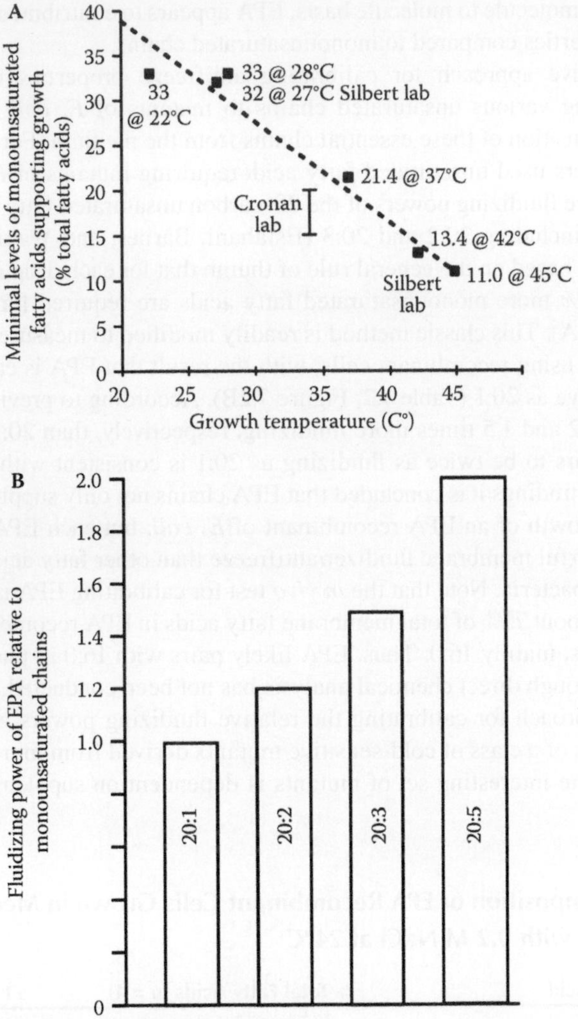


**FIGURE 7.2** An EPA recombinant of *E. coli* is used to calibrate the antiviscosity properties of EPA in a living cell. (A) Minimal growth requirement for monounsaturated fatty acids in *E. coli* increases as temperatures drop. These data from several laboratories serve as a standard curve for calibrating *in vivo* activity of EPA. (B) With this standard curve and data summarized in Table 7.2, it is calculated that one molecule of EPA working in a living cell provides roughly double the fluidizing power of monounsaturated chains such as 16:1. Data for 20:1, 20:2, and 20:3 are from the literature. DHA is not included in this study because DHA recombinants of *E. coli* dependent on DHA for growth are not yet available.

EPA for growth at 4°C. Further studies show that the specificity for EPA is not absolute since feeding some other polyunsaturated chains, such as arachidonic acid (20:4), also restores growth at 4°C. In contrast growth is poor when these mutants are fed monounsaturated chains. Recently Japanese scientists have used knockout mutants of specific EPA genes of *Shewanella* to confirm that EPA is required for growth of this organism at 4°C (Sato et al., 2008). These researchers conducted a comprehensive chemical analysis of phospholipid acyl pairing preferences showing that EPA often pairs in phospholipids with 16:1 (Kawamoto et al., 2009). This structure is expected to be one of the most powerful membrane antifreeze molecules in bacteria and helps explain why many natural isolates of *Shewanella* need only ~3% to 5% percent EPA as total membrane fatty acids in their bilayers. Note that about 75% of acyl chains in *Shewanella* membranes are unsaturated or branched (i.e., fluidizing), the opposite compared to membranes of EPA recombinant cells where saturated chains predominate. The point is that a small amount of EPA partnering in phospholipids with 16:1 and working in a membrane surface already "primed" against hardening might provide a suitable blend for membranes to support electron transport at temperatures around 0°C.

However, it is important to point out that the roles of EPA as a fluidizing chain remain controversial. For example, the authors of the above report present evidence against EPA as a fluidizing chain and instead highlight roles in cellular growth and development (Kawamoto et al., 2009). We believe that it is possible to reconcile both points of view in that there is clear evidence that EPA contributes fluidity essential for cellular growth, leaving open the possibility of more specific but unknown biochemical roles such as in cell division. Ultimately, we suggest that the fact that many DHA/EPA bacteria produce <5% DHA/EPA is not a sound argument against fluidizing roles for these chains. Instead we propose that DHA/EPA-producing bacteria must live by the DHA principle, specifically that too much DHA/EPA can be harmful, especially in this case where 75% of total chains are fluidizing. Recall that *Shewanella* does not produce cholesterol-like molecules to stabilize its membranes and instead appears to rely on large numbers of methyl-branched chains for this purpose. A mechanism to explain how relatively small amounts of DHA/EPA phospholipids might facilitate disruption of gel-phase membrane islands is described next.

## 7.4 SEEDING MODEL OF DHA IN DISRUPTING LIPID RAFTS

Molecular dynamic simulations clearly demonstrate that DHA as a lipid tail undergoes wild and rapid conformational swings that help to explain its ultralow melting point and provides insight on how DHA tails may act as seeds of liquefaction to disrupt lipid rafts. These models show DHA tails as being molecular contortionists in the membrane, rapidly extending and contracting in length and width, and striking unusual conformations such as helical and bent back shapes. However, dynamic simulations are as yet unable to capture the impact of DHA chains on the physical status of surrounding phospholipids. We envision the molecular motion of DHA lipids as destabilizing to the gel phase on account of the wild and rapid conformational shifts. For example, we envision several of DHA's conformations as being likely to interact with surrounding lipid tails in such a way that the weak interactions holding individual lipids to the gel phase may be disrupted. The rapid changes in conformation and rapid lateral motion may allow DHA to effectively plow through lipid rafts like a molecular icebreaker, in which case the antifreeze properties of DHA are tied also to the formation kinetics of "lipid microrafts" reforming in its wake. The stoichiometry of membrane lipids in psychrophilic bacteria may provide some insight on the sphere of influence for DHA lipids. For example, the presence of 10% to 17% DHA in *Moritella* suggest that at low temperature DHA lipids may only act to fluidize a handful of immediate neighbors, perhaps acting more to inhibit raft formation than disrupting rafts. At higher temperatures the radius of influence is likely to expand. We refer to this scenario as a "seeding" model, since relatively few molecules of DHA are envisioned to greatly amplify the proportion of surface maintained in the fluid state. This simple picture becomes more complicated when the disparity in phase transition temperatures between DHA phospholipids as icebreakers and other bulk phospholipids as icebergs becomes larger or smaller. Also, the presence of additional classes of lipids, especially cholesterol, adds further complexity to this picture. Nevertheless, the compact, flexible conformations of DHA chains appear to confer fluidity to the membrane. DHA might either break apart lipid rafts (icebreaker model) or inhibit their formation (antifreeze model), or both.

## 7.5 SUMMARY

There is increasing evidence that DHA/EPA chains contribute powerful antifreeze properties to the bilayers of cells ranging from deep-sea bacteria and cold-adapted mitochondria to rod cells of the human eye. We propose that this is a universal property of these phospholipid molecules that can be explained by their structures and resulting conformational dynamics. Doubling the number of DHA/EPA chains per phospholipid from one to two or pairing DHA/EPA with a second fluidizing chain such as 16:1 greatly decreases the phase transition temperatures and likely strongly increases antifreeze properties. A variety of other structures, such as di-18:$1_{n7}$ and di-polyunsaturated chains, also play roles similar to DHA/EPA. The widespread distribution and unique properties of low melting point phospholipids are drawing more attention to their important contributions to membrane biochemistry. A number of later chapters are devoted to this subject. At the molecular level, chemical modeling has led to pictures showing the exceptional conformational diversity of DHA chains, a property that might be shared with EPA, arachidonic acid (ARA 20:4), 22:5, and stearidonic acid (18:4). Virtually all of the conformations of DHA chains seen by chemical simulation display strong antibonding or antifreeze properties. One of the most exciting topics for future research involves the role of DHA working as a molecular icebreaker to disrupt lipid rafts.

## SELECTED BIBLIOGRAPHY

Brown, D. H. 2006. Lipid rafts, detergent-resistant membranes and raft targeting signals. *Physiology* 21:430–439.

Buchholz, F., and R. Saborowski. 2000. Metabolic and enzymatic adaptations in northern krill, *Meganyctiphanes norvegica*, and Antarctic krill, *Euphausia superba*. *Can. J. Fish. Aquat. Sci.* 57(Suppl. 3):115–129.

Esfahani, M., E. M. Barnes, and S. Wakil. 1969. Control of fatty acid composition in phospholipids of *Escherichia coli*: Response to fatty acid supplements in a fatty acid auxotroph. *Proc. Natl. Acad. Sci. U.S.A.* 64:1057–1064.

Fernandes, M. X., M. A. R. B. Castanho, and J. Garcia de la Torre. 2002. Brownian dynamics simulation of the unsaturated lipidic molecules oleic acid and docosahexaenoic acid confined in a cellular membrane. *Biochim. Biophys. Acta* 1565:29–35.

Frye, L. D., and M. Edidin. 1970. The rapid intermixing of cell surface antigens after formation of mouse-human heterokaryons. *J. Cell. Sci.* 7:319–335.

Gupte, S., E. S. Wu, L. Hoechli et al. 1984. Relationship between lateral diffusion, collision frequency, and electron-transfer of mitochondrial inner membrane oxidation-reduction components. *Proc. Natl. Acad. Sci. U.S.A.* 81:2606–2610.

Hagen, W., T. Yoshida, P. Virtue, et al. 2007. Effect of a carnivorous diet on the lipids, fatty acids and condition of Antarctic krill, *Euphausia superba. Antarct. Sci.* 19:183–188.

Haight, J. J., and R. Y. Morita. 1966. Some physiological differences in *Vibrio marinas* grown at environmental and optimal temperature. *Limnol. Oceanogr.* 11:470–474.

Kawamoto, J., T. Kurihara, K. Yamamoto, et al. 2009. Eicosapentaenoic acid plays a beneficial role in membrane organization and cell division of a cold-adapted bacterium, *Shewanella livingstonensis* Ac10. *J. Bacteriol.* 191:632–640.

Koynova, R., and M. Caffrey. 1998. Phases and phase transitions of the phosphatidylcholines. *Biochim. Biophys. Acta* 1376:91–145.

Mayzaud, P., P. Virtue, and E. Albessard. 1999. Seasonal changes in the lipid and fatty-acid composition of the euphausiid *Meganyctiphanes norvegica* from the Ligurian Sea. *Mar. Ecol. Prog. Ser.* 186:199–210.

McElhaney, R. N. 1974. The effects of alterations in the physical state of the membrane lipids on the ability of *Acholeoplasma laidlawii B* to grow at different temperatures. *J. Mol. Biol.* 84:145–157.

Munro, S. 2003. Lipid rafts: Elusive or illusive. *Cell* 115:377–388.

Overath, P., H. V. Schairer, and W. Stoffel. 1970. Correlation of *in vivo* and *in vitro* phase transitions of membrane lipids in *Escherichia coli. Proc. Natl. Acad. Sci. U.S.A.* 67:606–612.

Sato, S., T. Kurihara, J. Kawamoto, M. Hosokawa, S. B. Sato, and N. Esaki. 2008. Cold adaptation of eicosapentaenoic acid-less mutant of *Shewanella livingstonensis* Ac10 involving uptake and remodeling of synthetic phospholipids containing various polyunsaturated fatty acids. *Extremophiles* 12:753–761.

Seelig, A., and J. Seelig. 1977. Effect of a single *cis*-double bond on the structures of a phospholipid bilayer. *Biochemistry* 16:45–50.

Simons, K., and D. Toomre. 2000. Lipid rafts and signal transduction. *Nature Rev. Mol. Cell Biol.* 1:31–39.

Sinensky, M. 1974. Homeoviscous adaptation: A homeostatic process that regulates the viscosity of membrane lipids in *Escherichia coli. Proc. Natl. Acad. Sci. U.S.A.* 71:522–525.

Tanaka, T., S. Izuwa, K. Tanaka, et al. 1999. Biosynthesis of 1,2-dieicosapentaenosyl-*sn*-glycero-3-phosphocholine in *Caenorhabditis elegans. Eur. J. Biochem.* 263:189–194.

Watts, J. L. and J. Browse. 2002. Genetic dissection of polyunsaturated fatty acid synthesis in *Caenorhabditis elegans. Proc. Natl. Acad. Sci. U.S.A.* 99:5854–5859.

# 8 DHA as a Mediocre Permeability Barrier against Cations: *Water Wire Theory*

Hypothesis: DHA-enriched membranes are inherently leaky against $H^+$ and have evolved under conditions that minimize or counter the negative impact of such leaks.

The most fundamental role of membrane lipid chains is in forming a permeability barrier against protons, sodium, and other metabolites essential for cellular bioenergetics and growth. The importance of proton permeability is highlighted in the discovery by Peter Mitchell that energy stored in the form of proton electrochemical gradients across membranes represents the primary energy form powering cellular life (Mitchell, 1961; see Chapter 1 reference). Previously, this honor was given to ATP, which we consider secondary to proton chemiosmotic energy. Sodium in the form of chemiosmotic gradients also plays an important role in bioenergetics, but it is generally secondary to protons.

It is now believed that all fatty acid–based bilayers are inherently leaky or chemically permeable to protons (Figure 8.1) and that cells have evolved various fatty acid–based membrane defenses to enhance proton fidelity of their membranes. A theory linking fatty acid conformations and proton permeability of membranes was developed using chemically defined membrane vesicles and later bacteria and is referred to as homeoproton permeability adaptation—bacteria strive to adapt their membrane fatty acid composition to minimize proton leakage. It is now known that excessive proton leakage caused by high temperatures as well as hyperfluidizing or decoupling agents trigger cellular death.

Lipid membranes are far less permeable to sodium ions than to protons. This explains a strategy introduced in Chapter 5 regarding a major shift in bioenergetics of DHA-producing bacteria in which $Na^+$ replaces $H^+$ as the primary cation for storing electrochemical energy. It is important to recall that, even though membranes are several orders of magnitude less permeable to $Na^+$ than to $H^+$, the high levels of $Na^+$ surrounding many cells and organisms, including humans, can rapidly drive $Na^+$ into the cell. The staggering energy cost of $Na^+$ extrusion to the energy budget of red blood cells, mentioned previously, serves again as a classic example. The point we seek to emphasize here is that $H^+/Na^+$ leakage is a serious energy stress for cells, and the effective conservation of cellular energy has acted as a potent selective force driving the evolution of a battery of lipid-based defenses designed to prevent futile

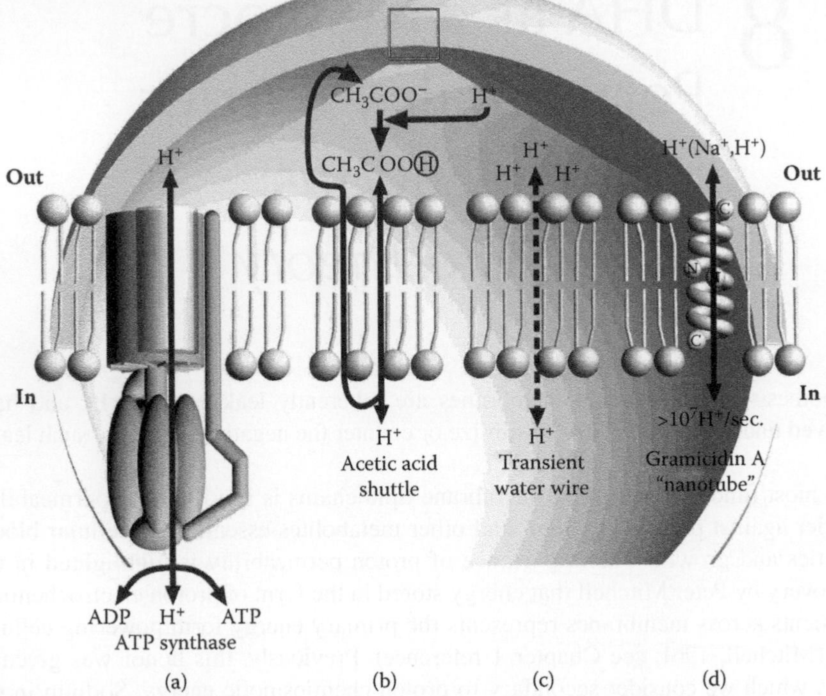


**FIGURE 8.1** Membrane lipids, including DHA/EPA, are essential for forming a molecular barrier against uncoupling of cation gradients. A variety of mechanisms are known to cause futile cycling of protons as follows (from left to right): leakage through critical membrane-bound enzymes such as ATP synthase; uncoupling by short-chain fatty acids capable of shuttling from one side of the bilayer to another; transient water wires emphasized here, and gramicidin A, the latter already introduced in Chapter 3 and discussed here as a model to explain the importance of water wires occurring in natural membranes. No matter the mechanism of loss, as chemiosmotic gradients are deenergized, the cell must invest metabolic energy to compensate for the loss.

cycling of these ions across the membrane. The focus here is on the structures and conformations of fatty acid chains, which contribute to or reduce membrane permeability toward cations. DHA/EPA appear to be extreme cases. It is important to keep in mind that the chemistry of attachment of DHA/EPA chains to their phospholipid headgroups, as well as environmental conditions such as temperature, also contribute important permeability properties to the bilayer.

## 8.1 A MEMBRANE-SPANNING NANOTUBE FORMED BY AN ANTIBIOTIC CREATES A MOLECULAR THREAD OF WATER THAT CONDUCTS PROTONS AT AMAZING RATES

Studies of the electrochemical properties of water, and later applied to membranes, revealed many years ago that protons somehow tunnel through ice at amazing speed.

About 50 years ago, the concept of proton tunneling via water wires was first applied to explain proton permeability properties of membranes. However, it wasn't until the early 1980s that proton tunneling, believed to be mediated by traces of water spontaneously entering the lipid region of membranes, was shown to occur in chemically defined membrane vesicles. Perhaps the biggest breakthrough in this field came from an unlikely source, studies of the molecular mode of action of energy uncoupling antibiotics used to fight diseases. As introduced in Chapter 3, these studies show that when about 21 molecules of water enter single file into the core of a membrane-spanning, nanotube-forming antibiotic called gramicidin A, a circuit is opened for conducting protons (Figure 8.2). It is now clear that it is the structure of chains or aggregates of water itself held in the core of gramicidin A that enables these fast-acting, proton-conducting pores in which protons appear to be transported along the molecular thread of water.

The antimicrobial activity of gramicidin A was first discovered during the era of massive screening of soil-borne organisms for antibiotics that, in nature, serve as chemical warfare agents used to gain competitive advantage against neighboring cells. The bacterium *Bacillus brevis* synthesizes and exports gramicidin A as a helical polypeptide monomer composed of 15 amino acid residues that inserts into the

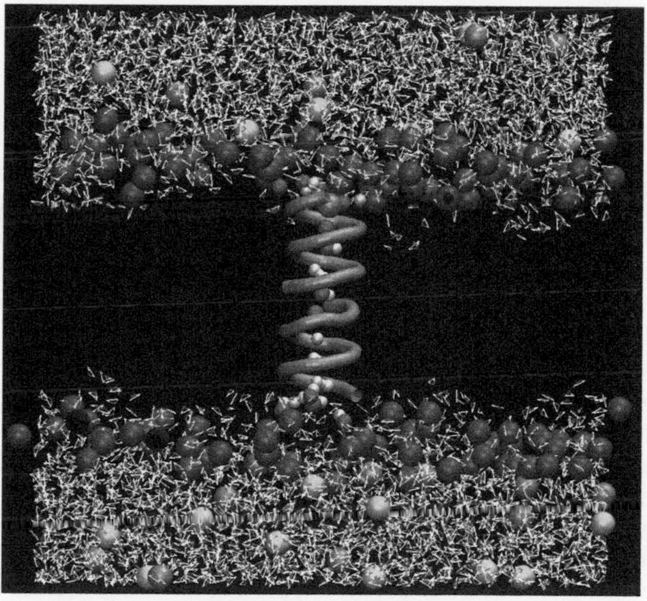

**FIGURE 8.2** (A color version of this figure follows p. 140.) Model of gramicidin A. This nanotube-forming peptide spans the membrane, holds a molecular thread of water in its core, and forms a conduit for rapid flux of protons. The structure and function of this membrane-spanning antibiotic have been studied in great detail and provide a clear picture of how water becomes toxic if allowed to permeate the membrane. Water wires forming spontaneously but infrequently in all membranes are proposed to act by the same mechanism, resulting in potentially catastrophic energy uncoupling. (Image generated by chemical simulation; courtesy of Serdar Kuyucak.)

membrane of a gram-positive target cell. A single monomer spans the distance of only one of the two leaflets forming the bilayer, requiring head to head dimerization with a second monomer to enable a full length nanotube. The core of each nanotube contains approximately 21 waters in single file (i.e., a water wire), a number sufficient to span across the membrane allowing "downhill" (i.e., high outside/low inside) proton flow. This catastrophic form of proton uncoupling short-circuits proton electrochemical gradients—in essence bleeding the cell of energy. The mature nanotubes formed by two monomers joined at their amino terminal ends are unstable within the bilayer and are constantly forming and disassociating, such that the average open time for a nanotube is about 1 second. However, with a large electrochemical gradient, one water wire can transport about 20,000 protons per open channel each millisecond, which is 1000 times faster than a single mobile cation carrier molecule such as valinomycin (see Chapter 3). Thus, water wires are an exceptionally fast mechanism for draining cation gradients compared to mobile carriers that work by diffusing from one side of the bilayer to the other. In contrast to diffusional carriers, gramicidin A nanotubes function in both physical states of the bilayer, the gel-phase and the liquid crystalline phase, although we are not aware that dimers become more stabilized when embedded in gel-phase lipids.

## 8.2 WATER WIRES LIKELY FORM SPONTANEOUSLY IN MEMBRANES

The development of water wire theory as applied to natural membranes began with the proton tunneling mechanism proposed to explain proton movement in ice. This mechanism was applied and tested in chemically defined membrane vesicles as well as bacterial membranes and was ultimately confirmed using gramicidin A nanotubes. Membrane bulking has been proposed as a defense, and here we generalize these observations to DHA membranes and further explore the broad ecological and applied implications.

The heart of water wire theory as applied to natural membranes involves two principles, one chemical and the other electrochemical. The first concerns refinement of the well-established chemical principle that water is largely excluded from or has little affinity for the lipid portion of the membrane. Chemical simulation models support the classic view that water seldom enters the interior of the membrane, but also show that water wires do form spontaneously, though infrequently. The second point is that, regardless of the low probability of water entering the bilayer, when a molecular thread of water does form, protons move downhill so rapidly that the cell's chemiosmotic energy store may be threatened. Perhaps the best validation of this model involves research on gramicidin A discussed above, where the molecular thread of water in the core of this antibiotic works as a molecular pore for sizing and transporting cations: $H^+ >>> Na^+ > K^+$. This mechanism of cation selectivity is clearly a property of the water wire pore of gramicidin A, and it is interesting to compare these data with cation permeability rates seen with chemically defined membrane vesicles used in biophysical experiments on permeability. The comparative rates for cations are remarkably similar, which is not only consistent with the presence of a water wire mechanism operating in membranes, but also suggests that the molecular thread of water surrounded by lipids has a similar sizing mechanism. Also recall that biochemical studies using bacterial membrane vesicles show that proton permeability is modulated by temperature. For example, substantially increasing temperature results in increasing proton leakiness that can be interpreted in terms of enhanced rates for water molecules entering hyperfluidized membranes. Importantly, cooling temperatures below the growth temperature of cells used to isolate vesicles also increases proton leakiness, which can also be envisioned to involve an increased probability of water wire formation. The latter data can be interpreted in terms of more water wires forming around defects in the membrane surface caused by "molecular cracks" perhaps at the edges of gel-phase lipid patches.

The mechanism of action of gramicidin A has been studied using both membrane vesicles and living bacterial cells (see Chapter 3). Typically, target cells are killed due to catastrophic uncoupling of cation gradients, but there are instances, as already introduced in Chapter 3, in which cells can be rescued and grow with high levels of gramicidin A in their bilayers. These cells have become dependent on gramicidin A as though water wires become essential for survival. In essence, a series of manipulations of the growth medium overcomes the lethal effects of proton conductance by the core water wire of gramicidin A. This involves the beneficial harnessing of water wires for transporting high levels of $K^+$ into the cell while artificially maintaining extracellular $H^+$ and $Na^+$ at levels compatible with intracellular needs.

Ecological and later genomic analyses provide a further validation of water wire theory. The ecological data involve isolation and physiological analysis of a bacterium living in a hot spring in New Zealand. This anaerobic isolate seemed to defy water wire theory for protons by growing at about 80°C, near record temperatures for cells producing fatty acid–based bilayers. The data are convincing that at these temperatures, so many water wires form spontaneously in fatty acid–based membranes that a system of bioenergetics based on proton energy gradients is no longer feasible (Figure 8.3). The membrane of this isolate was shown to support a system of energy transduction based on $Na^+$ as primary bioenergetic cation, which we argue is a direct evolutionary response related to energy stress from water wire formation. This requires evolutionary remodeling of the critical proton pump, ATP synthase working in reverse, to pump out the larger $Na^+$ ions entering the cytoplasm as primary cation in bioenergetics. It is proposed that once this adjustment in specificity of cation circulation is made, the sizing mechanism of water wires for cations works efficiently enough to protect $Na^+$ electrochemical gradients, in contrast to proton gradients, at this temperature. Evidence that these membranes are extremely leaky for protons is based on the observation that cells grow only at a narrow window of pH, near neutrality. Note the similarity to the case with gramicidin-inhibited *Streptococcus faecalis*, discussed in Chapter 3. Even more dramatic examples of $Na^+$ switching as a defense against water wires are discussed in Chapter 5 and Chapter 15.

Whereas ecological implications of water wire theory are highlighted here, it is important to note that many fundamental aspects of cation conductance remain to be discovered. For example, it has been proposed that enzymes such as proton pumps or proton-energized transporters harness water wire circuitry for moving protons at great rates to or from the membrane surface to enzyme sites situated in the interior

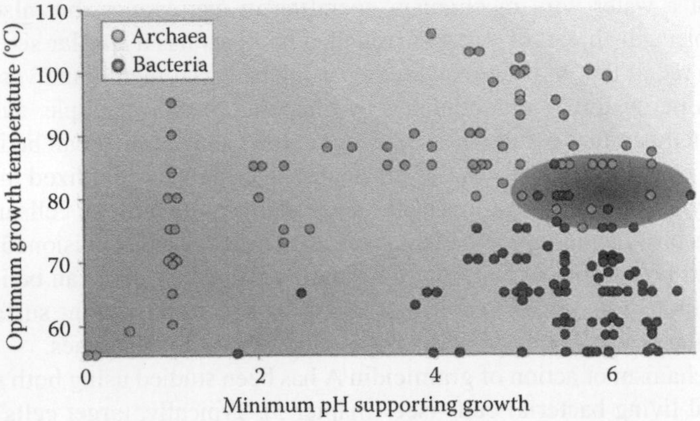


**FIGURE 8.3** Temperature and pH requirements for growth distinguish thermophilic bacteria and archaea: 72 archaeal species that represent 32 genera are included (lighty shaded dots), as are 107 bacterial species that represent 61 genera (dark shaded dots). Methanogenic archaea are excluded. Only the type strain for each species is shown, and an average temperature is given for species that have a range for the optimal growth temperature. The lower zone comprises environmental conditions to which bacteria are best adapted, the upper zone comprises conditions to which archaea are best adapted, and the ovular zone represents those conditions to which both archaeal and bacterial species are well adapted. (Data were compiled from Bergey's Manual of Systematic Bacteriology and from the primary literature. Reproduced from Valentine, D. L., *Nature Rev. Microbiol.,* 5:316–323, 2007. With permission.)

of the bilayer. Indeed, certain membrane enzymes handling protons are proposed to use water wires created within the peptide domains themselves to move protons. Truncated wires that span only one membrane leaflet might be used as a conduit for delivery of protons from the surface to a proton well or storage area, such as water aggregates confined in the protein or luminal region between monolayers. In addition to gramicidin A, a number of other peptide-based nanotubes, such as certain bacteriocins, likely involve water wires as a mechanism for cellular killing. An interesting case, in which water wires appear to be beneficial to yeast cells used in winemaking, is discussed in Chapter 16. Molecular roles of water wires in human cells are discussed in Chapters 13, 19, and 20.

## 8.3 DHA AND WATER WIRES

The benefits of DHA as powerful membrane antifreeze acting to enhance electron transport in phytoplankton and bacteria, and to disrupt lipid rafts in animal cells, are balanced against putative negative effects on proton permeability. Whereas there is convincing evidence that DHA-enriched membranes of mitochondria are prone to proton leakiness, molecular mechanisms are not clear. Indeed, there are many interesting questions that remain to be answered regarding the permeability properties of DHA-enriched membranes. For example, DHA enrichment of mitochondrial membranes is most prevalent among ectothermic organisms, leading to the speculation that cold temperatures might somehow reduce membrane permeability toward cations. However, we're not aware of any direct evidence on this point. The absence of DHA in most mitochondria of endothermic organisms might be explained on the basis of thermally induced changes weakening the permeability barrier against protons. However, there are exceptions to this rule involving DHA enrichment of mitochondria of fast muscles. As discussed in Chapter 12, fast muscles might tolerate a certain level of proton leakage, provided an excess supply of energy substrates is available.

Theoretical and experimental results consistent with proton conductance via a water wire mechanism in DHA-enriched bilayers are as follows:

- Water content in the interior of the bilayer increases with increasing chain unsaturation.
- Permeability of water and other solutes is increased.
- Relationship between DHA content and proton permeability is linear.
- DHA increases $H^+$ permeability of mitochondrial membranes.
- Loose lipid packing is predicted at the aqueous interface, resulting in penetration of water into the bilayer.

One of the simplest explanations of these data/properties is that water wires tend to form more readily in DHA-enriched bilayers. However, as discussed in Chapter 12, novel protective mechanisms may have evolved to conserve energy in DHA membranes.

## 8.4 FATTY ACID BULKING FOR "PLUGGING THE PROTON DIKE"

It has been proposed by Tom Haines (2001) that bulky lipid chains in the membrane, including cholesterol, hopanoids, and branched-chain fatty acids, prevent clusters of water (i.e., water wires) from forming spontaneously in the membrane. According to his theory, bulky fatty acid chains work directly as a defense against energy uncoupling of cation electrochemical gradients. Since bacteria place additional methyl groups and or cyclopropane groups on their lipid chains at different depths across the bilayer, systematically identifying these locations may provide clues as to where bilayers are most vulnerable to proton leakage, and ultimately provide a map of the route protons take to cross the bilayer. By this reasoning the locations of bulking agents likely identify the physical location of the rate-limiting step in proton leakage.

The artist's sketch in Figure 8.4 shows a side view of a membrane representing a composite picture of the location of extra bulky groups in membranes of *E. coli* and *Bacillus subtilus.* Note that *B. subtilis* adds extra methyl groups as methyl-branched fatty acids near the luminal region, in the form of *iso*-methyl branched chains. Methyl groups shown slightly further into the interior of the bilayer are derived from *anteiso*-branched fatty acids where the methyl group is attached one carbon further up the chain from the methyl end. Haines (2001) has proposed that methyl-branched fatty acids add extra bulk to close or pinch off water wires forming in or near the lumen. *E. coli* has evolved a different method of bulking, targeting cyclopropane groups to interior locations of fatty acid chains targeting sites of double

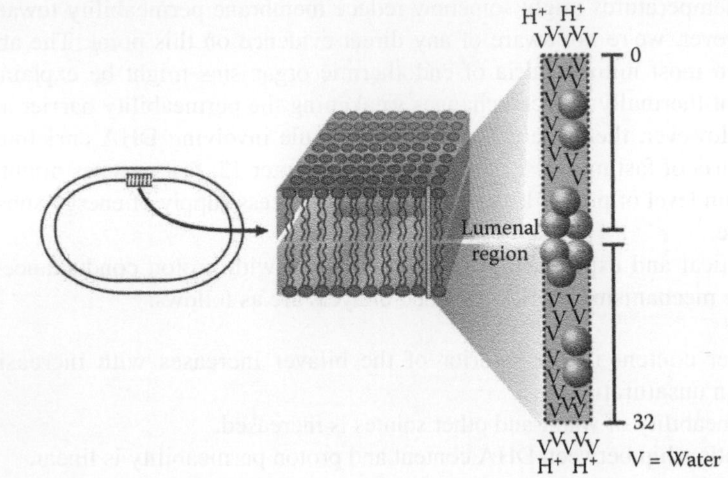


**FIGURE 8.4** Membrane bulking defenses used by bacteria provide a roadmap as to the locations where water wires are most effectively inhibited. Extra methyl groups donated by branched fatty acids and cyclopropane groups are shown as circles. The V-shaped structures are water molecules entering the lipid portion of the bilayer and initiating formation of water wires that are being pinched off by the increased bulk of the additional methyl groups. Data are based on fatty acid methylation patterns of *E. coli* and *B. subtilis*, and the concept is an extension of membrane bulking theory proposed by Tom Haines (2001).

bonds at positions $16{:}1_{n7}$ and $18{:}1_{n7}$. There are four possible locations for cyclopropane incorporation clustered in the interior regions, in contrast to the luminal region as seen with *B. subtilis*. Recall that cyclopropane groups start out as methyl groups and then undergo ring closure to form cyclopropane rings. Thus, a methyl/cyclopropane bulking pattern derived from these two bacterial membranes consists of eight locations across the bilayer involving bulking in the interior of each leaflet as well as luminal regions. We suggest that this methylation or cyclopropanation pattern serves to identify the most effective locations to inhibit formation of water wires. Based on this perspective, we suggest that water may penetrate more readily to several carbon lengths down from the membrane surface than to the midmembrane or luminal regions, consistent with proximity to aqueous solution. This is consistent with water wires forming a path for protons to cross the bilayer, with the greatest susceptibility to inhibition being in the midmembrane to luminal region.

The case histories of cyclopropane fatty acids as putative bulking agents are discussed in more detail next. Cyclopropane fatty acids (CFAs) are found in the phospholipids of membrane bilayers of many different bacteria. CFAs are produced by methylation of monounsaturated fatty acid phospholipids at an energy cost of multiple ATPs per molecule, and, as such, must have a function distinct from that of the precursor fatty acids. CFAs are derived from monounsaturated fatty acids esterified in membrane phospholipids and are believed to be essentially equivalent in maintaining the fluidity of the membrane as their monounsaturated precursors. It has been proposed that CFAs are simply more chemically stable analogs of monounsatured precursors, produced primarily in response to oxidative stress. While this is undoubtedly true, it is likely not the sole function of these molecules. For example, *S. faecalis* mutants lacking CFAs are more susceptible to the toxic effects of NaCl in the media than the CFA-producing parent strain.

CFA levels in wild-type *E. coli* strains are positively correlated with tolerance to large, rapid decreases in pH, which might help explain the role of CFAs. The fatty acid profile shown in Figure 8.5 (upper right) is typical of *E. coli* cells starved for energy or entering the dormant stage of the life cycle. Note the major peaks for CFAs, 17:0 CFA/19:0 CFA. The lower-right profile shows the absence of cyclopropane groups in a mutant used as control and blocked for the methylating enzyme

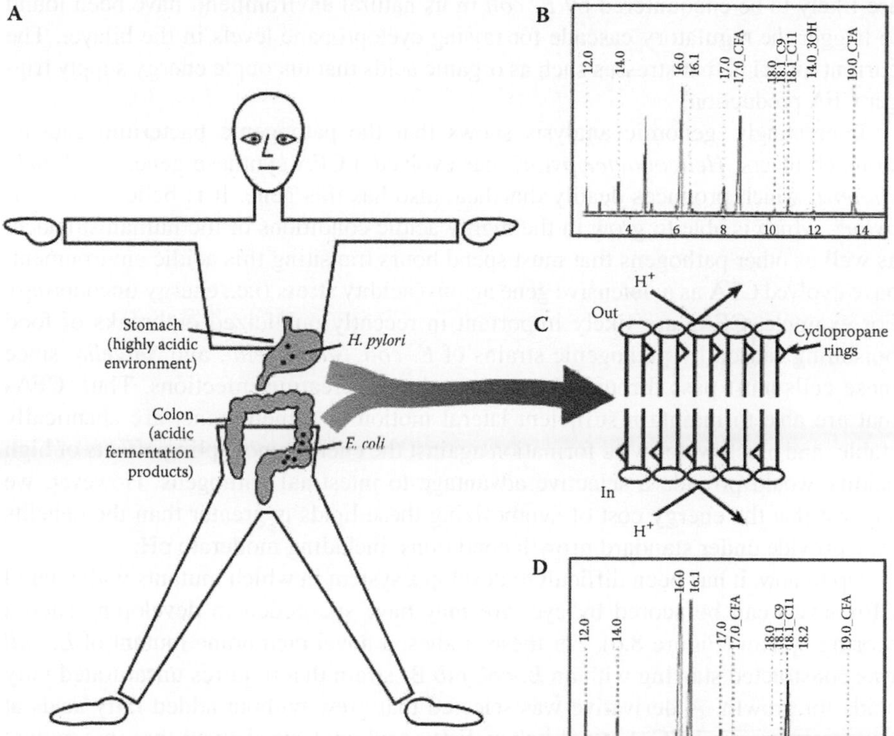


**FIGURE 8.5** Membrane bulking defenses used by some human intestinal pathogens. (A) and (C) Intestinal pathogens, such as foodborne derivatives of *E. coli*, *Salmonella*, and *Shigella* and the common ulcer-forming bacterium *Helicobacter pylori* likely combat energy uncoupling caused by high acidity in the stomach by bulking with cyclopropane groups. (B) The fatty acid profile represents wild-type *E. coli* grown under acidity stress, compared to the profile in panel (D) in which cyclopropane synthesis is blocked by mutation. *E. coli* has evolved a signaling system that senses and responds to acidity stress by replacing double bonds with cyclopropane groups. Note that 16:1 is not converted to its cyclopropane derivative in a mutant compared to wild type where most of the 16:1 peak is shifted to a peak labeled 17:0 CFA.

(mutant kindly provided by John Cronan). In this case peaks for the precursors of CFA, 16:1 and 18:1, predominate.

To learn more about the ecological importance of this mechanism in nature, we have studied the levels of CFA membranes of a wide range of enteric bacteria freshly isolated and typed at a local hospital. All of the isolates displayed high levels of cyclopropane derivatives after overnight incubation in a medium lacking energy sources, with some strains converting >90% of their total *cis*-double bonds to CFA groups. As mentioned above, based on mutant analysis, the cyclopropane group seems to play a major role in defense against acidity ($H^+$), raising the possibility of the direct role of cyclopropane rings as proton shields in the bilayer. This is consistent with water wire theory. Interestingly, proton energy uncouplers such as acetic acid, which are likely to be encountered by *E. coli* in its natural environment, have been found to trigger the regulatory cascade for raising cyclopropane levels in the bilayer. The current model is that stresses such as organic acids that uncouple energy supply trigger CFA production.

Interestingly, genomic analysis shows that the pathogenic bacterium causing stomach ulcers, *Helicobacter pylori*, has evolved a CFA synthase gene, and *Vibrio cholera,* which produces deadly diarrhea, also has this gene. It is believed that *H. pylori,* which is able to grow in the highly acidic conditions of the human stomach, as well as other pathogens that must spend hours transiting this acidic environment, have evolved CFA as a defensive gene against acidity stress (i.e., energy uncoupling). For example, CFAs are likely important in recently publicized outbreaks of food poisoning caused by pathogenic strains of *E. coli*, *Salmonella,* and *Shigella*, since these cells must pass through the stomach before creating infections. Thus, CFAs that are able to maintain sufficient lateral motion of a membrane, are chemically stable, and block water wire formation against the energy uncoupling effects of high acidity would provide a selective advantage to intestinal pathogens. However, we suggest that the energy cost of synthesizing these lipids is greater than the benefits they provide under standard growth conditions, including moderate pH.

Up to now, it has been difficult to develop a system in which mutants with altered CFA levels can be scored by eye. We may have succeeded in developing such a scoring system (Figure 8.6). For these studies, a novel membrane mutant of *E. coli* was constructed starting with an *E. coli fab* $B^-$ strain that requires unsaturated fatty acids for growth. A derivative was selected that grew without added fatty acids at temperatures of >30°C, but not below. Fatty acid analysis showed that this mutant produced a hypofluidized membrane of fixed (i.e., independent of temperature) composition—around 32% total monounsaturated fatty acids, predominantly $18{:}1_{n7}$. This mutant was found not only to be cold sensitive, but also salinity sensitive. However, addition of crystals of a variety of slightly acidic compounds to the center of the plate resulted in a halo of growth. Membranes of cells isolated in the growth region contain high levels of cyclopropane fatty acids that might protect these cells simultaneously against both mild acidity stress and salinity stress. Effects of acidity on $Na^+$ transporters have not been ruled out; however, acidity is expected to decrease $Na^+$ efflux. We suggest that the hypofluidized membrane of this mutant may be permeable to toxic levels of $Na^+$ with slight acidity triggering sufficient upregulation of

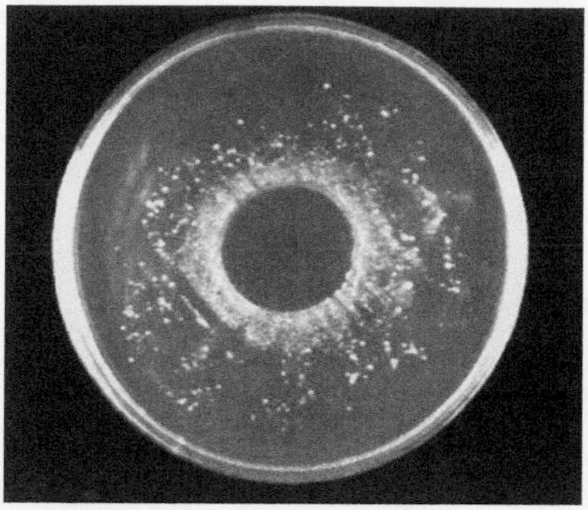

**FIGURE 8.6** Cyclopropane groups might also serve as a membrane defense against $Na^+$ toxicity. A derivative of *E. coli fab* $B^-$ failed to grow when incubated on Luria broth at 30°C. Preliminary experiments showed that cells are sensitive to salinity present in this medium, which is overcome by feeding unsaturated fatty acids. Surprisingly, addition of a few crystals of moderately acidic molecules such as ascorbic acid placed at the center of the plate also overcomes salinity stress and creates a halo of growth. Sublethal amounts of acidity in their environment appear to rescue these cells by signaling strong upregulation of levels of cyclopropane groups, perhaps working simultaneously as a defense against salinity and acidity.

CFA as a "bulking defense" against $Na^+$ to allow these cells to grow in the presence of relatively high levels of salinity.

## 8.5 SUMMARY

Initial data that chemically defined membrane vesicles are leaky against protons seemed to defy Mitchell's concept of the central role played by membranes in maintaining proton electrochemical gradients essential in bioenergetics. This issue has now been resolved with the conclusion that membranes without specialized lipid-based defenses are indeed leaky against protons. Water wires forming in the core of gramicidin A nanotubes and other lipophilic peptide uncouplers are known to cause rapid proton conductance, and this mechanism serves as a model for proton leakiness of natural membranes. Numerous triggers observed to cause cation leakage seem to be explained by water wire theory. These triggers include heat, cold, lipophilic solvents, defective protein-generated holes, excessive unsaturation, chain shortening, and detergents. Studies with vesicles also show that $Na^+$ is orders of magnitude less permeable than $H^+$, and this important property has now been found to be harnessed by certain bacteria that, in order to overcome chronic energy stress caused by proton leakage, evolve to use $Na^+$ gradients for energy storage. Evidence is accumulating that bulky fatty acids work as defenses against futile cycling of $H^+/Na^+$. These studies highlight the importance of fatty acid structure to membrane permeability and bioenergetics, key parameters in understanding how organisms interact with their environments. The conservation of proton gradients appears so important that cells have evolved a battery of membrane-based defenses, some of which are now coming to light. The permeability properties of DHA membranes can be explained partially by a water-wire mechanism with favorable conformations of DHA being dependent on low temperature and high pressure, as well as the configuration of the chemical linkage between DHA and its headgroup.

## SELECTED BIBLIOGRAPHY

Andersen, O. S., R. E. Koeppe II, and B. Roux. 2005. Gramicidin channels. *IEEE Trans. Nanobiosci.* 4:10–20.

Brogden, K. A. 2005. Antimicrobial peptides: Pore formers or metabolic inhibitors in bacteria? *Nat. Rev. Microbiol.* 3:238–250.

Chang, Y.-Y., and J. E. Cronan. 1999. Membrane cyclopropane fatty acid content is a major factor in acid resistance of *Escherichia coli*. *Mol. Microbiol.* 33:249–259.

Chou, T. 2004. Water alignment, dipolar interactions, and multiple proton occupancy during water-wire proton transport. *Biophys. J.* 86:2827–2836.

Grogan, D. W., and J. E. Cronan. 1997. Cyclopropane ring formation in membrane lipids of bacteria. *Microbiol. Mol. Biol. Rev.* 61:429–441.

Haines, T. H. 2001. Do sterols reduce proton and sodium leaks through lipid bilayers? *Prog. Lipid Res.* 40:299–324.

Jenski, L. J., and W. Stillwell. 2001. The role of docosahexaenoic acid in determining membrane structure and function: Lessons learned from normal and neoplastic leukocytes. In *Fatty Acids: Physiological and Behavioral Functions.* ed. D. I. Mostofsky, S. Yehuda, and N. Salem, Jr., 41–62. Totowa, NJ: Humana Press.

Nagle, J. F., and H. J. Morowitz. 1978. Molecular mechanisms for proton transport in membranes. *Proc. Natl. Acad. Sci. U.S.A.* 75:298–302.

Paula, S., A. G. Volkov., A. N. Van Hoek, T. H. Haines, and D. W. Deamer. 1996. Permeation of protons, potassium ions, and small molecules through phospholipid bilayers as a function of membrane thickness. *Biophys. J.* 70:339–348.

Roux, B., and M. Karplus. 1994. Molecular dynamics simulations of the gramicidin channel. *Annu. Rev. Biophys. Biomol. Struct.* 23:731–761.

Smart, O. S., J. M. Goodfellow, and B. A. Wallace. 1993. The pore dimensions of gramicidin A. *Biophys. J.* 65:2455–2460.

Speelmans, G., B. Poolman, T. Abee, and W. N. Konings. 1993. Energy transduction in the thermophilic anaerobic bacterium *Clostridium fervidus* is exclusively coupled to sodium ions. *Proc. Natl. Acad. Sci. U.S.A.* 90:7975–7979.

Stillwell, W., L. J. Jenski, F. T. Crump, and W. Ehringer. 1997. Effect of docosahexaenoic acid on mouse mitochondrial membrane properties. *Lipids* 32:497–506.

Venable, R. M., and R. W. Pastor. 2002. Molecular dynamics simulations of water wires in a lipid bilayer and water/octane model systems. *J. Chem. Phys.*116:2663–2664.

# 9 DHA/EPA Membranes as Targets of Oxidative Damage

Hypothesis: DHA/EPA-enriched membranes are a source of oxidative stress to the cell because they are readily oxidized, and cells have therefore evolved a variety of protective mechanisms against lipoxidation of these bilayers.

DHA membranes in mammals turn over rapidly and are converted to potentially toxic molecules following chemical oxidation by molecular $O_2$ (i.e., lipoxidation). These reactions occur in the dark but are accelerated in light. Thus, DHA/EPA molecules are not always beneficial for cells and indeed can be considered a double-edged sword in terms of benefits versus risks, especially for aerobic organisms. Evidence summarized in this chapter emphasizes the risks inherent in incorporating DHA/EPA into the membrane. We suggest that these substantial risks must be counterbalanced by extraordinary benefits to account for the prevalence of DHA and EPA in the biosphere.

The distribution in the biosphere of cells with DHA/EPA membranes provides clues as to the risks associated with these molecules:

- *Niche specialization*—DHA/EPA-producing bacteria, the simplest cells forming these chains, are rare in nature and are often isolated from the gastrointestinal tracts of marine animals and anoxic sediments, niches low in $O_2$ and where oxidative stress is minimized.
- *Abstinence*—Many organisms/cells, including *E. coli* and wine yeast, avoid incorporating or synthesizing DHA/EPA chains.
- *Marine bias*—Land plants do not incorporate or synthesize DHA/EPA, in contrast to marine plants that produce copious amounts. A simple explanation of this distribution pattern is that marine plants experience a net benefit from these chains, in contrast to land plants where risks exceed benefits (Figure 9.1).
- *Migration and environmental manipulation*—The nematode *C. elegans* has evolved dual behavioral systems that help to avoid lipoxidation of its EPA enriched bilayers. These mechanisms involve taxis toward subambient $O_2$ levels (usually in dark environments). In addition, social feeding behavior communally lowers $O_2$ levels.
- *Physiological adaptations*—Hummingbirds and certain long-distance flyers have evolved novel mechanisms, including torpor as well as migration and life at high elevation, to protect their DHA membranes. A natural

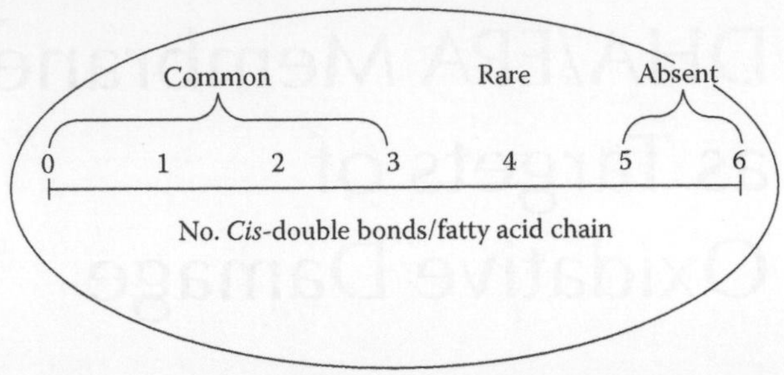


**FIGURE 9.1** Unsaturation ladder showing a maximum of four double bonds in fatty acids of land plants. Lipoxidation chemistry can be used to help explain why terrestrial plants limit the numbers of double bonds in their membranes compared to marine plants. That is, EPA and DHA, with five and six double bonds, while widespread in marine plants, are proposed to be harmful to land plants (see Chapter 17). In contrast, benefits outweigh risks in marine plants, which produce prodigious amounts of DHA/EPA for their membranes. The pendulum of benefit-risk in land plants is believed to swing into negative territory beyond the third rung of the unsaturation ladder. The benefits of DHA/EPA in marine plants are discussed in Chapter 11.

doping mechanism in migratory birds is seemingly used to provide maximum benefit while minimizing duration of exposure.

- *Selective cellular targeting*—Humans target DHA to rod cells, neurons, and sperm and away from most other kinds of cells. Rod cells and neurons have evolved sophisticated repair, recycling, and biochemical defenses to limit damage to their membranes. Sperm may avoid lipoxidation because of low $O_2$ levels in the female reproductive tract (discussed in Chapter 13).

We propose that chemical oxidation and toxicity of associated lipoxidation products strongly modulates the distribution pattern of DHA/EPA-enriched membranes in the biosphere, as well as the physiology and behavior of organisms that incorporate these lipids in their membranes.

## 9.1 BRIEF CHEMISTRY OF LIPOXIDATION OF DHA/EPA MEMBRANES

The chemical stability of biomembranes in an oxygenated environment is dependent to a large degree on the number of double bonds in the fatty acid chains of phospholipids (Figure 9.2). The chemical principles governing lipid oxidation are now understood in detail and are readily applied to membranes. For example, omega-3–enriched membranes are subject to chemical oxidation in which molecular oxygen attacks the double bonds of these fatty acids. The oxidation of unsaturated fatty acids is a spontaneous chemical process and as such will occur wherever the proper conditions exist, from a bottle of cooking oil to a cellular membrane. Polyunsaturated fatty acids are far more susceptible to oxidation than monounsaturated fatty acids, with linolenic acid ($18{:}2_{n6}$) being oxidized at a rate more than 10-fold greater than oleic acid ($18{:}1_{n9}$). The large increase in oxidation rate is due to the methylene-interrupted

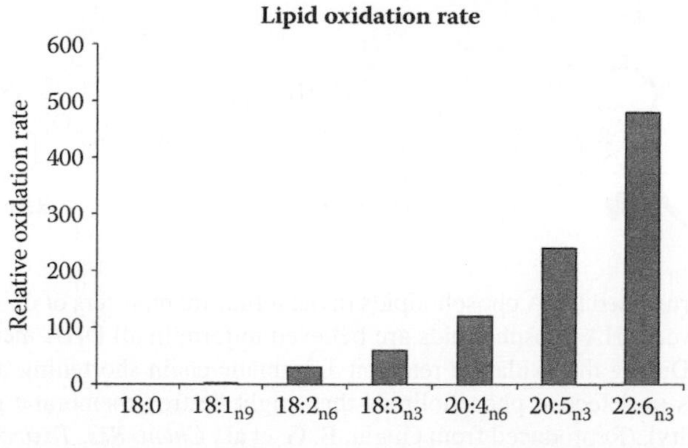


**FIGURE 9.2** Lipoxidation rates depend to a large degree on the number of double bonds in the fatty acid chains of phospholipids. DHA is the most readily oxidized fatty acid in nature and wherever $O_2$ is present yields chemical breakdown products that create defective membranes and are also toxic to the cell. (Courtesy of Luke Hillyard.)

structure of the double bonds of fatty acids that allow for resonance stabilization of the free radical intermediate that is formed as through radical chain autooxidation (as opposed to photooxidation). For every double bond past two per fatty acid, the rate of oxidation is approximately doubled; thus, the rate of oxidation of DHA is ~480 times the rate of oxidation of $18{:}1_{n9}$, and the rate for EPA is estimated to be about half of this value. Various environmental factors, such as increased temperature, trace metals (e.g., iron), oxygen concentration, and especially irradiance, can greatly increase oxidation rates. Rates of photooxidation are much higher as compared with autooxidation (dark). For example, the rate of photooxidation, such as that which occurs in plants, rod cells, and so forth, of $18{:}1_{n9}$ is ~30,000 times that of autooxidation, and photooxidation of $18{:}3_{n3}$ occurs ~900 times faster than does autooxidation.

Thus, as discussed by an expert (Mitchell, 1961; see reference in Chapter 1) in this field, photooxidation differs from autooxidation in several important aspects: (1) it involves reaction with singlet oxygen atoms produced from molecular oxygen by light interacting with a sensitizer, such as chlorophyll; (2) it is not a free radical process as with autooxidation; (3) it displays no induction period; (4) it is unaffected by antioxidants normally capable of inhibiting autooxidation, but is inhibited by singlet oxygen quenchers, such as carotene; (5) it gives products that are similar in type, but not identical in structure to those obtained by autooxidation; and (6) as already mentioned, the reaction occurs more quickly than autooxidation, especially for monounsaturated fatty acids, such as $18{:}1_{n9}$, and its rate is related to the number of double bonds, rather than the number of methylene groups between double bonds.

Several products of lipid oxidation are known to play important biological roles. For example, organic peroxides are common intermediates in autooxidation of unsaturated fatty acids. These molecules belong to a class of metabolites called reactive oxygen species, which behave as potent toxicants and are implicated in aging, as well as cancer and other diseases. Organic peroxides produced by lipid oxidation contribute to the cellular pool of reactive oxygen species, triggering the oxidative stress cascade. Because the chemistry of oxidative stress involves a chain reaction,

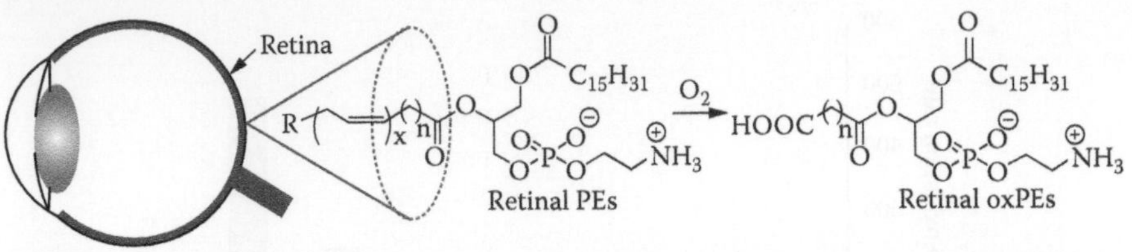


**FIGURE 9.3** Truncated DHA phospholipids in the retina are reporters of oxidative damage. Oxidatively cleaved DHA phospholipids are believed to form in all DHA membranes in the presence of $O_2$. During the oxidation reaction a dramatic chain shortening or cleavage can occur and results in defective phospholipids that might destroy membrane processes (e.g., cation permeability). (Reproduced from Gugiu, B. G. et al., *Chem. Res. Toxicol.*, 19:262–271, 2006. With permission.)

the formation of organic peroxides is self-perpetuating and can lead to severe cellular damage, including generation of mutations in DNA, and ultimately trigger programmed cellular death (i.e., apoptosis).

Another product arising from oxidative damage of DHA- or EPA-enriched membranes is truncated phospholipids (Figure 9.3). In this structure the oxidized fatty acid chain is cleaved, yielding a dramatically shorter fragment, which remains attached to its headgroup. Advances in analytical chemistry for detecting oxidation products present in only trace amounts in membranes has led to the identification of truncated phospholipids in virtually all oxygen-exposed membranes containing unsaturated fatty acids. This validates the general concept that wherever DHA is present in the membranes of aerobic cells, oxidative damage of membranes is inevitable. Indeed, methods are now sufficiently sophisticated as to detect membrane damage triggered by the smoking of a single cigarette. The "hormonal" roles of oxidatively damaged lipids remaining in the bilayer are being actively investigated, and new functions are being reported. Truncated phospholipids may also reduce membrane fidelity with respect to cations, with the reasoning as follows.

1. Owing to their lipophilic nature, truncated phospholipids arising from membrane lipoxidation remain within the membrane itself.
2. The truncated fatty acids are physically shorter than other membrane lipids, effectively creating a gap in the critical region between the middle of the leaflet and the lumen; this arrangement is likely to increase water wire formation as discussed in Chapters 7 and 16. Truncation of other polyunsaturated phospholipids besides DHA yields a hierarchy of fatty acid chain lengths, and we suggest this general class of molecules is energy uncoupling.
3. Because of the rapid electrochemical drain associated with water wires, a small number of oxidatively truncated phospholipids may be sufficient to uncouple both proton gradients in mitochondria and sodium gradients in the plasma membrane.

Truncated phospholipids are indeed toxic and have been found to trigger apoptosis in animal cells. Cells have evolved repair lipases as defenses for removing these damaged molecules from the membrane. One of the most interesting experiments involves the linkage between truncated phospholipids and apoptosis in animal cells. The key findings are that the truncated phospholipids work as triggers for apoptosis, whereas transgenic cells overexpressing repair lipases are not subject to apoptosis. The most efficient mechanism of repair involves the presence of repair lipases targeted to both sides of the membrane where these enzymes can apparently access and remove damaged phospholipids efficiently from either membrane leaflet. Repair phospholipases generate another class of potentially energy uncoupling phospholipids called lyso glycerophosphorylcholine. This molecule has strong detergent properties also capable of damaging membrane architecture and likely generating water wires.

It is now clear that the large database on mechanisms of chemical lipid oxidation of unsaturated fatty acids has applications in understanding the effect of oxidative damage on membranes of DHA/EPA-enriched cells. Next, the focus shifts to documenting the oxidation risk associated with of omega-3 fatty acids across a hierarchy of life forms as viewed from an ecological perspective. Later chapters deal with specific examples of how oxidative damage is amplified to harm (Chapters 17–20) and in one case benefit (Chapter 21) the cell or organism. A comparative approach involving cells ranging from EPA recombinants of *E. coli* to rod cells of the eye is used to estimate the oxidative risk associated with DHA/EPA chains.

## 9.2 DHA MIGHT BE TOXIC TO *E. Coli*

*E. coli* does not synthesize DHA and, while routinely exposed to these chains in the human diet, has not evolved an uptake system. This is surprising since *E. coli* is able to incorporate virtually all other unsaturated fatty acids common in the human diet. We have recently developed a sensitive growth assay using *E. coli* in which an unsaturated fatty acid–requiring mutant is used as a tester strain able to "detect" the incorporation of even small amounts of DHA into phospholipids. The tester strain was derived from an *E. coli* fab B$^-$ mutant supplied by John Cronan that normally requires relatively large amounts of monounsaturated fatty acids to enable sufficient membrane motion to support metabolism (roughly 50% of total fatty acids). However, the tester strain constructed for these studies already produces most, but not all, of its unsaturated fatty acids required for growth (i.e., about 25% to 30% of total fatty acid chains). Because the levels of unsaturation in this membrane are set by mutations only slightly below the threshold for growth, uptake of only a few percent of exogenous fatty acids tips the balance permitting growth. Temperature also modulates growth of this strain.

In control experiments, we were able to show that EPA stimulates growth and is incorporated into membrane phospholipids at maximum levels of only about 3% total membrane fatty acids. This level of EPA is far below quantities usually required for supplying bulk fluidity and points both to the precarious state of the membrane in the tester strain as well as to the benefits of EPA in providing motion to the membrane. In side-by-side experiments in which EPA supports growth, DHA is not active in supporting growth. There are at least two possible mechanisms to explain these data, lack of uptake or toxicity.

The pattern of DHA uptake in the tester strain has been analyzed and found to be very weak, on the order of less than 1% of total membrane fatty acids. Recall that free fatty acids must transit two membrane barriers before entering the cytoplasm, and movement of fatty acids across the outer membrane requires a non–energy-dependent, fatty acid–specific porin for long chains, such as DHA/EPA. Transiting the inner membrane is believed to occur by a chemical flip-flop process with attachment to highly water-soluble coenzyme A molecules, essentially pulling the oily chains out of the inner leaflet and into the cytoplasmic acyl-SCoA pool. We are not aware that the substrate specificity of either the porin or long chain acyl CoASH synthetase has been tested against DHA, with either step being capable of inhibiting significant uptake. Thus, lack of uptake may explain why DHA does not promote growth of the tester strain.

However, toxicity cannot be ruled out as a possible mechanism. The highest levels of incorporation of DHA, which Calgene scientists first reported in the literature for *E. coli* or its recombinants, are only 1% to 2%. Later, researchers found a 5% value for DHA recombinants, but these cells synthesize essentially a wild-type membrane composition, which might help mask any DHA toxicity. We're not aware of any studies on the viability of *E. coli* with DHA-enriched membranes, so the toxicity question remains open. Toxicity might occur by two separate mechanisms, oxidative damage as emphasized here versus membrane defects arising from conformational changes, as discussed in Chapter 17.

## 9.3 GROWTH OF EPA RECOMBINANTS OF *E. Coli* INDICATES $O_2$ TOXICITY *IN VIVO*

A negative relationship between ambient levels of oxygen and growth of the EPA recombinant of *E. coli* has been demonstrated using a classic microbiological experiment called the deep-stab method. This involves inoculating EPA recombinant cells from a fresh colony using a long needle that is stabbed into tubes partially filled with solid medium. This deposits inoculant along the length of the tube from top to bottom, in what amounts to an oxygen gradient. $O_2$ levels reaching individual cells along the deep-stab are modulated by masses of cells growing along the inoculation tract and consuming $O_2$, with the highest $O_2$ experienced at the surface. Results are that cellular growth below the surface first becomes visible after ~24 hours at 18°C and then rapidly intensifies. However, growth of the recombinant at the agar surface is delayed up to several days, compared to the rapid growth of wild-type cells used as control. This is not expected and is consistent with some inhibitory or toxic effects of oxygen on cellular growth.

EPA recombinant cells have also been found to be more resistant to oxidants such as hydrogen peroxide, $H_2O_2$. Our preferred interpretation of these data is that EPA phospholipids present in recombinant membranes are sensed as oxidants, causing upregulation of natural enzymatic defenses against reactive oxygen species in general. In other words, EPA chains in the recombinant may behave like a general oxidative stressor signal, bolstering a battery of oxidative defenses. In this case the threat is from within; that is, the highly unsaturated chains of EPA are synthesized inside the cell and then incorporated into the membrane. The take-home lesson from studying recombinants is that EPA is likely a serious oxidative threat but is at least partially managed by upregulation of the conventional umbrella of oxidative defenses present in *E. coli*. However, another simple microbiological experiment discussed in Chapter 18 shows that EPA recombinant cells become highly vulnerable to photooxidative killing, which can be explained as overpowering "dark" defenses against damage caused by EPA lipoxidation products. It is interesting to speculate that DHA is an even more powerful oxidant than EPA and is capable of triggering oxidative death of *E. coli*.

## 9.4 YEAST (*SACCHAROMYCES Cerevisiae*) SYNTHESIZES ONLY MONOUNSATURATED CHAINS AND FEEDING THESE CELLS POLYUNSATURATED CHAINS CAN BE TOXIC

Yeast synthesizes only the most oxidatively stable class of unsaturated fatty acid chains, mainly monounsaturates, such as $18{:}1_{n9}$. Mutants blocked at the level of fatty acid desaturation can be forced to incorporate a variety of unsaturated chains, including EPA and DHA. However, a considerable period of time is required for "adaptation" to DHA. Incorporation of other polyunsaturated chains, specifically 18:2 and 18:3, has been studied in detail, with the conclusion that these chains might do more harm than good, at least under certain conditions. For example, cells with polyunsaturated lipids in their membranes are killed by oxidants such as $Cu^{2+}$ at concentrations that have little permanent effects on wild-type cells. There is convincing evidence that within a few minutes after treatment with copper, yeast/fungal cells, especially those with polyunsaturates in their bilayers, become permeabilized at the plasma membrane as measured by $K^+$ leakage. Thus, results with yeast are consistent with a mechanism in which $Cu^{2+}$-induced membrane damage generates water wires that uncouple critical cation gradients at the level of the plasma membrane.

## 9.5 C. *Elegans* PRODUCES EPA AND SEEKS OR CREATES LOW $O_2$ ENVIRONMENTS

This tiny roundworm, about 1 mm long, has a transparent body that exposes its EPA-enriched membranes, including 302 neurons, to visible and ultraviolet radiation. Thus these neurons and membranes in general are targets of photooxidative damage due to their high unsaturation levels. It has been established that EPA plays an important role in *C. elegans*, apparently at the level of neuronal efficiency. Obviously, the need for EPA in its neurons would expose this organism to both light and dark oxidative damage of membranes, leading to oxidative stress. *C. elegans* produces up to 34% of its total membrane fatty acids as EPA, with a significant fraction as di-EPA phospholipids. These structures stand out in being among the most unsaturated membrane components documented in nature (i.e., ten double bonds) and are expected to be highly vulnerable to oxidative damage. *C. elegans* has evolved an umbrella of biochemical mechanisms against oxidative damage and, in addition, has evolved novel behavioral patterns that likely reduce lipoxidation. *C. elegans* moves away from ambient oxygen and seeks environments under soil detritus with roughly one-third of ambient oxygen concentrations; also note that worms in this niche are simultaneously protected from photooxidative damage. If there is no area of low oxygen concentration to be found, the worms will congregate and communally lower the oxygen tension through their communal respiration. Thus, *C. elegans* uses both conventional biochemical defenses as well as migratory and communal mechanisms to avoid atmospheric levels of $O_2$. We hypothesize that these behaviors evolved to minimize lipoxidation of EPA that can trigger cellular death; we further suggest that incorporation of EPA into membranes of *C. elegans* must provide a strong selective advantage in order to force these fundamental behaviors.

## 9.6 BIRDS

Hummingbirds incorporate high quantities of DHA into their mitochondrial membranes, seemingly to maximize energy output for flight. These birds have also evolved several mechanisms to help protect their DHA-enriched membranes from lipoxidation. These mechanisms include an energy-conserving sleep state known as torpor, as well as migratory patterns that minimize oxygen exposure.

Hummingbirds require two to three times their body weight in food per day and conserve energy at night by entering into a state of deep sleep called torpor. In the torpid state energy consumption and fluxes of $O_2$ to electron transport chains are drastically lowered, which allows birds to survive the night without feeding, while protecting their DHA-enriched membranes. Hummingbirds lack specialized heat-saving feathers, so body temperatures are also lowered sharply. A secondary benefit of a lowered body temperature is predicted to be a lower rate of lipoxidation of DHA membranes. Interestingly, species of hummingbirds adapted to life at high elevations, where the partial pressure of $O_2$ is considerably less than 20 kPa, are reported to display decreased rates of mutations in their mitochondrial genomes. This datum is consistent with lipoxidation as a factor in production of DNA-damaging reactive oxygen species.

Many hummingbirds are migratory. For example, small hummingbirds arrive in Southeast Alaska in mid-April after a long migration from southern climates. It is believed that the maximum distance between refueling stops is about 600 miles. Once in Alaska, birds are often faced with a dearth of early flowers. Temperatures, moderated by the maritime climate, vary with 8–9°C at night to about 15–19°C in the daytime during spring and summer. Why have hummingbirds evolved such an odd pattern of migration? Clearly this behavior must be beneficial, but how? The picture that emerges is that hummingbird flight, which requires the most sophisticated, but oxidation-prone mitochondrial membrane structure in nature (see Chapter 12), essentially forces these birds to occupy what amounts to restricted environmental niches. That is, hummingbirds seem to seek environments where low nighttime temperatures and reduced $O_2$ partial pressures work together with this bird's lower nocturnal respiration to minimize damage to DHA membranes essential for rapid flight.

Another perspective on lipoxidation in birds comes from studies of migratory birds. Natural doping with DHA/EPA is a term used by molecular ecologists who study the relationships between long over-ocean migrations of shorebirds and their food preferences. By gorging on DHA/EPA-rich Fundy mud shrimp during a 10- to 20-day refueling stop in Nova Scotia, semipalmated sandpipers migrating from the Arctic then travel nonstop and over the ocean to South America, a distance of about 4000 miles. During their crucial stopover on the food rich mud flats of the Bay of Fundy, sandpipers store oil in bulk (i.e., almost half of their weight) as energy food but also remodel membranes of their flight muscles using DHA/EPA present in their diet (see Chapter 12). This natural doping of their membranes with DHA/EPA might be a necessary adaptation for sustaining long flights, but it is only a temporary fix. That is, once reaching their wintering grounds in South America, it is unlikely that such DHA/EPA-rich food is available. Thus, after a spike in membrane levels of DHA/EPA, a drop likely follows. The pulse of DHA/EPA can be envisioned as an avoidance mechanism, especially if more saturated fatty acids are used to replace DHA/EPA no longer needed in such high levels for adaptation to their new winter environment.

## 9.7 HUMANS: RHODOPSIN DISK MEMBRANES ARE HIGHLY ENRICHED WITH DHA, OXIDIZE RAPIDLY, AND REQUIRE CONTINUOUS RENEWAL

Rhodopsin membrane disks (RMD) are targets of photooxidation during waking periods and subject to autooxidation during sleep. Owing to their high DHA content, it is not surprising that these specialized bilayers are unstable and require continuous renewal. For example, the life span of a newly minted membrane entering the bottom of the stack (see Figure 9.4) and exiting at the apex is about 10 days. That is, roughly 100 aged disks are removed from the top of the stack daily, being replaced with an equal number on the bottom. The biochemistry and cell physiology of renewal of RMD is extremely elaborate, with spent disks being shed and taken up by pigmented epithelial cells acting as phagocytes. Within pigmented epithelial cells, disks become surrounded by fragment cell membranes to form inclusions called phagosomes. The phagosomes are subsequently degraded within the pigmented epithelial cells, with undamaged DHA being salvaged and used for new RMD synthesis. Phagocytosis of spent RMD is accelerated as expected when the disks in the intact eye are exposed to excessive light, with too much light causing permanent damage or even blindness. Interestingly, the life span of RMD in frogs' eyes is significantly increased, which is consistent with known effects of cool temperatures in lowering chemical oxidative damage to DHA membranes. In addition to membrane damage, critical proteins in rhodopsin disks are also subject to photooxidation, but this interesting subject is not discussed any further here.

Thus, the environmental conditions of warm temperatures, light, and high fluxes of $O_2$, combined with the high DHA composition of rod cells, make these bilayers ideal "substrates" for lipid oxidation. The retina has evolved several defense mechanisms that protect against oxidative damage. Conventional defenses include high levels of superoxide dismutase found in photoreceptor outer segments. However, antioxidants are the main line of defense against autooxidation in the retina. Vitamin E appears

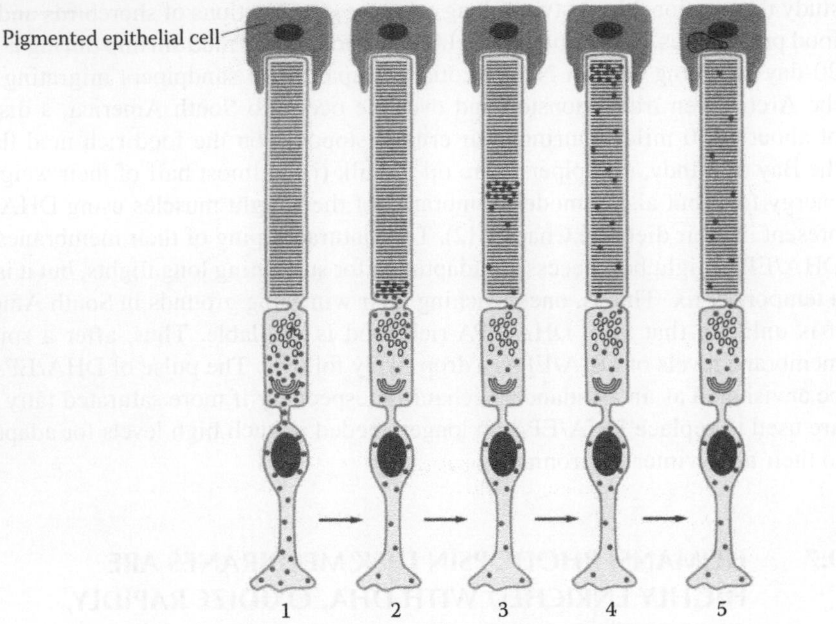


**FIGURE 9.4** DHA-enriched, rhodopsin disk membranes of rod cells decay rapidly and require constant regeneration. Newly minted rhodopsin disks (second panel from left) concentrate radioactive amino acid tracers that label rhodopsin and other membrane proteins. In about 10 days the pulse of radioactivity moves from the bottom 100 disks to the top 100 disks where damaged disks are recycled. We favor a model in which DHA oxidation is at the heart of membrane damage. (From Alberts et al., 1994. Reproduced by permission of Garland Publishing.)

to be located strategically in RMD and protects DHA from oxidation. Incorporation of vitamin E into DHA-enriched membrane vesicles significantly reduces damage to DHA. The presence of ascorbate and reduced glutathione in the retina suggests that they too may serve as retinal antioxidants. Even though the retina has evolved a battery of conventional defenses against oxidative damage, the presence of high levels of DHA appears capable of overwhelming these systems. This necessitates the need for additional protection in the form of active remodeling and highly sophisticated membrane renewal mechanisms as discussed above. Over time or through environmental insults, defenses might begin to deteriorate, resulting in serious damage to vision; that is, at high light intensities the risks associated with DHA may outweigh the benefits.

## 9.8 SUMMARY

Any comprehensive picture regarding the roles of DHA and EPA in the membrane must take into account chemical lipoxidation of these fatty acids. Lipoxidation damages the integrity of the bilayer and has the power to trigger oxidative death. Based on concepts developed above and pursued in later chapters, it seems likely that oxidatively truncated derivatives of DHA- and EPA-phospholipids behave as powerful energy uncouplers. The paucity of DHA in most human cells is consistent with a toxicity mechanism. Finally there is the idea that the proposed extraordinary benefits of DHA in specialized cells and organs must be considerable to compensate for the potential toxicity.

## SELECTED BIBLIOGRAPHY

Alberts, B., D. Bray, J. Lewis, M. Raff, K. Roberts, and J. D. Watson. 1994. *Molecular Biology of the Cell,* 3rd ed. New York: Garland Publishing, Inc.

Avery, S. V., N. G. Howlett, and S. Radice. 1996. Copper toxicity towards *Saccharomyces cerevisiae*: Dependence on plasma membrane fatty acid composition. *Appl. Environ. Microbiol.* 62:3960–3966.

Bleiweiss, R. 1998. Slow rate of molecular evolution in high elevation hummingbirds. *Proc. Natl. Acad. Sci. U.S.A.* 95:612–616.

Frankel, E. N. 2005. *Lipid oxidation*. Oily Press Lipid Library. Vol. 18. Bridgwater, UK: P. J. Barnes & Associates.

Giusto, N. M., S. J. Pasquaré, G. A. Salvador, P. I. Castagnet, M. E. Roque, and M. G. Ilincheta de Boschero. 2000. Lipid metabolism in vertebrate retinal rod outer segments. *Prog. Lipid Res.* 39:315–391.

Gray, J. M., D. S. Karow, H. Lu, et al. 2004. Oxygen sensation and social feeding mediated by a *C. elegans* guanylate cyclase homologue. *Nature* 430:317–322.

Gugiu, B. G., C. A. Mesaros, M. Sun, X. Gu, J. W. Crabb, and R. G. Salomon. 2006. Identification of oxidatively truncated ethanolamine phospholipids in retina and their generation from polyunsaturated phosphatidylethanolamines. *Chem. Res. Toxicol.* 19:262–271.

Gunstone, F. 1996. *Fatty Acid and Lipid Chemistry.* Glasgow, U.K: Blackie Academic and Professional.

Orikasa, Y., T. Nishida, A. Yamada et al. 2006. Recombinant production of docosahexaenoic acid in a polyketide biosynthesis mode in *Escherichia coli. Biotechnol. Lett.* 28:1841–1847.

# Section 4

## Cellular Biology of Omega-3s and Other Membrane Lipids

In this section we consider the roles of DHA and other lipids in membranes from a selection of important cells and organelles: bacteria, chloroplasts, mitochondria, and human sperm. Starting with bacteria, several case histories are developed that highlight the contributions of membrane fatty acids to cellular bioenergetics.

Bacteria are masters at adapting their membrane fatty acid composition to fit their environmental needs and serve as an important model to understand the complexities of membrane structure-function. For example, evidence is accumulating that energy is at the heart of membrane adaptation in bacteria, with applicability to mitochondria and chloroplasts as well.

DHA or EPA are present in mitochondria of cold-adapted animals such as krill and marine fish. It is proposed that DHA/EPA serve as an adaptation to combat energy stress, which is created by a variety of conditions and circumstances. This concept is expanded to explain the benefits of DHA in other mitochondria, such as the fast muscles of the hummingbird breast, which are subject to energy stress created by flight.

Light-harvesting membranes of chloroplasts contain the largest quantities of polyunsaturated fatty acids found in the biosphere. We suggest that light-harvesting efficiency of photosynthetic membranes is dependent on the presence of this vast pool of polyunsaturated phospholipids, which facilitate long distance electron transport. Photosynthesis in marine plants, but not land plants, is often dependent on DHA/EPA phospholipids.

Sperm cells spend an important stage of their life cycle traveling in the female reproductive tract, which is a hostile and potentially lethal environment. To survive, sperm cells have made numerous adaptations, including enriching their tail membranes with DHA. We consider the role of energy stress to understand the molecular roles of DHA in sperm membranes and in doing so develop a new concept on the roles of DHA in mechanically stressed bilayers.

# 10 Bacteria: *Environmental Modulation of Membrane Lipids for Bioenergetic Gain*

Hypothesis: Bacteria actively modulate the fatty acid content of their membrane for optimizing production and conservation of energy.

In this chapter the general properties and evolutionary history of DHA/EPA are generalized to include several other classes of fatty acids important in membranes of bacteria. As background recall that bacteria live in diverse environments and their membrane lipid chains play crucial roles in adaptation and survival. Thermodynamic ecology is the study of how bacteria adapt to energy sources in their environments. Bacteria are well known to blend different fatty acids into their bilayers to match their environment and growth needs; we argue that such changes are driven purely by the bioenergetic needs of the cell.

Several case histories described here highlight the importance of specific membrane fatty acids in bacterial ecology. In the field of bacterial membrane lipid compositions, the series edited by Ratledge and Wilkinson covering fatty acid compositions of major groups of bacteria remains the key reference (e.g., see Wilkinson 1988). What stands out from this large database is the great variety of membrane fatty acids synthesized by bacteria, in contrast to the great similarity in composition among evolutionarily related species. In other words, membrane fatty acid compositions among different isolates of the same species and among related genera are remarkably predictable. Indeed, fatty acid/membrane lipid compositions are often used as important markers in identifying organisms in pure/mixed cultures or microbial communities found in natural environments. For example, lipids extracted from microbial mats forming around marine methane gas seeps were found by fatty acid analysis to have the characteristic spectrum of the common bacterial fatty acid palmitoleic acid (i.e., 16:1). However, closer analysis shows that *cis*-double bonds normally in the n7 (i.e., $16{:}1_{n7}$) position in most bacteria were found to be distributed in different positions along the chain. This composition matches structures obtained from pure cultures of methane-oxidizing bacteria and helps marine microbial ecologists identify the importance of methane-oxidizing bacteria in this natural microbial community. Clearly, these cells have evolved fatty acid elongases/dehydrases, which insert double bonds in this distinct location, a property missing from most other strains.

Pioneering researchers such as John Cronan have identified the core biochemical pathways used by *E. coli* and other bacteria to synthesize their membrane fatty acids and provided important clues in understanding fatty acid structure-function (Cronan and Rock, 1996). A long-range goal in this field is to develop a predictive model for explaining in biochemical language why bacteria synthesize or incorporate different fatty acids into their bilayers. We are not aware of previous attempts to decipher the membrane fatty acid "code" in a generalized way. Some assumptions and rules that we apply in developing a generalized membrane code are as follows:

- Each and every change in membrane fatty acid composition has a biochemical explanation and confers advantage for bacteria in their natural environments.
- At least three essential biochemical parameters must be considered in deciphering membrane composition in bacteria: viscosity, cation permeability, and oxidative stability (note that the latter is most relevant to oxygenated environments).
- We apply the general concept that biochemical functions of membrane lipid chains are determined by their structure.

## 10.1 METHYL-BRANCHED FATTY ACIDS AS MEMBRANE BULKING AGENTS

Many bacteria living in diverse or extreme environments have evolved to synthesize or incorporate methyl-branched fatty acids into their membranes. Branched chains are believed to serve multiple biochemical roles, including bulking agents against proton leakage, and promoting fluidity while exhibiting oxidative stability. This ideal set of properties must be balanced against the complex and energy intensive pathways needed to synthesize these chains, a fact that may help to explain their absence in membranes of most plants and higher animals. As already discussed in Chapter 8, additional bulky methyl groups placed at or near the methyl end of the chain can be pictured to strengthen the proton permeability barrier by way of suppressing formation of the water-based conduits. It has been hypothesized that these additional methyl groups appended to otherwise normal fatty acid chains work at the molecular level by contributing extra density or bulk to the bilayer and thus pinching off the conducting mechanism against protons. Bacteria producing methyl-branched fatty acids are able to fine-tune the structure of their methyl-branched chains by modulating chain length as well as the position of the methyl group near the end of a chain. Note also that isoprenoidal lipid tails of Archaea can be viewed as being highly methylated, consistent with their low permeability to protons.

The important pathogen, *Listeria monocytogenes*, is widely publicized as a major foodborne disease threat. As summarized in Figure 10.1, this organism appears to improve its energy efficiency and perhaps its pathogenic prowess by differentially incorporating methyl branched chains in its membrane lipids when moving back and forth between two extremes in temperature: its free-living lifestyle on refrigerated food (i.e., 2°C to 4°C) versus its human pathogenic state (i.e., 37°C). It is now clear that the extreme shift in temperatures from 2–4°C $\rightarrow$ 37°C requires that these cells restructure their membrane fatty acid compositions to reflect these dramatic

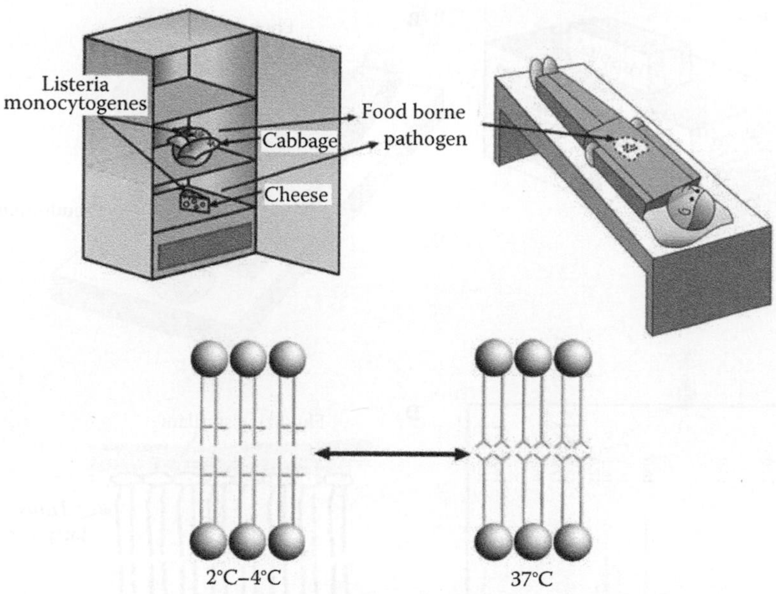


**FIGURE 10.1** Altering structures of membrane fatty acids allows an important foodborne pathogen to lead a double life: refrigerator (2°C) to human host (37°C). *Listeria monocytogenes* is able to grow and contaminate food held at refrigerator temperatures, aided by modification to their membranes including shorter chain fatty acids and relocation of methyl groups, shown on the lower left. Modifying fatty acid chains as an adaptation for life as a human pathogen (lower right) then requires incorporating longer fatty acid chains and methylation at the luminal region, presumably to counter acidity stress associated with transport through the stomach as well as adaptation to growth at body temperature (37°C).

changes in temperature. It is predicted that these changes are necessary to maintain fundamental proton permeability properties allowing these cells to adapt to such different temperatures. Of course, this is not good news for humans, who depend on the refrigerator for food preservation. Indeed this pathogen seems to have "evolved along with the refrigerator," now becoming a major threat to health.

## 10.2 TRANS FATTY ACIDS PLAY MULTIPLE BENEFICIAL ROLES

*Trans* fatty acids are considered a health hazard when consumed in high quantities in the human diet but are beneficial to bacteria living in certain niches subject to rapidly changing conditions. The ecological significance of *trans* fatty acids in a hospital setting is summarized in Figure 10.2. In this scene, representing a typical hospital operating room, a nurse is seen using a powerful disinfectant, such as toluene, to sterilize the surface of the operating table against ubiquitous hospital contaminants belonging to *Pseudomonas* species. Many sterilants are believed to target the membrane itself, essentially dissolving into and destroying vital permeability barrier properties of the bilayer. General sterilants/disinfectants are designed to instantaneously kill pathogens, but after continuous use bacteria are known to become increasingly resistant

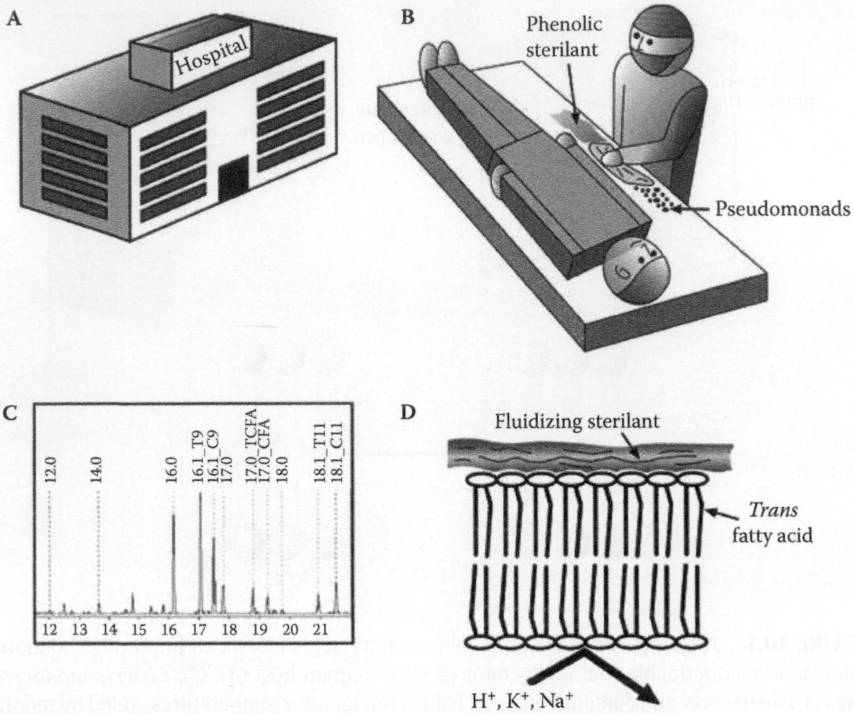


**FIGURE 10.2** *Trans* fatty acids thought to be harmful for human health but beneficial to some bacteria. Bacteria often live in environments (A and B) where rapid shifts in temperature or exposure to membrane-altering sterilants are common. In these cases, bacteria have evolved an enzyme called *cis-trans* isomerase that rapidly alters the isomeric content of the membrane (C, note that T = *trans*) thus changing membrane architecture to confer more energy-conserving properties to the bilayer.

and can cause serious secondary infections in compromised patients. It is difficult to imagine how a cell might defend itself against instantaneous flooding of its environment with such rapid-acting toxic chemicals, but pseudomonads have apparently evolved just such a defensive mechanism to protect their membranes. This involves a constitutive enzyme always present in the cytoplasm and designed to rapidly switch the *cis*-configuration of double bonds of existing monounsaturated fatty acid chains of membrane phospholipids to the *trans* isomer. Although the number of double bonds remains the same, the fixed conformation caused by the *cis* double bond and contributing to sensitivity becomes less exaggerated in the *trans* configuration (i.e., less fluidizing). This conformational change is believed to counter the hyperfluidizing effect of the sterilizing solvent as the *trans* structure contributes to increased viscosity. In other words, the *trans* conformation of these chains is predicted to help seal the membranes of these cells against catastrophic futile cycling/uncoupling of their primary chemiosmotic gradients, analogous in some ways to the incorporation of cholesterol into the bilayer.

*Cis-trans* isomerization as a protective mechanism against hyperfluidizing chemicals presumably evolved in bacteria long before the first application of sterilants by humans. This idea is based on the finding of *trans* fatty acids in bacteria isolated from marine ecosystems subject to rapid fluctuations in temperatures, such as caused by eddies of cold water mixing with warmer water. Indeed, the strain used to generate the fatty acid profile of Figure 10.2 was isolated from marine sediment, although its levels of *trans* fatty acids seems representative of hospital isolates. The favorable thermodynamics of *cis* → *trans* chemistry works in favor of rapid remodeling of existing membranes necessary in the face of apparently catastrophic proton uncoupling caused by sterilants or rapid shifts in temperature; that is, the cell can enact this change rapidly with no significant input of energy.

Another example of *cis-trans* isomerization of fatty acids comes from ruminant animals, such as cows and sheep (Figure 10.3). Indeed, this mechanism might result in health benefits to humans who consume dairy or meat products from free-grazing animals where the product of isomerization, conjugated linolenic acid (CLA), is produced and proposed to possess beneficial bioactive properties. CLA biosynthesis in the rumen involves a mutualistic strategy for protecting syntrophic archaea, an essential class of ruminant microorganisms. Recall that cool-weather plants often used as forage by ruminants contain high levels of polyunsaturated fatty acids such as linoleic acid (18:2). These highly unsaturated membranes are a necessary plant adaptation to cool growth temperatures. There is evidence that methanogenic archaea, carrying out

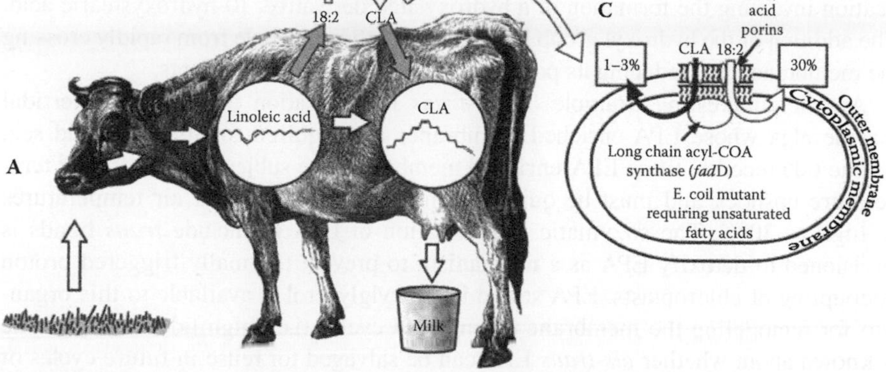


**FIGURE 10.3** Evolution of an efficient ruminant archaea-bacteria symbiosis depends on a novel enzyme capable of detoxifying polyunsaturated fatty acids. Archaeal cells (i.e., methanogens) as essential nutritional symbionts in ruminant animals are believed to be killed by an excess of polyunsaturated fatty acids derived from forage plants and released during the digestive process. Free fatty acids are proposed to behave as potent energy uncouplers against archaea. The dairy cow shown in the illustration grazes on cool-weather grasses that release free, polyunsaturated fatty acids that are toxic to archaea. Bacteria also living in the rumen produce an enzyme called linolenic acid *cis-trans* isomerase, which protects archaea by converting 18:2 to the far less toxic conjugated linoleic acid (CLA). As an added bonus, CLA incorporated into animal products is believed to confer health benefits to humans.

a critical step of balancing hydrogen gas in ruminant metabolism, are vulnerable to energy uncoupling caused by these free fatty acids. Free fatty acids, including 18:2 and others, are formed from plant matter by phospholipases acting in rumen fluid, and tend to build up when the diet contains an abundance of cool-weather grasses. Methanogens seem to lack fatty acid detoxifying genes themselves and are thus sensitive to free fatty acids as toxicants. This concept is based on studies of anaerobic waste digesters—which are similar in many ways to the rumen—where methanogens repeatedly fail to adapt to shock loads of free fatty acids. The current working model is that free fatty acids, such as 18:2, chemically absorb into archaeal membranes and act to shuttle protons in a futile cycle across the membrane. Thus, free fatty acids are believed to cause energy stress in archaea by bleeding their proton chemiosmotic gradient. Die-off of methanogenic archaea ultimately disrupts the hydrogen balance in the rumen, causing bloating, which can be fatal. Certain members of the ruminant bacterial community other than archaea are believed to produce and contribute *cis-trans* isomerases, enzymes that effectively detoxify chains such as linoleic acid by changing their conformation (i.e., 18:2 all *cis* to the *cis-trans* conformation). This structural modification apparently renders these molecules less toxic for archaea. It is known from model studies with *E. coli* that *trans* isomers are weakly taken up unless strains are genetically adapted, and we have repeated this experiment using CLA. However, we are not aware of inhibitor studies with archaea in which CLA is shown to be less toxic than its precursor 18:2, though we consider this is a reasonable inference. Since the enzyme CLA isomerase is not active against monounsaturated fatty acids such as oleic acid ($18{:}1_{n9}$) common in animal forage, other mechanisms to detoxify these chains have likely evolved. For example, there is evidence that oleic acid detoxification may involve a chemical modification involving the formation of a hydroxylated derivative, 10-hydroxystearic acid. The addition of the hydroxyl group likely prevents this molecule from rapidly crossing the membrane, thus reducing its potential for draining proton gradients.

Another interesting example of *cis-trans* isomerization involves an intertidal marine alga whose EPA-enriched membranes are required for life in a cold sea. As the tide recedes, these EPA-enriched membranes are subjected to dramatic temperature upshock and must be quickly remodeled to function at air temperatures as high as 30°C. The enzymatic isomerization of EPA to include *trans* bonds is envisioned to detoxify EPA as a mechanism to prevent thermally triggered proton uncoupling of chloroplasts. EPA stored in triacylglycerol is available to this organism for remodeling the membrane during cold cycles (i.e., high tide), though little is known about whether *cis-trans* EPA can be salvaged for reuse in future cycles of membrane remodeling.

## 10.3 *CIS*-VACCENIC ACID CONFORMATION ENABLES ENERGY-TRANSDUCING MEMBRANES DEPENDENT ON CHOLESTEROL-LIKE MOLECULES

*Cis*-vaccenic acid ($18{:}1_{n7}$) is a close chemical relative of the more common oleic acid, $18{:}1_{n9}$. However, the sole double bond in *cis*-vaccenate is shifted slightly away from the middle of the chain and closer to the methyl end, compared to oleic acid. This shift of double bond position has the effect of significantly decreasing the antifreeze properties of these chains. However, two of these chains are commonly incorporated into a phospholipid to form di-*cis*-vaccenate in which both fatty acid tails are $18{:}1_{n7}$. The antifreeze or viscosity properties of di-$18{:}1_{n7}$ have already been discussed in Chapter 7. The main conclusion is that in the di-unsaturated configuration, this phospholipid contributes powerful antifreeze properties, helping to explain its distribution in many bacteria such as cold-adapted *E. coli*. Because of the geometric tilt of phospholipids at the membrane surface, $18{:}1_{n7}$ chains located in the *sn*-1 position on the glycerol backbone are expected to extend significantly deeper into the opposing leaflet; this configuration and the longer effective chain length of $18{:}1_{n7}$ might add thickness and bulk to the membrane, perhaps improving proton fidelity by inhibiting water wire formation near the luminal region of the bilayer.

An interesting ecological story is emerging regarding bacteria that produce membranes in which virtually all membrane chains are *cis*-vaccenate (Table 10.1). These natural isolates include strictly respiratory and photosynthetic bacteria, which share the common property of living in ecological niches where their essential respiratory substrates for generating proton energy are either perpetually in short supply, or there is selective pressure for maximizing energy efficiency (e.g., soybean root nodule bacteria). The case history of one class of *cis*-vaccenate-producing bacteria, class II methane oxidizers, is shown in Figure 10.4. These ubiquitous bacteria are shown living in the waters of a flooded rice paddy where $O_2$ is limiting for growth. Their other substrate, methane gas, is produced by methanogens lower down in the anaerobic zone and is in plentiful supply.

Although methane is plentiful, a paucity of available $O_2$ causes energy stress. We hypothesize that these bacteria have adapted their membranes as a mechanism to overcome energy stress and adapt to these environmental niches. The highly monounsaturated bilayers produced by these methanotrophic bacteria seem to stand at the crossroads between moderately fluidized membranes such as those of *E. coli* and often highly unsaturated mitochondrial membranes. The low $O_2$ levels typical of these environments may prevent rapid oxidation of the unsaturated lipid tails, and

**TABLE 10.1**
***Cis*-Vaccenate Bacteria**

| Name | % $18{:}1_{11}$ |
|---|---|
| *Blastobacter* sp. | 88 |
| *Sappia tellulatum* | 93 |
| *Hymphomicrobium* sp. | 86 |
| *Methylobacterium organophilum* | 93 |
| *Methylocystis minimus* | 96 |
| *Methylosinus trichosporium* | 88 |
| *Methylocystis parvus* | 84 |
| *Methanomonas methanooxidans* | 90 |
| *Bradyrhizobium* sp. (root nodule symbiont) | 91 |
| *Nitrobacter winogradskyi* | 92 |
| *Rhodobacter capsulatus* | 92 |
| Antarctic marine bacterium S-7 | 89 ($18{:}1_9$) |

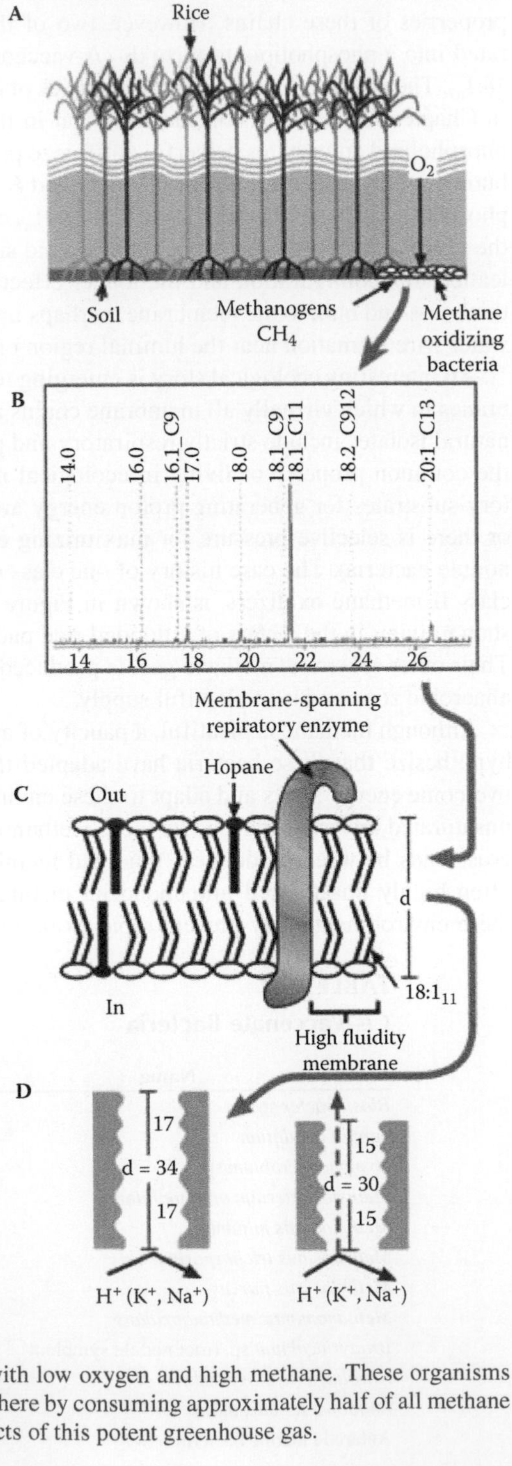


**FIGURE 10.4** Methane oxidizing bacteria (shown here in a rice field) have evolved an unusual membrane structure believed to enhance energy production in their oxygen-limited environment. Cells that use methane, called methanotrophs, are ubiquitous in nature, and many produce a membrane comprised almost entirely of *cis*-vaccenate ($18{:}1_{n7}$) (see Table 10.1). This class of membranes displays low melting temperature and rapid lateral motion beneficial for energy production and is stabilized from oxidation by their localization to environments with low oxygen and high methane. These organisms serve an important function in the biosphere by consuming approximately half of all methane produced, thus reducing radiative impacts of this potent greenhouse gas.

the high fluidity afforded by the membrane may facilitate activity of the key energy-generating enzyme in methanotrophic bacteria—the membrane-bound methane monooxygense. The low $O_2$ levels may also help to explain the inclusion of only monounsaturated fatty acids in the bilayer of these bacteria, since additional oxygen would be required for the synthesis of polyunsaturated fatty acids. In fact, the lack of available oxidant may be the reason type II methanotrophs use the serine pathway for carbon fixation, in contrast to the less energy-intensive RuMP pathway used by type I methanotrophs. The serine pathway involves the incorporation of two molecules of methane to one molecule of $CO_2$, in contrast to the RuMP pathway that uses only methane-derived carbon. We suggest it is advantageous for type II methanotrophs to expend extra ATP to fix $CO_2$, rather than to use limited $O_2$, which provides greater net energy acquisition if reserved only for catabolism.

Bacteria producing membranes with di-monounsaturated lipids first came to light during a survey and study of the role of cholesterol-like molecules called hopanoids in bacteria (Rohmer, Bouvier-Navé, and Ourisson, 1984), although at the time a linkage between hopanoid-producing strains and high membrane monounsaturation was not recognized. An unrelated evolutionary study aimed at identifying possible bacterial precursors of mitochondria led to this same set of bacteria, later shown to synthesize this unique membrane structure. Finally, a literature net that we cast in searching for terrestrial bacteria that produce highly fluidized bilayers led to this same group of bacteria. In our survey, we did not expect to find EPA or DHA bacteria because of the rarity of these lipids in terrestrial ecosystems.

Note that the membrane unsaturation rule (i.e., high unsaturation leads to elevated proton leakiness), long a central dogma in the field of membranes, seems to be broken in the case of bacteria such as methanotrophs in which virtually all of the chains are unsaturated. Fatty acid analysis shows that in certain bacteria this bilayer is composed almost entirely of phospholipids in which both acyl chains are $18{:}1_{11}$ (Table 10.1). The presence of this membrane structure in methanotrophs is puzzling because earlier studies of *E. coli* membranes of similar composition showed these bilayers to be highly leaky for $K^+$ and presumably protons. We have also found, using mutants of *E. coli* selected for high *cis*-vaccenate production, that *E. coli* tolerates up to about 68% $18{:}1^{11}$ at 37°C before losing viability or becoming heat sensitive (unpublished data). How do bacteria solve their apparent membrane hyperfluidity and proton leakage problems with membranes in which >90% of lipid chains are $18{:}1_{n7}$? Genomic analysis of *Methylococcus capsulatus* (Bath), a type I marine methanotroph characterized as having relatively high levels of monounsaturated chains in its bilayer, seems to answer this question. *M. capsulatus* has evolved at least 11 genes for synthesis of methyl cholesterols and hopanoids, including an alternative, mevalonate-independent pathway starting with glyceraldehyde-3-phosphate. It has been proposed that the highly unsaturated membranes of methanotrophs such as *M. capsulatus* (Bath) (71% $16{:}1_9$ at 30°C compared to >90% $18{:}1_{11}$ for type II methotrophs) require stabilization by incorporation of hopanoids or sterols. Also, recall that methanotrophs often synthesize extensive intracytoplasmic membranes to house their methane-oxidizing apparatus (i.e., a much more complex intracellular organization of membranes compared to *E. coli*), which may require cholesterol-like chains for stability. In addition to cholesterol-like additives, *M. capsulatus* has evolved three additional types of membrane adaptations discussed previously, likely relevant to its lifestyle as follows: (1) addition of cyclopropane groups, (2) *cis-trans* isomerization, and (3) positioning double bonds away from the center of the fatty acid chain, which reduces fluidity. For example, $18{:}1_{n7}$ is believed to be only about 70% as fluidizing compared to $18{:}1_{n9}$ because of the shift in the double bond away from the center of the chain. It has been proposed that cholesterol-like molecules of bacteria work by blocking spontaneous leakage of $H^+/Na^+$ essential for maintaining electrochemical gradients.

Thus, from a bioenergetic perspective *cis*-vaccenate–producing bacteria appear to "have their cake and eat it, too." That is, these cells are expected to harness their low melting point phospholipids (i.e., di-$18{:}1_{n7}$) to maximize collisions among components of the electron transport chain, in essence enhancing energy production. However, this low-viscosity membrane is predicted to serve as a mediocre barrier against $H^+/Na^+$. The membrane fidelity for cations might be partially compensated due to increased thickness of the bilayer contributed by these 18-carbon long chains, and through the incorporation of sterols/hopanoid. It is interesting to speculate that the union of high unsaturation with cholesterol-like molecules represents an evolutionary milestone in terms of maximizing rates of both energy production and energy conservation in energy-stressed bacteria. From an evolutionary perspective there are striking similarities between this class of bilayers and those of mitochondria and chloroplasts. Indeed, *cis*-vaccenate membranes seem to represent a missing link between bacteria and organelles, and in some ways we consider them to be something of a "poor man's mitochondria."

## 10.4 BACTERIA MIGHT PRODUCE PLASMALOGENS AS $H^+/Na^+$ BLOCKERS

Plasmalogen is the generic name for an important class of phospholipids introduced in Chapter 6 and found in animals, in some bacteria, and in plants in which the ester linkage of acyl chains to the glycerol backbone, typical of many phospholipids, is replaced in one position by an ether linkage (Figure 10.5). Fatty acid chains in ether linkage are generally located at the *sn*-1 position of the glycerol moiety paired with conventional ester-linked chains attached at the adjacent *sn*-2 position. The chemistry, biochemistry, and function of this interesting class of glycerol phospholipids have been studied in detail mainly in animals. There is general agreement regarding the chemistry and biosynthesis of plasmalogens, whereas there is no generally accepted theory to explain their function in membranes, where they appear to play multiple roles and function alongside ester-linked phospholipids. The emphasis here is on a reductionist idea starting with Bacteria/Archaea and working up to eukaryotic cells. This approach seems justified because of the presence of plasmalogens in single-celled organisms such as *Clostridium butyricum* (i.e., butyric acid producer) and various bacteria fermenting lactic acid as substrate.

Ether-linked lipids have been found to occur across all four life forms, Archaea, Bacteria, plants, and animals, pointing toward a fundamental role, such as bioenergetics. For example, the fermentative bacterium, *C. butyricum*, contains significant amounts of ethanolamine plasmalogens. The presence of plasmalogens in *C. butyricum* provides an important clue toward function that might apply to primary and secondary energy-transducing membranes in general. These cells produce so much butyric acid, as fermentation product, that the pH drops rapidly to the point of killing vegetative cells that have not entered the spore stage. Thus, these cells create their own energy stress because butyric acid is known to be a potent proton energy uncoupler. Recall that *C. butyricum* belongs to a class of bacteria that depend on proton circulation for pH homeostasis and as fuel for energizing nutrient uptake pumps, but not for primary energy production. *C. butyricum* produces monounsaturated and cyclopropane chains for fluidity, with the cyclopropane chains presumably doubling as a defense against acidity. However, the acidity stress/energy stress caused by the generation of butyric acid might require additional membrane-based defenses to increase the robust nature of its membranes against energy uncoupling driven by either butyric acid or water wires. Recall that in model membranes, ether-linked lipids have been shown to decrease ion permeability and lower the phase transition temperature of membrane bilayers compared to their diacyl counterparts. *C. butyricum* appears to be able to modulate membrane function by multiple mechanisms. These include altering the ratio of ether versus acyl type ethanolamine phospholipids in synchrony with environmentally induced changes in the degree of lipid unsaturation of the membrane. The experiments with bacteria indicate that the substitution of plasmenylethanolamine for phosphotidylethanolamine in biomembranes is another example of a defense to conserve proton electrochemical gradients (i.e., conserve energy) at the membrane level.

Thus, plasmalogens might have evolved in bacterial membranes as a defense against cation leakage. Ether-linked phospholipids are known to be more

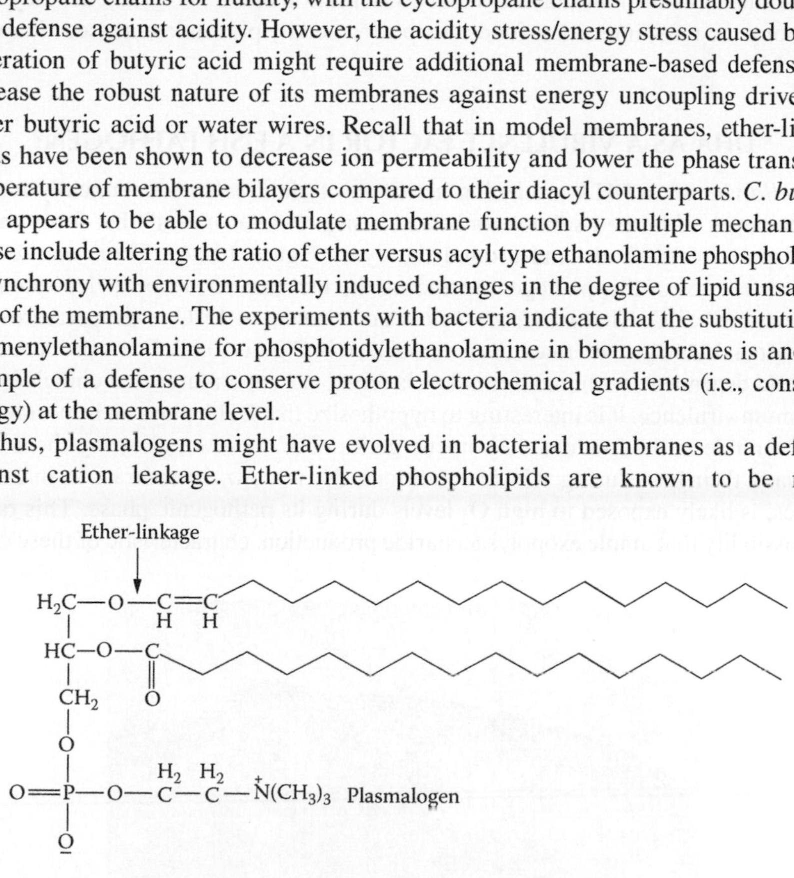


**FIGURE 10.5** New roles proposed for membrane plasmalogens as an adaptation against acidity stress. Plasmalogens, a class of ether-linked phospholipids, are found in membranes of butyric acid bacteria as well as lactic acid fermenting bacteria. We suggest that these "streamlined" phospholipids tighten membrane architecture and work as an energy-saving membrane-based defense against proton energy uncoupling and acidity stress. Thus, bacteria thriving in the presence of normally toxic levels of organic acids adapt not only their membrane fatty acid composition, but also the nature of chemical linkages between acyl chains and their headgroups. (Courtesy of Luke Hillyard.)

streamlined and are believed to be more tightly packed because of loss of the protruding keto group (=0) during formation of the ether versus ester bond. We hypothesize that tight packing of headgroups in essence decreases the numbers of water wires forming spontaneously in the membrane and may also serve to prevent diffusion of butyric or other organic acids across the membrane. It is further hypothesized that energy conservation is the elemental driving force behind evolution of ether-linked phospholipids. That is, conservation of gradients of protons, sodium, potassium, and so forth are equivalent to producing more energy for cellular functions and thus provide a selective advantage. Recall that in the case of plasma membranes of animal cells that conservation of $Na^+/K^+$ gradients may also be among the most important functions performed by plasmalogens (see Chapters 6 and 13).

## 10.5 DHA AS A VIRULENCE FACTOR IN A FISH PATHOGEN?

*Moritella viscosa*, a DHA-producing marine bacterium, is an economically important pathogen of farmed salmon and cod causing "winter ulcers" that form on the scaled parts of the body and around the eyes (Figure 10.6A). There are several unusual features of this pathogen, including the fact that virulence is dependent on temperatures <7°C. That is, *M. viscosa* is not virulent at 15°C, a temperature supporting maximum growth rates. As summarized in Figure 10.6B, DHA production is also cold dependent, with maximum levels found at temperatures that coincide with maximum virulence. It is interesting to hypothesize that DHA plays an important but unknown role in the fundamental bioenergetics of *M. viscosa*, allowing these cells to invade their host during winter conditions. *M. viscosa*, in contrast to intestinal species, is likely exposed to high $O_2$ levels during its pathogenic phase. This raises the possibility that ample exopolysaccharide production, characteristic of these cells,

**FIGURE 10.6A** DHA levels in an important fish pathogen correlate with virulence. (A) Winter ulcers in salmon are caused by *Moritella viscosa*. (Photo courtesy of Poppe Trygve.)

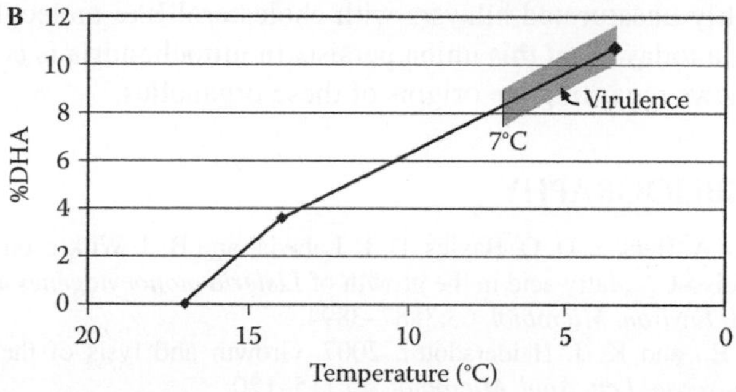


**FIGURE 10.6B** (B) *Moritella viscosa* was grown on marine agar at three different temperatures and analyzed for DHA content. Note that *M. viscosa* is virulent only at temperatures ≤7°C, the temperatures that coincide with peak membrane enrichment with DHA.

is actually an $O_2$ shield for protecting their DHA-enriched bilayers. To test this idea we have constructed high polysaccharide mutants of an EPA isolate that display a simultaneous upregulation in EPA levels. Finally, there are clues that DHA-enriched membranes are best suited for $Na^+$ bioenergetics in contrast to $H^+$ bioenergetics, raising the question of the importance of $Na^+$ bioenergetics in pathogenicity of this organism (see Chapter 5).

## 10.6 SUMMARY

Progress is being made in deciphering why bacterial cells choose specific fatty acid structures for their bilayers. These studies support the generalization that fatty acid structures and conformational dynamics play crucial roles in the molecular ecology of bacteria. We suggest that at least three related parameters form the biochemical basis of a predictive membrane code for bacteria. These involve fatty acid structures and dynamics that contribute important bioenergetic properties to the bilayer, including cation permeability, freedom of motion, and oxidative stability. Environmental signals as well as chemical interactions among acyl chains and their headgroups are also suggested to modulate fatty acid motion as well as conformational dynamics. The net effect is to generate a dynamic lipid molecular architecture most appropriate for a specific environment. DHA chains are found in relatively few bacteria, and these cells appear to be highly specialized toward conditions in the marine world. Data suggest that harnessing DHA chains in energy-transducing bilayers in bacteria involves balancing risks, including enhanced lipoxidation and a mediocre proton permeability barrier, against the potential benefits of high membrane fluidity and rapid electron transport. Plasmalogen phospholipid structure is hypothesized to tighten membrane architecture and enhance acidity stress tolerance in bacterial membranes. This mechanism might apply to critical animal cells such as neurons and cardiac muscle cells where the ether-linked phospholipids are abundant but where their mode of action is poorly understood. Finally, studies of fatty acid conformation-function in bacterial membranes provide a new perspective of major steps of membrane evolution. For example, the selective advantage of combining highly unsaturated bilayers with cholesterol-like molecules is evident in many bacteria today. That this union persists in mitochondria is consistent with evolutionary views regarding the origins of these organelles.

## SELECTED BIBLIOGRAPHY

Annous, B. A., L. A. Becker, D. O. Bayles, D. P. Labeda, and B. J. Wilkinson. 1997. Critical role of anteiso-$C_{15:0}$ fatty acid in the growth of *Listeria monocytogenes* at low temperatures. *Appl. Environ. Microbiol.* 63:3887–3894.

Benediktsdóttir, E., and K. J. Heidarsdóttir. 2007. Growth and lysis of the fish pathogen *Moritella viscosa. Lett. Appl. Microbiol.* 45:115–120.

Cronan, J. E., and C. O. Rock. 1996. Biosynthesis of membrane lipids. In *Escherichia and Salmonella: Cellular and Molecular Biology*. F. C. Niedhardt et al., Eds. Washington, D.C.: ASM Press.

Davis, M.-T. B. and D. F. Silbert. 1974. Changes in cell permeability following a marked reduction of saturated fatty acid content of *Escherichia coli* K-12. *Biochim. Biophys. Acta* 373:224–241.

Goldfine, H. 1984. The control of membrane fluidity in plasmalogen-containing anaerobic bacteria. In *Biomembranes*, Vol. 12, 349–377, ed. M. Kates and L. A. Manson. New York: Plenum Press.

Hanson, R. S., and T. E. Hanson. 1996. Methanotrophic bacteria. *Microbiol. Rev.* 60:439–471.

Holtwick, R., F. Meinhardt, and H. Keweloh. 1997. *Cis-trans* isomerization of unsaturated fatty acids: Cloning and sequencing of the *cti* gene from *Pseudomonas putida P8. Appl. Environ. Microbiol.* 63:4292–4297.

Okuyama, H., N. Okajima, S. Sasaki, S. Higashi, and N. Murata. 1991. The *cis/trans* isomerization of the double bond of a fatty acid as a strategy for adaptation to changes in ambient temperature in a psychrophilic bacterium, *Vibrio* sp. strain ABE-1. *Biochim. Biophys. Acta* 1084:13–20.

Paltauf, F. 1994. Ether lipids in biomembranes. *Chem. Phys. Lipids* 74:101–139.

Rohmer, M., P. Bouvier-Navé, and G. Ourisson. 1984. Distribution of hopanoid triterpenes in prokaryotes. *J. Gen. Microbiol.* 130:1137–1150.

Ward, N., Ø. Larsen, J. Sakwa, et al. 2004. Genomic insights into methanotrophy: The complete genome sequence of *Methylococcus capsulatus* (Bath). *PLoS Biol.* 2:1616–1628.

Weber, F. J., S. Isken, and J. A. M. de Bont. 1994. *Cis/trans* isomerization of fatty acids as a defense mechanism of *Pseudomonas putida* strains to toxic concentrations of toluene. *Microbiology* 140:2013–2017.

Wilkinson, S. G. 1988. Gram-negative bacteria. In *Microbial lipids,* Vol. 1, 299–457, ed. C. Ratledge and S. G. Wilkinson. London: Academic Press.

Zheng, W., M. L. Wise, A. Wyrick, J. G. Metz, L. Yuan, and W. H. Gerwick. 2002. Polyenoic fatty acid isomerase from the marine alga *Ptilota filicina*: Protein characterization and functional expression of the cloned cDNA. *Arch. Biochem. Biophys.* 401:11–20.

# 11 Chloroplasts: *Harnessing DHA/EPA for Harvesting Light in the Sea*

Hypothesis: DHA/EPA-enriched membranes of chloroplasts enhance photosynthetic electron transport and increase primary productivity of the oceanic food web.

Harvesting light in frigid ocean water such as Antarctica requires specialized chloroplast membranes not found in terrestrial plants. We suggest this is why marine phytoplankton are the major producers of DHA/EPA in the biosphere. DHA and EPA produced by marine phytoplankton and cycled throughout the marine food web are the major source of these molecules finding their way into the human diet. Whereas the term "fish oils" is generally applied to this class of fatty acids, many of these chains are made by marine plants for the purpose of photosynthesis.

The following rough estimate of annual production of DHA/EPA by eukaryotic marine plants highlights the importance of these fatty acids in the biosphere and sets the stage for discussions here. The following assumptions are used to arrive at a production estimate of 200 to 400 million metric tons of DHA+EPA annually:

- 15 to 30 billion metric tons of carbon fixed annually by eukaryotic marine algae (out of a global total of about 75 to 100 billion metric tons).
- ~10% of carbon converted to membrane fatty acids (2 to 4 billion metric tons per year, accounting for hydrogen and oxygen incorporation).
- ~10% of fatty acids are DHA+EPA (0.2 to 0.4 billion metric tons per year)

We further assume that a significant fraction of DHA/EPA ends up in photosynthetic membranes.

A biochemical explanation for the importance of DHA/EPA in photosynthesis in the marine world is rooted in research carried out by Australian scientists studying chloroplasts of land plants more than two decades ago (Anderson and Andersson, 1982). Also, pioneering research by Japanese scientists sheds light on the importance of membrane fatty acid unsaturation in photosynthesis (Wada et al., 1990; Tasaka et al., 1996). For example, the latter workers used knockout mutations of desaturases to show that growth of cyanobacteria at cool temperatures is dependent on polyunsaturated photosynthetic membranes. Interestingly, cyanobacteria are now recognized as playing an important role in primary productivity in the marine world. However, photosynthetic membranes in these cells lack DHA/EPA, which might help to explain the predominance of these cells in tropical marine waters. Indeed, cyanobacteria such as *Prochlorococcus* seem restricted to growth above about 15°C in seawater, in contrast to cold-adapted eukaryotic algae, which are enriched with DHA/EPA. A brief look at how light is trapped by photosynthetic membranes sets the stage for understanding the molecular roles of DHA and EPA in marine chloroplasts.

Photosynthetic membranes of chloroplasts (Figure 11.1) of both marine and land plants house the photochemical machinery needed to harvest light. It can be calculated that if all photosynthetic membranes were "stitched together," they would form a continuous film surrounding the entire globe. Chloroplast membranes are the forefront of bioenergy production on earth, powering synthesis of food, fiber, and fuels for mankind and driving the biogeochemical cycles of the planet. Evolutionarily, plants also provided raw material for generation of fossil fuels, produced $O_2$ to drive respiration, and have moderated the greenhouse effect by scrubbing atmospheric $CO_2$. The numbers of chloroplasts working in the biosphere today are prodigious and are about equally distributed between terrestrial and marine realms.

As introduced in Chapter 7, DHA/EPA-enriched phytoplankton blooms in the ocean surrounding Antarctica support about 750 million metric tons of krill. According to a rule of thumb of biological oceanography, only about one carbon out of every ten fixed by phytoplankton ends up in biomass of krill. Thus, it is estimated that roughly $10^6$ metric tons of DHA/EPA are produced in this ecosystem ($7.5^9$ metric tons of phytoplankton carbon at 1% DHA + EPA) to support the krill population. A significant proportion of these chains is likely localized to chloroplast membranes essential in harvesting light. The prodigious numbers of chloroplasts required to energize the Antarctic food web are adapted to carry out photosynthesis and photosynthetic electron transport at temperatures in the range –2°C to 4°C. We propose that this is where DHA and EPA are most needed.

The primary role of photosynthetic membranes of chloroplasts is to facilitate production of ATP and reducing power needed for $CO_2$ fixation. The inner membrane of chloroplasts is packed with the molecular machines needed for harnessing light energy, splitting water and forming ATP, and reducing power. The overall process involved in forming ATP is called photosynthetic electron transport (Figure 11.2). The components of the photosynthetic electron transport chain and the ATP synthesizing apparatus consist of four membrane complexes, photosystem II (PSII), photosystem I (PSI), cytochrome $b_6f$ complex, and ATP synthase ($CF_1$-$CF_0$). Electrons are transferred from water to the central soluble electron carrier nicotinamide adenine dinucleatide phosphate (NADPH). During electron transfer, energy from photochemically excited electrons is harnessed by cytochrome $b_6f$ working as a proton pump to help generate a proton electrochemical gradient—high in the lumen and low in the stroma. This proton gradient is ultimately used for the synthesis of ATP by ATP synthase. Plastoquinone (PQ) is a small lipophilic electron carrier linking large enzyme complexes. Plastocyanin (PC) is also an electron carrier, but unlike plastoquinone, is highly soluble in water. Fd is ferredoxin and FNR is the enzyme ferredoxin—NADP reductase. Note that the electron transport complexes as well as plastoquinone are housed in the membrane, which in the case of marine chloroplasts is often enriched with DHA or EPA.

Because chloroplasts and photosynthetic membranes are so important in both marine and terrestrial ecology and agriculture, there is active research in this area. However, there are many important basic and applied questions yet to be answered.

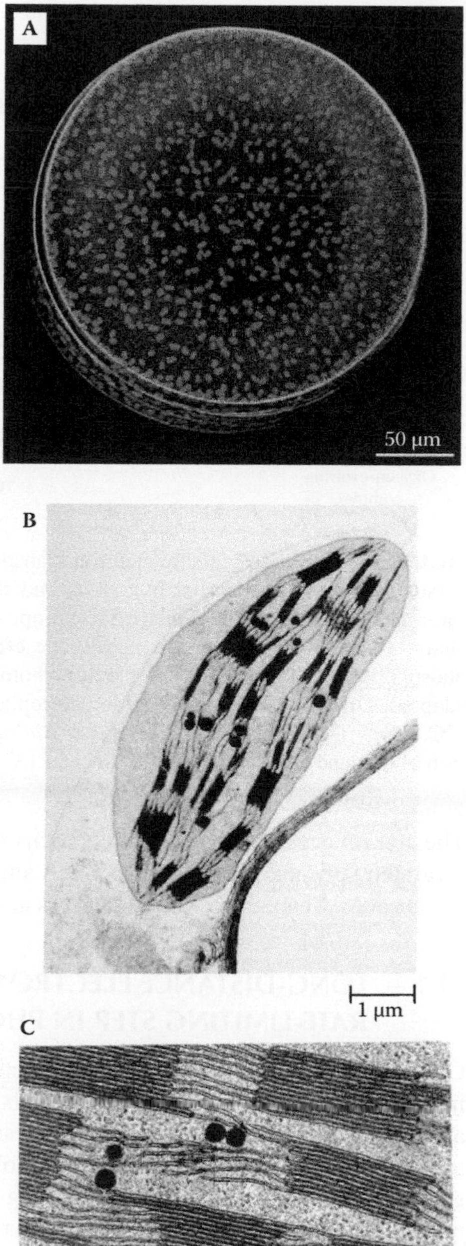


**FIGURE 11.1** (A color version of this figure follows p. 140.) Chloroplast membranes: a vast polyunsaturated membrane surface has evolved in plants to harvest light for primary production. (A) Individual chloroplasts are visible in a marine diatom. The image is a confocal laser scanning micrograph showing the frustule (stained with PDMPO) and the chloroplasts (autofluorescence) of a live specimen of the planktonic diatom species *Coscinodiscus wailesii*. (Courtesy of Jan Michels) (B) Photosynthetic membranes in a single chloroplast of a land plant. (C) High, resolution electron micrograph showing stacks of chloroplast membranes. Several adaptations aimed at increasing energy output are apparent in chloroplasts. These include increasing total photosynthetic membrane surface area by raising the density of chloroplasts per cell, packing in more membranes per individual chloroplast or inserting more photosynthetic electron transport chains into each membrane. We suggest that in marine plants a fourth adaptation involving DHA/EPA has occurred that provides great benefit in harvesting light in the sea. [(B) and (C) Courtesy of W. Wergin ARS.]

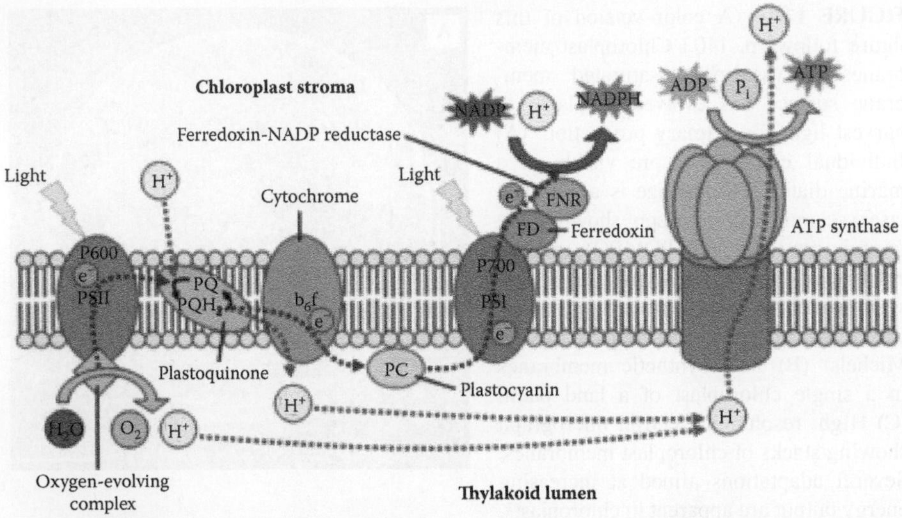


**FIGURE 11.2** Photosynthetic electron transport chains of chloroplasts of marine plants versus land plants are similar in content and function, and both are embedded in polyunsaturated membranes. These bilayers are proposed to drive high rates of collisions along the chain, essentially increasing photosynthetic efficiency. The harnessing of polyunsaturated phospholipids as a mechanism to increase photosynthesis seems to reach a climax with cold-adapted, DHA/EPA-enriched marine chloroplasts operating at temperatures from −2°C to 2°C. (Copyright Alberts et al., *Molecular Biology of the Cell,* 2008. Reproduced by permission of Garland Science/Taylor & Francis, LLC.)

The central question raised here concerns the fundamental roles played by polyunsaturated fatty acid chains such as DHA and EPA in phospholipids of photosynthetic membranes. Chapters 17 and 18 address applied aspects of this question.

## 11.1 LONG-DISTANCE ELECTRON TRANSPORT AS A RATE-LIMITING STEP IN PHOTOSYNTHESIS

Chloroplasts of plants have evolved to function in a great diversity of habitats, including different light intensities, salinity levels, and temperatures, to name a few. These are examples of powerful environmental signals that can generate energy stress at the level of the membrane itself. Various solutions have evolved to maximize energy generation in plants and include increasing the levels of photosynthetic catalysts per unit membrane or raising the total numbers of chloroplasts per cell. We argue that the combination of energy stress and competition among plants is a potent selective force to maximize energy yields, and we therefore hypothesize that chloroplast membranes are highly adapted for this purpose. Inherent in this hypothesis is the argument that lipids such as DHA/EPA incorporated into such membranes are the optimal lipids for this purpose.

Long-distance electron transport seems to be the weak link for photosynthetic electron transport in both marine and terrestrial chloroplasts. That is, the transport of electrons for long distances through the membrane seems to limit efficiency of photosynthetic electron transport in chloroplasts and serves as the rate-limiting step in photosynthesis. Long-distance electron transport was defined some years ago by experts in this field and is rooted in the uneven localization or spacing between electron donor enzyme complexes and recipients forming the electron transport chain. Careful analysis of localization patterns of components of the electron transport chain show an unexpected lateral heterogeneity, as summarized in Table 11.1. For example, photosystem II (PSII) is localized primarily in the stacked membrane regions of the photosynthetic membrane, whereas PSI and ATP synthase are almost exclusively localized in the unstacked or stroma-exposed membrane regions. The cytochrome $b_6f$ complex of the electron transport chain is distributed evenly throughout the membrane regions. The physical separation of the photosystems necessitates mobile electron carriers that shuttle reducing equivalents between the spatially separated membrane complexes. A closer look at Figure 11.2 depicts the electron carrier plastoquinone shuttling electrons only a short distance between PSII and $Cytb_6f$. However, a much wider lateral chasm often separates these complexes and helps define long distance electron transport in terms of a relatively long shuttle distance that must be covered by plastoquinone (Figure 11.3). This journey is made more difficult because plastoquinone is confined to travel laterally from donor to acceptor via the lipid portion of the membrane, a more viscous milieu compared to water. Plant biochemists studying the kinetics of photosynthetic electron transport have found that the frequencies of collisions between reduced plastoquinone and cytochrome $b_6f$ appear to be rate limiting for chloroplast energy production in nature. Thus, long distance electron transport can be considered a weak link in photosynthesis, and we suggest that the evolution of DHA/EPA phospholipids in photosynthetic membranes is a response to this weakness.

## 11.2 SPEEDING UP LONG-DISTANCE ELECTRON TRANSPORT

Over a billion years of evolution plants have evolved chloroplasts that are highly efficient at photosynthetic electron transport. For example, several mechanisms have evolved to speed up rates of photosynthetic electron transport. One case history involves the water-soluble electron carrier plastocyanin, which seems to speed

**TABLE 11.1**
**Photosynthetic Components Are Not Evenly Distributed in Chloroplast Membranes, Indicating the Need for Long-Distance Electron Transport**

| Component | Localization in Grana Stacks (%) | Photosynthetic Membranes Stroma-Exposed (%) |
|---|---|---|
| PSII | 85 | 15 |
| PSI | 10 | 90 |
| Cytochrome $b_6f$ complex | 50 | 50 |
| ATP synthase | 0 | 100 |
| Plastocyanin | 40 | 60 |

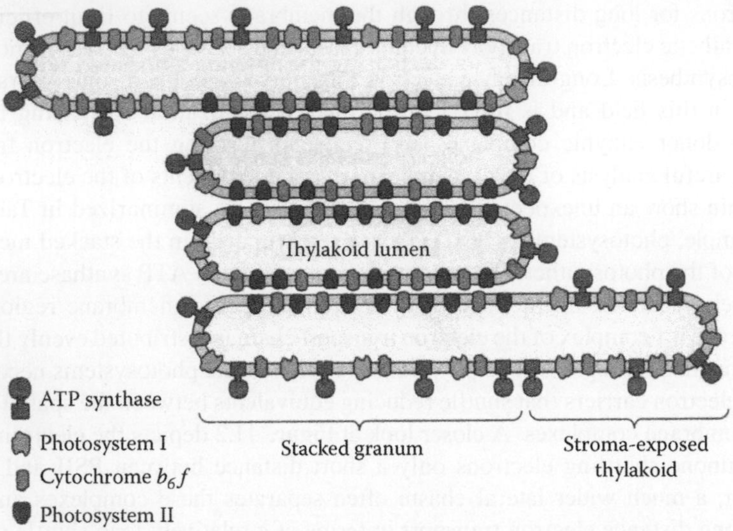


**FIGURE 11.3** Artist's sketch showing uneven distribution of components of the photosynthetic electron transport chain in chloroplast membranes. Plastoquinone (not shown) is a lipophilic electron carrier that often must travel long distances to link electron donors and electron acceptors along the chain. This defines long distance electron transport and highlights the need for DHA/EPA especially in chloroplasts operating near 0°C. (Reproduced from R. Malkin and K. Niyogi. In *Biochemistry and molecular biology of plants*, B. B. Buchanan, W. Gruissem, and R. Jones, Eds., 568–628, 2000. Rockville, MD: American Society of Plant Physiologists. With permission.)

up rates of electron transfer between cytochrome $b_6f$ feeding electrons to PSI. That is, the relatively long lateral distances separating components of the photosynthetic electron transport chain can be covered more rapidly via diffusion of plastocyanin through the water phase in contrast to lipophilic carriers moving in membranes. We hypothesized earlier that DHA/EPA phospholipids work by enhancing collisions between the lipophilic carrier plastoquinone and components of the photosynthetic electron transport chain. This might be especially important in marine chloroplasts whose membranes are often subject to the hardening effects of cool temperatures.

The fatty acid composition of a major component of the inner membrane of plant chloroplasts is summarized in Table 11.2. Photosynthetic membranes contain several lipid classes and species found nowhere else in the plant cell. The most abundant species is the galactolipid monogalactosyldiglyceride (MGDG), often making up almost 50% of the total membrane lipids. Digalactosyldiglyceride along with MGDG makes up a vast majority of the membrane surface. Minor constituents include the sulfolipid, sulfoquinovosyldiglyceride, phosphatidylcholine, phosphatidylglycerol, and phosphatidylinositol.

The roles of galactolipids are described next, and for this discussion, photosynthetic membranes of chloroplasts are subdivided into two classes, marine and terrestrial. We chose this separation because the inner membrane of marine chloroplasts is often highly enriched with DHA/EPA in contrast to land plants, whose membranes

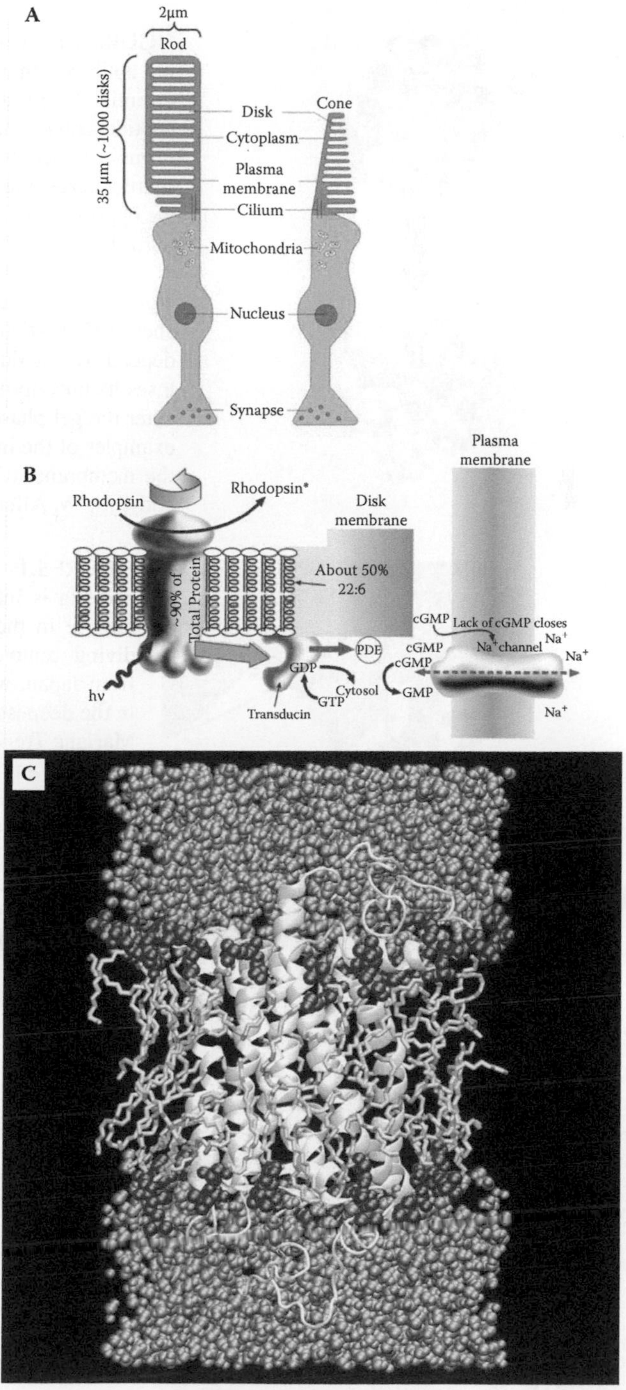


**FIGURE 2.4** Vision depends on DHA-enriched membranes that house the rapidly moving, light-sensing protein rhodopsin that triggers the visual cascade. (A) Perception of dim light occurs using strategically placed membrane disks acting as microantennae in the outer segment of rod cells in the eye. Approximately 1000 DHA-enriched membrane disks act as the light-sensing apparatus. (B) DHA-enriched membranes enable rhodopsin to rotate and move laterally at rapid speed. This facilitates collisions between rhodopsin and its G-protein transducin, firing the visual cascade. (C) Rhodopsin is shown to be surrounded by DHA phospholipids in an image generated by chemical simulation. In addition to driving motion of rhodopsin, DHA phospholipids might also be involved in the biochemistry of rhodopsin. [(A) Courtesy of Frank Muellerand (C) courtesy of Scott Feller.]

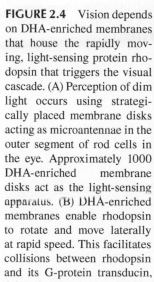

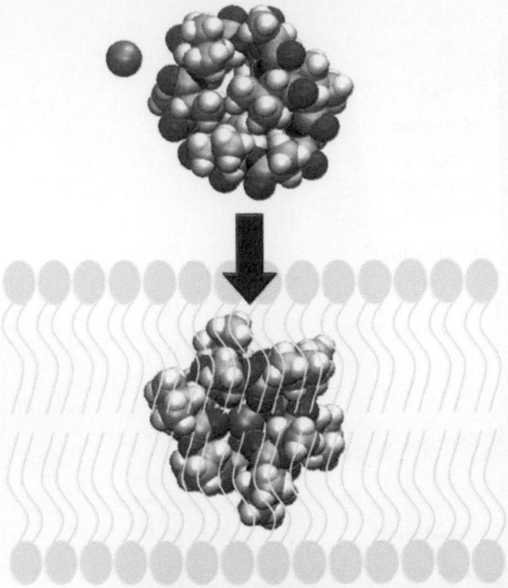

**FIGURE 3.1** Valinomycin is a membrane-acting antibiotic that kills bacteria by uncoupling essential $K^+$ gradients. This lipophilic, cyclic peptide enters the membrane of target cells, forms a molecular cage around a $K^+$ ion, and shuttles across the bilayer, releasing the $K^+$ ion on the opposite surface. This mode of action transports $K^+$ across the membrane in the direction of the chemiosmotic gradient, effectively deenergizing the $K^+$ gradient and wasting cellular energy. Of particular interest here, valinomycin depends on the fluidized state of the bilayer and loses its uncoupling capacity when membranes enter the gel phase. This is one of the simplest examples of the importance of the fluid state of the membrane. (Courtesy of Borislava Bekker and Toby W. Allen, personal communication.)

**FIGURE 5.1** Discovery of DHA-producing bacteria is intimately linked to the search for life in the deep ocean. (A) The deep-diving remotely operated vehicle, *Kaiko*, from Japan, was designed to search for life in the deepest part of the ocean, such as the Mariana Trench (>30,000 feet in depth). (B) Sediment samples taken by *Kaiko's* robotic arm contained DHA-producing bacteria, confirming the discovery in 1986 by DeLong and Yayanos, who first described the isolation of DHA-containing bacteria from water samples taken from the deep ocean. (Photos courtesy of JAMSTEC.)

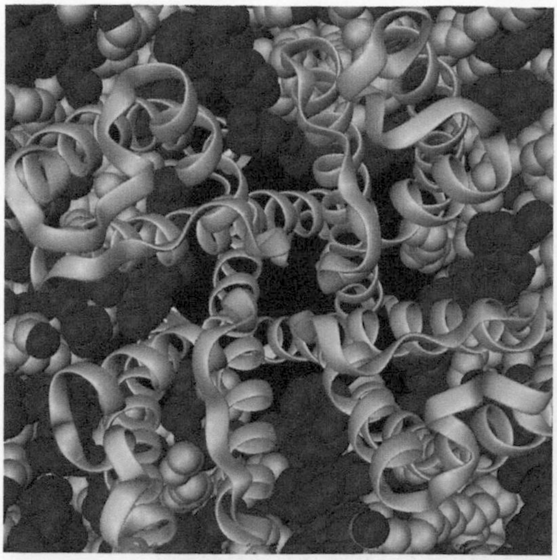

**FIGURE 6.3** Model of a $K^+$ channel (KcsA) in yellow shows a close association with DHA phospholipids (red). Antirafting properties of DHA might help maintain neuronal membrane proteins in motion, such as already described above for the G-protein–mediated receptors. Alternatively, a more direct role in enzyme catalysis can be envisioned. (Courtesy of Igor Vorobyov, Scott Feller, and Toby W. Allen, unpublished data.)

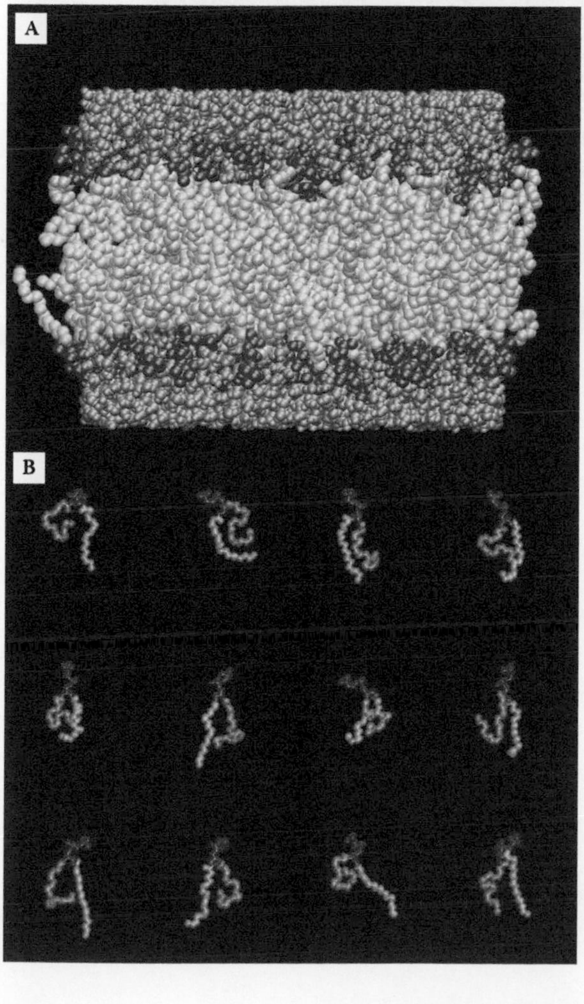


**FIGURE 7.1** DHA chains display powerful membrane antibonding properties. (A) DHA-enriched membrane bilayer generated using chemical simulation techniques (gold chains are DHA and greenish chains are 18:0). (B) Extraordinary conformations of DHA seen in this series of images are consistent with the powerful membrane antifreeze properties attributed to these highly unsaturated chains. DHA chains are shown as golden, 18:0 as greenish, and the head group as red. (Images courtesy of Scott Feller, and generated for this book by Matthew B. Roark, both of Wabash College.)

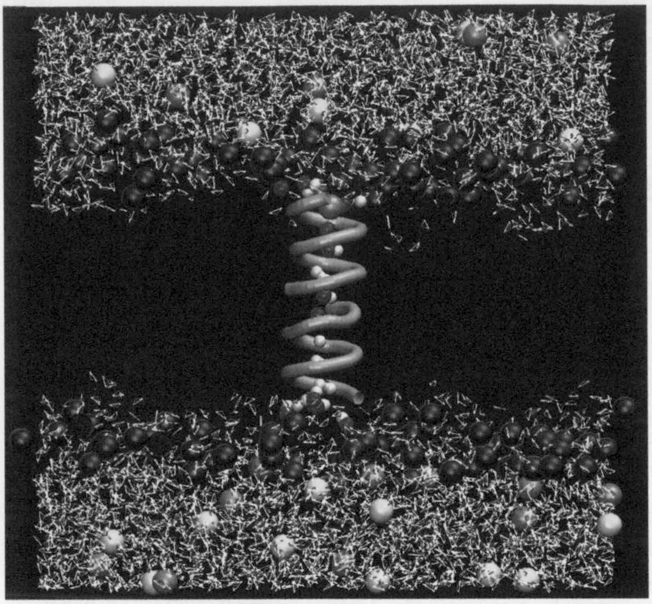

**FIGURE 8.2** Model of gramicidin A. This nanotube-forming peptide spans the membrane, holds a molecular thread of water in its core, and forms a conduit for rapid flux of protons. The structure and function of this membrane-spanning antibiotic have been studied in great detail and provide a clear picture of how water becomes toxic if allowed to permeate the membrane. Water wires forming spontaneously but infrequently in all membranes are proposed to act by the same mechanism, resulting in potentially catastrophic energy uncoupling. (Image generated by chemical simulation; courtesy of Serdar Kuyucak.)

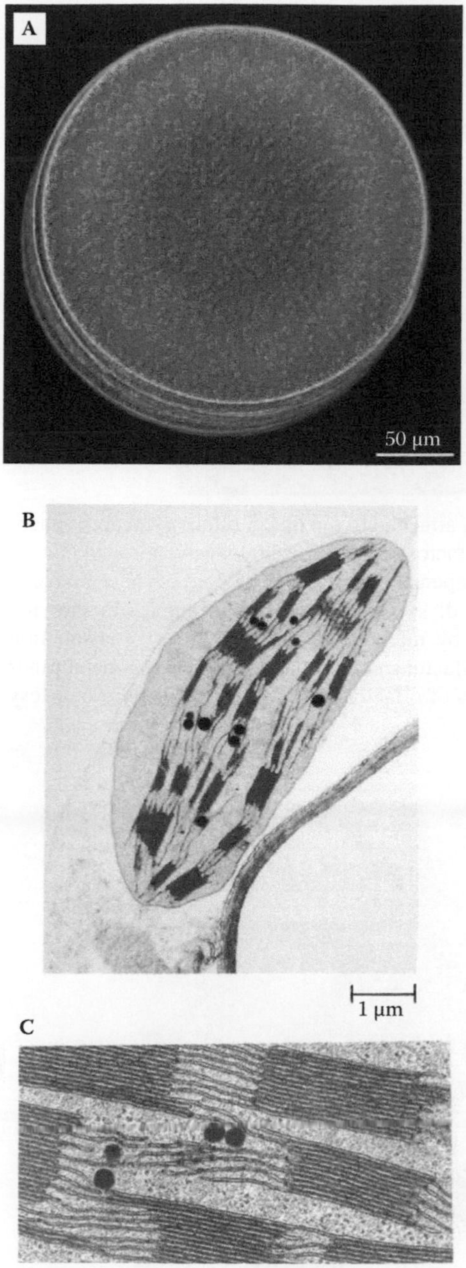


**FIGURE 11.1** Chloroplast membranes: a vast polyunsaturated membrane surface has evolved in plants to harvest light for primary production. (A) Individual chloroplasts are visible in a marine diatom. The image is a confocal laser scanning micrograph showing the frustule (stained with PDMPO) and the chloroplasts (autofluorescence) of a live specimen of the planktonic diatom species *Coscinodiscus wailesii.* (Courtesy of Jan Michels.) (B) Photosynthetic membranes in a single chloroplast of a land plant. (C) High-resolution electron micrograph showing stacks of chloroplast membranes. Several adaptations aimed at increasing energy output are apparent in chloroplasts. These include increasing total photosynthetic membrane surface area by raising the density of chloroplasts per cell, packing in more membranes per individual chloroplast, or inserting more photosynthetic electron transport chains into each membrane. We suggest that in marine plants a fourth adaptation involving DHA/EPA has occurred that provides great benefit in harvesting light in the sea. [(B) and (C) Courtesy of W. Wergin ARS.]

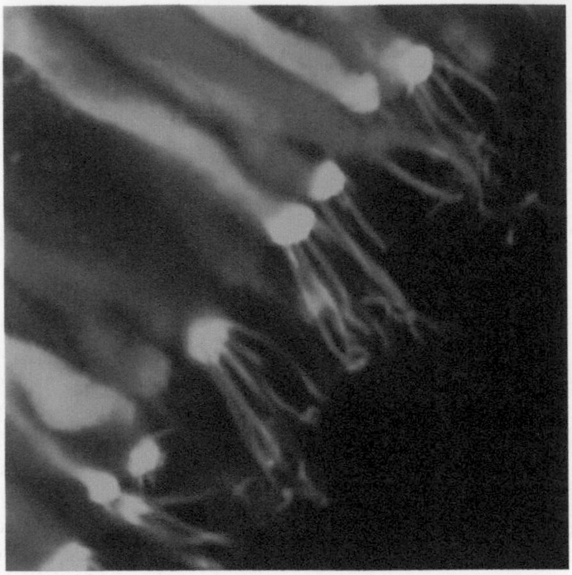

**FIGURE 13.3** Membranes of olfactory cilia attached to the tips of olfactory neurons share properties with sperm tails. Membranes of olfactory cilia are excitatory in nature and house olfactory receptors, whose mode of action depends on the physical state of these bilayers. Olfactory membranes along with rhodopsin disks serve as striking examples in showing the essential biochemical roles contributed by membranes in neurosensory transduction systems. Recall that hundreds of different olfactory receptors present in these membranes combine to create the sense of smell. (Photos of olfactory cilia of larval *Xenopus* courtesy of Ivan Mangini.)

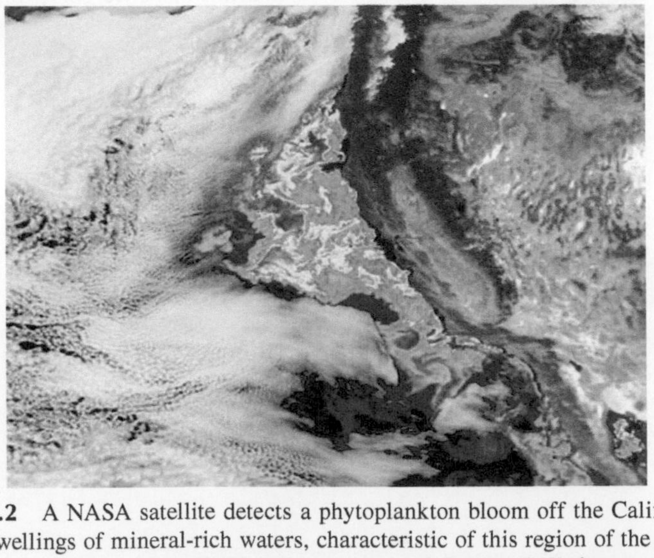

**FIGURE 17.2** A NASA satellite detects a phytoplankton bloom off the California coast. Periodic upwellings of mineral-rich waters, characteristic of this region of the coast, result in large blooms of algae. One of the greatest salmon fisheries of all time once flourished here, with enormous runs navigating up the Sacramento River and its tributaries each year to spawn. (Image courtesy of the *SeaWiFs* Project, NASA/Goddard Space Flight Center, and ORBIMAGE.)

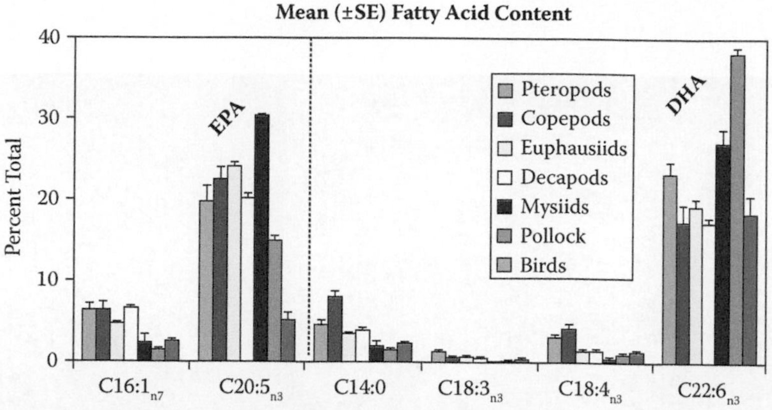


**FIGURE 17.5** Distribution of DHA/EPA in a hierarchy of animals of the eastern Bering Sea. Data are arranged with the diatom-indicating FA on the left of the vertical line and the small cell-indicating FA to the right of the line. Pteropods = *Limacina heliciana;* Copepods = *C. marshallae, N. cristatus, M. pacifica, E. bungii;* Euphausiids = *T. raschii, T. inermis, T. spinifera, T. longipes*, furcilia; Decapods = three unidentified taxa, two zoea, one megalopes; Mysiids = Unidentified; Birds = two common murres (*Uria aalge*), three short-tailed shearwaters (*Puffinus tenuirostris*). Summary of data from all taxa averaged over all stations. This survey of animal life around the Pribilof Islands shows the importance of DHA/EPA in marine animals starting with zooplankton and ending with pollock and birds. Note that whereas DHA/EPA are roughly equally distributed among zooplankton, DHA predominates further up the chain. Also note the high total levels of DHA plus EPA in these animals. Recall that DHA/EPA are an important energy food for animals but also play important structural roles in membranes of mitochondria and neurons. We hypothesize that DHA/EPA are needed to build efficient neurosensory membranes in both endothermic and ectothermic marine animals. (Courtesy of J. M. Napp, L. E. Schaufler, G. L. Hunt, and K. L. Mier, unpublished data.)

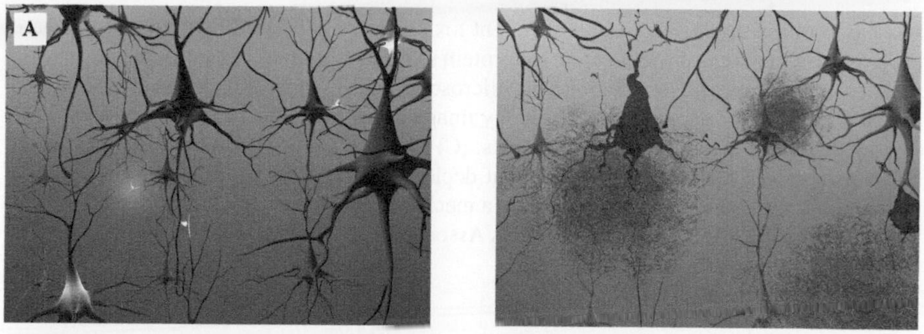


**FIGURE 20.1** Loss of DHA from axonal membranes during Alzheimer's disease might be a cause rather than effect. (A) At left: healthy neurons maintain their DHA-enriched axonal membranes through an active and sophisticated repair and remodeling system for removing damaged DHA. DHA is replenished through the diet and via biosynthesis. With age the delicate balance needed for maintaining membrane structural homeostasis might slowly shift such that net DHA levels begin to drop. As discussed in the text, a steep drop of DHA is characteristic of AD and might be a cause rather than an effect. At right: loss of DHA in neurons has been implicated in defective protein processing and generation of toxic peptides such as β-amyloid shown in this diagram. However, a direct linkage between β-amyloid and

*(Continued on next page)*

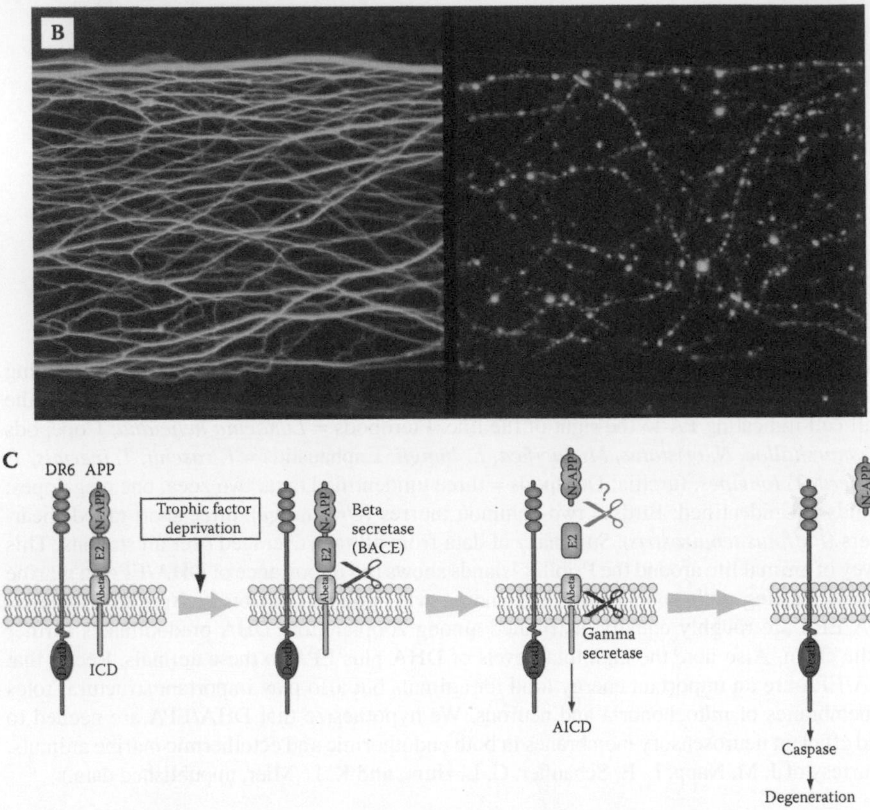


AD has not been established. (B) A different toxic peptide generated by defective protease processing of the same amyloid precursor protein (APP) acts as a suicide molecule destroying sensory connections. At left: fluorescent microscopy image of healthy sensory connection of neurons. At right: fluorescent microscopy image of neurons after treatment with N-APP, displaying destruction of neural connections. (C) Model proposed for role of N-APP in AD by scientists at Genentech. We propose that depletion of DHA in neural membranes might result in increased production of N-APP by a mechanism similar to that described in the text for β-amyloid. [(A) Courtesy of Alzheimer's Association. (B) and (C) Courtesy of M. Tessier-Lavigne and Genentech Inc.]

are devoid of these chains. For example, analysis of total fatty acids of the marine phytoplankton *Amphidinium carterae* shows that 25% of the chains are DHA and 14% EPA. It has been clearly established that individual lipids such as MGDG in photosynthetic membranes of various marine chloroplasts are often enriched with highly unsaturated fatty acids, with four to six double bonds per chain. Perhaps the most dramatic aspect of the molecular architecture of photosynthetic membranes involves the fact that almost all of the fatty acid chains in major lipid classes, such as MGDG, are unsaturated (Table 11.2). For example, 91% of the fatty acid chains in MGDG in the algae *Ochromonas danica* are unsaturated, 88% being polyunsaturated. The main conclusion from studies of fatty acid compositions of marine chloroplasts is that photosynthetic membranes are often highly enriched with polyunsaturated chains. From a biochemical perspective it is important to know that much of the surface area of the inner membranes of marine chloroplasts is composed of the most highly unsaturated membrane lipids in nature, presumably contributing powerful antifreeze properties.

The same mechanism applies to terrestrial chloroplasts, except unsaturated chains with one to three double bonds predominate. Nevertheless, di-polyunsaturated or di-unsaturated lipid structures are characteristic of photosynthetic membranes of many kinds of terrestrial plants. This class of phospholipids is hypothesized to be a powerful fluidizer of the inner chloroplast bilayer. Indeed, the total pool of di-unsaturated species of membrane lipids found in photosynthetic membranes of chloroplasts in nature dwarfs the levels found in all other cells combined. We hypothesize that the

**TABLE 11.2**
**Phospholipids of Photosynthetic Membranes of Chloroplasts**

| | Plant Chloroplasts (Fatty Acid Composition in Monogalactosyl Diglyceride) | | | |
|---|---|---|---|---|
| **Fatty Acids (% Total in MGDG)** | **Green Algae (*Chlorella vulgaris*)** | **Diatoms (*Navicula muralis*)** | **Red Algae (*Porphyra yezoensis*)** | **Spinach** |
| 16:0 | 5 | 3 | 13 | |
| 16:1 | 2 | 9 | | |
| 16:2 | 19 | 14 | | |
| 16:3 | 10 | 27 | | 30 |
| 18:0 | 2 | 2 | | |
| 18:1 | | | 4 | 1 |
| 18:2 | 17 | | 1 | 1 |
| 18:3 | 45 | 2 | | 67 |
| 20:1 | | | 1 | |
| 20:2 | | | 1 | |
| 20:3 | | | 3 | |
| 20:4 | | | 2 | |
| 20:5 | | 42 | 74 | |
| % Total unsaturated Chains in MGDG | 93 | 94 | 86 | 99 |

*Note:* Phospholipids with galactosyl headgroups are predicted to contribute both potent membrane antifreeze properties and proton fidelity to the bilayer.

evolution of low melting point membrane lipids in chloroplasts across so many plant types can be explained on the basis of "turbocharging" of photosynthetic electron transport chains to maximize energy production.

Data from studies on the mobility of electron carriers support such a "collision model." For example, the motility of plastoquinone in protein-free liposomes appears to be roughly 100 times faster compared to photosynthetic membranes. The large reduction in plastoquinone mobility in the native membrane as compared to the liposome bilayer is due to a hindering of free diffusion by the presence of relatively large amounts of intrinsic membrane proteins, such as electron transport complexes. This concept is supported by the finding that a 10-fold reduction in quinone mobility occurred in a membrane in which membrane proteins occupied 16% to 26% of the interfacial surface area, compared to 0% in liposomes. In stacked photosynthetic membranes, protein complexes may occupy as much as 50% of the membrane volume. Researchers in this field proposed a model in which electron transfer by plastoquinone is restricted by the presence of a fluctuating network of intrinsic membrane proteins. Since these results were published, the concept of long-distance electron transport as a rate-limiting step in photosynthesis has been validated (Malkin and Niyogi, 2000). We propose that chloroplast membranes containing DHA/EPA phospholipids evolved as a mechanism to overcome both the long distances and tortuosity for electron transport in photosynthesis.

## 11.3 A DELICATE BALANCING ACT BETWEEN PROTON PERMEABILITY, MOTION, AND OXIDATIVE STABILITY

It seems clear that photosynthetic membrane lipids are responsible simultaneously for at least three critical biochemical properties in chloroplasts: motion, as already discussed; oxidative stability, discussed in Chapters 9, 17, and 18; and permeability, discussed in Chapter 8. According to the current working model, photosynthetic membranes of chloroplasts have evolved membrane lipids with dual unsaturated fatty acid chains for the purpose of increasing rates of electron transfer. This membrane adaptation is regarded as a fundamental step of photosynthetic energy transduction in chloroplasts of plants. Both marine and terrestrial chloroplasts share this need, with marine chloroplasts requiring molecules that provide even more lateral motion, such as DHA/EPA. As discussed in Chapter 8 and Chapter 18, these chains are considered as being far too photooxidatively unstable to provide a net benefit for terrestrial plants. Photooxidation by direct sunlight is likely to be much greater in terrestrial plants compared to marine plants, where high-energy sunlight is modulated in the overlying water. Species of di-polyunsaturated membrane lipids found in terrestrial chloroplasts are relatively highly fluidizing while being far less susceptible to photooxidation compared to marine chloroplasts, and they clearly serve their purpose effectively.

One of the most puzzling aspects of the molecular architecture of photosynthetic membranes of chloroplasts concerns the apparent absence of membrane lipid-based defenses against proton leakage such as found in bacteria. Recall that the vast majority of di-unsaturated phospholipids in nature occur as sugar lipids found in the photosynthetic membrane of chloroplasts. In marine chloroplasts, these galactolipids often contain fatty acids such as DHA/EPA with four to six double bonds. These membranes likely have poor cation fidelity. We hypothesize that galactose headgroups work by tightening the permeability barrier against protons perhaps through stabilizing effects of hydrogen bonding among headgroups, as seen in archaeal membranes (see Chapter 2). Alternatively, these highly unsaturated membranes may enable substantially greater energy production than is lost to cation infidelity, providing a net energy benefit compared to other structures.

## 11.4 SUMMARY

Chloroplast membranes housing photosynthetic electron transport chains are fundamental for life as we know it. Photosynthesis is an amazingly sophisticated system and pushes membrane biochemistry to its limits. Simply increasing the total number of chloroplasts per cell or cramming more and more electron transport chains into membrane surfaces has its limits. Harvesting of dim light in the relatively cool marine world seemingly requires adaptation, involving maximization of collisions among components of the electron transport chain. DHA and EPA are proposed to speed up rates of electron transport in marine chloroplasts. The immense benefit of DHA/EPA chains is again tempered by their high permeability and oxidative instability, likely explaining why land plants often use di-unsaturated chains with two to three double bonds per chain for the same roles. A mechanism involving glycolipid headgroup interactions may improve the permeability properties of DHA/EPA-enriched membranes. Attempts to develop DHA-producing land plants are discussed in Chapter 18.

## SELECTED BIBLIOGRAPHY


Anderson, J. M., and B. Andersson. 1982. The architecture of photosynthetic membranes: Lateral and transverse organization. *Trends Biochem. Sci.* 7:288–292.

Blackwell, M., C. Gibas, S. Gygax, D. Roman, and B. Wagner. 1994. The plastoquinone diffusion coefficient in chloroplasts and its mechanistic implications. *Biochim. Biophys. Acta* 1183:533–543.

Harwood, J. L. 1998. Membrane lipids in algae. In *Lipids in photosynthesis: Structure, function and genetics,* ed. P.-A. Siegenthaler and N. Murata, 53–64. Dordrecht: Kluwer Academic Publishers.

Lavergne, J., J.-P. Bouchard, and P. Joliet. 1992. Plastoquinone compartmentation in chloroplasts. II. Theoretical aspects. *Biochim. Biophys. Acta* 1101:13–22.

Malkin, R., and K. Niyogi. 2000. Photosynthesis. In *Biochemistry and molecular biology of plants*, ed. B. B. Buchanan, W. Gruissem, and R. Jones, 568–628. Rockville, MD: American Society of Plant Physiologists.

Napp, J. M., L. E. Schaufler, G. L. Hunt, Jr., and K. L. Mier. 2006. Summer food web structure in the eastern Bering Sea: Fatty acid composition of plankton, fish, and seabirds around the Pribilof Islands. Presented at PICES 15th Annual Meeting, October 13–22, 2006, Yokohama, Japan.

Siegenthaler, P-A., and N. Murata, eds. 1998. *Lipids in photosynthesis: Structure, function, and genetics*. Dordrecht: Kluwer Academic Publishers.

Singh, S. C., R. P. Sinha, and D.-P. Häder. 2002. Role of lipids and fatty acids in stress tolerance in cyanobacteria. *Acta Protozool.* 41:297–308.

Tasaka, Y., Z. Gombos, Y. Nishiyama, et al. 1996. Targeted mutagenesis of acyl-lipid desaturases in *Synechocystis*: Evidence for the important roles of polyunsaturated membrane lipids in growth, respiration and photosynthesis. *EMBO J.* 15:6416–6425.
Vargas, C. A., R. Escribano, and S. Poulet. 2006. Phytoplankton food quality determines time windows for successful zooplankton reproductive pulses. *Ecology* 87:2992–2999.
Wada, H., Z. Gombos, and N. Murata. 1990. Enhancement of chilling tolerance of a cyanobacterium by genetic manipulation of fatty acid desaturation. *Nature* 347:200–203.

# 12 Mitochondria: *DHA-Cardiolipin Boosts Energy Output*

Hypothesis: The incorporation of DHA cardiolipin in mitochondrial membranes increases energy output and enables extremes in physiology, including flight of hummingbirds, fast heartbeats of small mammals, rattle muscles of snakes, swimming by cold-water fishes, and long migratory flights of birds.

There is increasing evidence that DHA-enriched mitochondria are the most effective power sources for fast muscles such as hummingbirds' flight muscles. It is well established that hummingbirds have undergone a series of evolutionary changes required to achieve wing beats of about 80 per second while hovering and up to about 200 per second during courtship. These adaptations include (1) a relatively large and powerful heart and circulatory system needed to deliver $O_2$ (Figure 12.1A) to (2) densely packed mitochondria in muscle cells (e.g., up to 35% total volume). A third adaptation (3) involves a doubling of inner membrane cristae (Figure 12.1B), theoretically resulting in a doubling of ATP output. The simplest explanation for these data is that the demands of rapid flight create an intense energy need or stress in flight muscles that has led to a series of adaptations aimed at increasing power output. Thus, it is not surprising that ATP production by mitochondria of hummingbird muscles is believed to set the standard in nature for the maximum energy output by this organelle. The finding of DHA phospholipids in breast muscle provides an important clue about a possible fourth adaptation essential for fast flight in hummingbirds and lays the groundwork for discussion here.

Since the roles of DHA as a pacemaker in respiration in whole animals have been reviewed in 1999 by Hulbert and Else, our focus here shifts toward a molecular explanation of how the conformational dynamics of DHA might be harnessed to boost energy production in mitochondria of fast muscles. Fast muscles include heart muscle cells of small rodents such as miniature voles (~1100 beats/min) and hummingbirds (~1360 beats/min). The goal here is to explain the need for DHA phospholipids in these cells.

## 12.1 MAKING A CASE FOR DHA-CARDIOLIPIN IN FAST MUSCLES

What do mitochondrial membranes of oysters, cold-adapted fishes, hummingbirds, long-distance flying shorebirds, and rattlesnakes have in common? One answer is DHA-cardiolipin. Cardiolipin is an important membrane constituent targeted to the inner mitochondrial membrane of these animals. Recall that this membrane houses

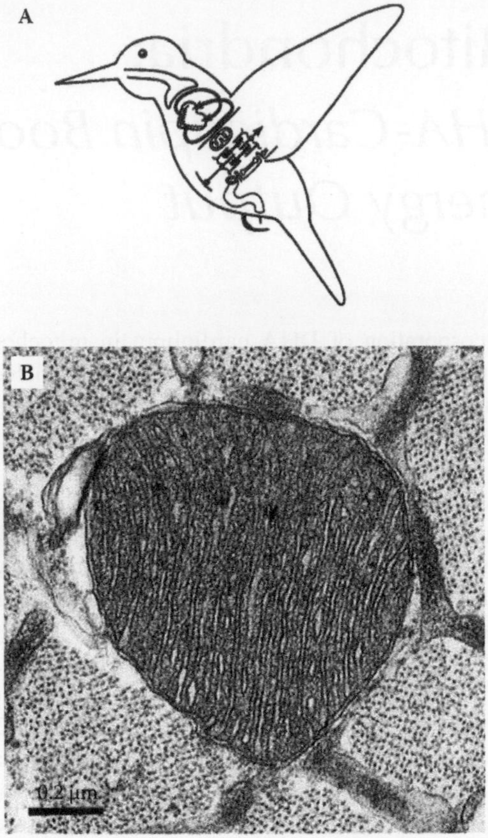


**FIGURE 12.1** Diagram of a hummingbird in flight. Several major adaptations in bioenergetics are necessary for powering high-speed wing beats. (A) A robust cardiovascular system is needed for supporting high rates of mitochondrial respiration in flight muscles. (B) Electron micrograph showing high density of mitochondrial membranes (cristae), thought to approximately double ATP production. A third adaptation (not shown) involves increasing the density of mitochondria per flight muscle cell toward a theoretical limit (i.e., about 35% to 50% total cellular volume). A possible fourth adaptation involving DHA phospholipids is discussed in the text. [Panel (A) reproduced from Suarez, R. K., *J. Exp. Biol.*, 201:1065–1072, 1998. With permission. Panel (B) micrograph courtesy of O. Mathieu-Costella.]

the electron transport chain and also functions as a critical permeability barrier against futile cycling of protons. Cardiolipin is characterized by the presence of four acyl chains attached by ester linkage to dual headgroups. DHA-cardiolipin with up to four DHA chains has been reported in oysters and abalone (Figure 12.2). In clams a cardiolipin species with two EPA and two DHA chains has been identified. In contrast, such highly unsaturated classes of cardiolipin are largely absent in mitochondria of humans, with one possible exception described in Chapter 21. However, 18:2-cardiolipin with four 18:2 chains is common in heart mitochondria of mammals. The focus here is on the biochemical roles of DHA-cardiolipin with an eye toward understanding the roles of this unusual phospholipid in bioenergetics of fast muscles.

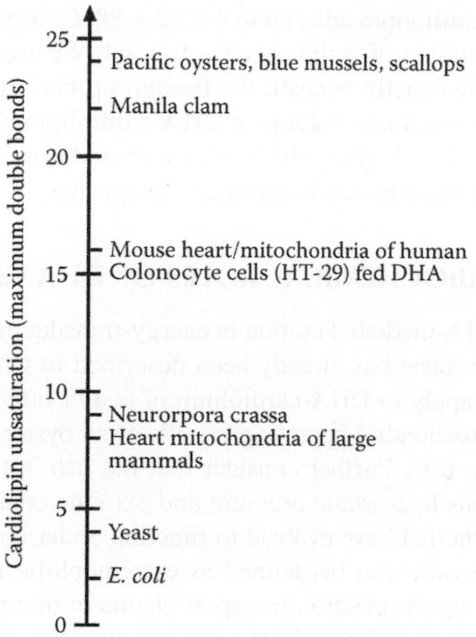


**FIGURE 12.2** Cardiolipin unsaturation ladder. Number of double bonds per cardiolipin molecule varies from one to twenty-four across life forms from bacteria to large mammals to Pacific oysters. DHA phospholipids perhaps as cardiolipin are proposed as an energy booster in membranes of mitochondria in fast muscles. Cardiolipin is synthesized in mitochondria and targeted to the inner membrane of this organelle. DHA cardiolipin is hypothesized to contribute multiple important biochemical properties essential for energizing fast muscles.

There are several extraordinary structural aspects of DHA-cardiolipin that point toward novel functions. The first concerns the presence of four DHA chains, two attached to each conjoined headgroup. We're not aware that the conformations of DHA chains in cardiolipin have been analyzed by chemical simulation. This leaves conformational dynamics of DHA as seen in diacylglycerides as the best model available for interpreting molecular dynamic properties of DHA-cardiolipin. A second novel structural feature with potential biological consequences involves the covalent joining of two conventional phospholipids together via headgroup-headgroup linkage. This might have significant functional ramifications where specific interactions between headgroups are needed for bioactivity. Conjoined headgroups of cardiolipin also carry double negative charges that might strengthen ionic bonding with oppositely charged regions of membrane proteins (see Chapter 3). DHA-cardiolipin is also expected to be subject to rapid rates of lipoxidation. Another feature of DHA-cardiolipin that sets it apart from more saturated species of cardiolipin involves what might be considered the overall bulkiness of this molecule. To visualize this concept imagine an aerial view on a single DHA-cardiolipin molecule in a membrane. The conjoined headgroups help identify the target. Four DHA chains are tethered to the base of the headgroups and lie largely out of view, being underneath the headgroups. The main point is that if one assumes a normal membrane thickness (e.g., 32 C-C bonds) as yardstick then the total number of carbons tucked beneath the twin headgroups of DHA-cardiolipin adds up to 4 × 22 = 88. Compare this to cardiolipin of *E. coli* where the number of carbons is 4 × 16 = 64. We are not claiming that all 88 carbons of DHA lie exactly beneath the headgroup but, rather, that the greater density of carbon atoms in a unit volume of DHA-cardiolipin membrane might have biological meaning such as bulking the bilayer with additional carbon atoms while simultaneously contributing powerful antifreeze properties.

## 12.2 POSSIBLE BIOCHEMICAL ROLES OF DHA CARDIOLIPIN

The importance of DHA-mediated motion in energy-transducing membranes including bacteria and chloroplast has already been described in Chapters 5 and 11, and these concepts might apply to DHA-cardiolipin of fast or otherwise energy-starved muscles. Consider mitochondria operating in cells of an oyster or fish living at temperatures approaching 0°C. Further consider that the fish is forced to move while subject to the enormous hydrostatic pressure and extreme cold characteristic of the deep sea. Mitochondria that have evolved to function under conditions of cold temperature and high pressure can be defined as extremophilic mitochondria. Recall that enzymes making up the electron transport chains of mammalian mitochondria have been calibrated to move laterally at great speeds when tested at physiological temperatures (e.g., 37°C). However, at 0°C motion is expected to stall based on the predictions discussed in Chapter 7. We hypothesize that DHA-cardiolipin works as powerful membrane antifreeze to maintain motion of components of the electron transport chain of mitochondria of cold-adapted animals, but that it may also confer a simultaneous permeability benefit compared to more saturated forms of cardiolipin.

Does the concept of DHA-cardiolipin as antifreeze also apply to fast muscles such as hummingbird breast? Obviously, mitochondria operating at above 40°C in hummingbird breast muscle seem far removed from their counterparts in the deep sea. In the marine environment the primary threat is believed to be excessive portions of the membrane in the gel phase leading to a lack of energy production and enhanced futile ion cycling—both forms of energy stress caused by the cold. However, there is a common stress shared by both classes of mitochondria when viewed from the perspective of energetics. Breast muscle cells can be considered to be energy stressed because of the demands of fast flight. That is, maximum performance of the organism as a whole is limited by output of these muscles, thereby putting their mitochondria under constant energy stress by pushing their limits. Thus, energy stress is passed down to the level of mitochondria, which must meet the needs of muscle cells in energizing the exceptionally rapid wing beats needed for hummingbird flight. According to this scenario, DHA-cardiolipin might play identical roles in mitochondria of both endothermic and ectothermic organisms—optimizing energy output by optimizing motion of components of the electron transport chain. But this is only part of the story, and below we further explore two concepts as to how the structure of DHA-cardiolipin might also act to prevent energy loss in the form of futile ion cycling.

As discussed previously, preventing energy loss at the membrane level is equivalent to increasing energy output. Even though mitochondria are believed to have evolved from single-celled prokaryotes, these organelles seem to lack putative ion-shielding phospholipid structures used by free-living bacteria. It can be argued that membrane lipid defenses against proton leakage are no longer needed by mitochondria because of their intracellular location, which might "buffer" these organelles from energy stresses, common in the bacterial world (e.g., pH and salinity). However, we suggest that membranes of mitochondria operating at temperatures typical of endotherms face a potentially significant energy-uncoupling problem that cannot be ignored and if corrected can effectively increase ATP production by mitochondria. This led us to the idea that cardiolipin in the inner mitochondrial membrane might work as a defense against proton leakage.

We propose three principles about proton fidelity in the inner mitochondrial membrane that lead to further consideration of cardiolipin:

- Proton energy uncoupling via membrane-based water wires is unavoidable in the inner mitochondrial membrane.
- Membrane defenses against energy uncoupling have evolved in the inner mitochondrial membrane.
- Like bacteria, mitochondria harness structural properties of their fatty acids and phospholipids as a defense against water wires.

Based on these principles we explore cardiolipin as a possible candidate for enhancing proton fidelity. Cardiolipin is targeted to and present in significant levels in the inner mitochondrial membrane, and it clearly contributes to permeability properties of the bilayer. The question is what role it plays. There are several chemical features of DHA-cardiolipin that might distinguish permeability properties of this phospholipid from conventional DHA phospholipids as follows:

- Increased density of carbon atoms contributed by DHA chains perhaps works as a bulkiness mechanism or as a deterrent against water wires.
- Joining two headgroups together might hinder water from penetrating the bilayer.
- More powerful antifreeze/antirafting properties are predicted for DHA-cardiolipin as a mechanism to minimize lipid rafts and diminish water wire formation believed to occur at the interface between liquid and gel-phase lipids.
- Owing to its larger size and extensive ionic bonding, chemical flip-flop rates would be expected to be significantly lower, thus reducing levels of any flip-flop–mediated proton uncoupling.

As discussed in Chapter 3, the negative charges on the headgroups of cardiolipin might be harnessed as molecular glue for generating more efficient supramolecular aggregates of electron transport chains. Currently this is the most widely accepted theory linking cardiolipin and bioenergetics. An alternative view as to the mechanism of cardiolipin places a greater emphasis on the headgroup. From this perspective the tight internal hydrogen bonding, the negative charge on the phosphorus group, and the ions thought to associate with the carboxyl groups may serve to repel adjacent lipids, or at least to prevent strong ordering in the lipid portion of the membrane, thereby increasing the potential for lateral membrane motion by loosening head group packing and weakening interactions between adjacent lipid tails. The headgroup of cardiolipin would thus provide much of the fluidity for mitochondria. In this model, the conformational dynamics of DHA takes on a greater significance as the numerous conformations that extend radially outward from beneath the cardiolipin headgroup provide the space filling or bulking needed to prevent excessive water wire formation facilitated by loose packing of headgroups. Such a space-filling role for DHA in cardiolipin can also be applied to other polyunsaturated chains that share the fundamental properties of achieving numerous stable radial conformations compared to saturated chains. We suggest that this general property of polyunsaturated lipid tails may explain why they are incorporated extensively in cardiolipin throughout the biosphere. Interestingly, the conformational dynamics of the various mono- and polyunsaturated chains suggest a cardiolipin permeability hierarchy, with DHA > EPA >> other polyunsaturated chains >> saturated chains. This hierarchy is dependent on the dynamics of radial conformations of the lipid tails and is virtually opposite the permeability hierarchy of standard phospholipids.

## 12.3 CARDIOLIPIN CASE HISTORIES

We next compare the bioactivity of cardiolipin across a range of different organisms with an eye toward understanding function in fast muscles. The first case history involves explaining the needs for cardiolipin in lactic acid bacteria where current theories on the roles of cardiolipin do not seem to apply. Lactic acid bacteria clearly produce cardiolipin, which raises the possibility of some new roles for this class of phospholipids that remain to be discovered. Interestingly, recent genomic analysis has revealed a "hidden" respiratory chain in a lactic acid bacterium, a system that might depend on cardiolipin as molecular glue, a previously proposed mechanism. However, on balance it seems unlikely that the major role of cardiolipin in lactic acid bacteria is as glue to meld electron transport chains together, since these chains are usually not functional. These cells actively ferment high levels of lactose in milk to produce prodigious amounts of lactic acid—hence their name. The rapid drop in pH caused by accumulation of lactic acid denatures milk proteins in the cheese-making process and in the production of yogurt. However, such low pH, while beneficial to the cheese maker, is toxic to lactic acid bacteria and subjects these cells to acidity stress—high outside proton concentrations tend to drive protons across the bilayer in an energy-uncoupled fashion. Lactic acid bacteria have evolved a cyclopropane system, presumably as a membrane-bulking defense against acidity, but might require additional mechanisms of protection against such high levels of lactic acid. We hypothesize that cardiolipin in this anaerobic cell protects proton electrochemical gradients essential for survival in the presence of high levels of lactic acid acting as a proton energy uncoupler. One mechanism by which this might happen involves the protonation of the cardiolipin headgroup at low pH, which according to the above model might reduce repulsion between headgroups and potentially tighten membrane architecture during acidification.

The second case history involves the roles of cardiolipin in yeast. Recall that polyunsaturated fatty acids are not usually present in cardiolipin of yeast whose mitochondria contain fatty chains with only one double bond. However, mutants blocked in synthesis of monounsaturated chains accept a hierarchy of chains carrying from one to six double bonds (i.e., 18:1 → 22:6) to build their mitochondrial membranes. The absence of polyunsaturated cardiolipin in *S. cerevisiae* can best be explained in terms of the selective advantages to *S. cerevisiae* of having oxidatively stable monounsaturated lipids in its membranes. The main evidence supporting this idea is that *S. cerevisiae* cells fed polyunsaturated fatty acids become highly sensitive to oxidative stress and killing, as already discussed in Chapter 9.

Since many other kinds of yeast/fungi, such as *Neurospora crassa*, produce polyunsaturated bilayers, to their advantage, we decided to search for a window in their growth cycle in which yeast cells might also benefit from polyunsaturation. A DNA recombinant approach was used to construct, introduce, and express the fungal Neu (*Neurospora*) 111, Δ12 desaturase gene in *S. cerevisiae.* The Δ12 desaturase recombinant produces high levels of polyunsaturated chains (e.g., 18:2, 16:2) when grown in shake cultures using the respiratory substrate galactose. At first, comparison of the growth rates between the recombinant producing polyunsaturated membranes and a control cell line, in which the Δ12 desaturase gene carried on a plasmid vector is replaced by the *LacZ* gene of *E. coli* as a control, showed no differences. However, this picture changes significantly when recombinant and control cells are exposed to cooler growth temperatures (Figure 12.3). In this experiment, growth is carried out not in liquid culture, but in colonies; recall that cells in the interior of the colony might be energy starved due to $O_2$ deprivation. Growth rates approximately doubled in 18:2/16:2 recombinant colonies grown at 24°C and more than quadrupled at 17°C. Fatty acid analysis of these cells derived from colonies revealed an enrichment of about 5% total fatty acids with 18:2/16:2, considerably lower than levels observed in cells grown under aerated conditions where fatty acid desaturases are saturated with $O_2$ needed as substrate.

A literature survey to see whether growth benefits of polyunsaturation in yeast have been reported previously led to data that appear to explain our results. It was found that partial mutants targeted to electron transport complexes were difficult to score compared to a wild-type control when cells are grown in maximally aerated liquid medium. However, growth differences between partial mutants and wild type were easily scored using a colony growth assay such as we employed. This is consistent with the idea that partial defects in the electron transport chain presumably resulting in energy stress are somehow amplified during colony growth. Recall that cells growing on galactose are largely dependent on electron transport (i.e., mitochondria) for energy production. This observation raises the possibility that polyunsaturated chains might be incorporated into cardiolipin and perhaps work by enhancing mitochondrial energy production above wild-type levels. Although it is rare to improve on Mother Nature, there appears to be a narrow thermal window in yeast in which polyunsaturation swings from risk to net benefit, as measured by faster growth rates. However, the tracing of the observed growth effect to the level of 18:2-cardiolipin and its putative effect as an energy booster in yeast mitochondria remains to be accomplished.

Case history three involves research on thermal regulation of cardiolipin unsaturation in fungi (Schlame, Brody, and Hosteter, 1993). Once again conventional views of cardiolipin functions do not satisfactorily explain these data. Unlike *S. cerevisiae*, cardiolipin of fungi such as *N. crassa* is enriched with polyunsaturated chains, and this provides an opportunity to explore how environmental signals such as temperature modulate unsaturation levels. Data obtained with *N. crassa* point

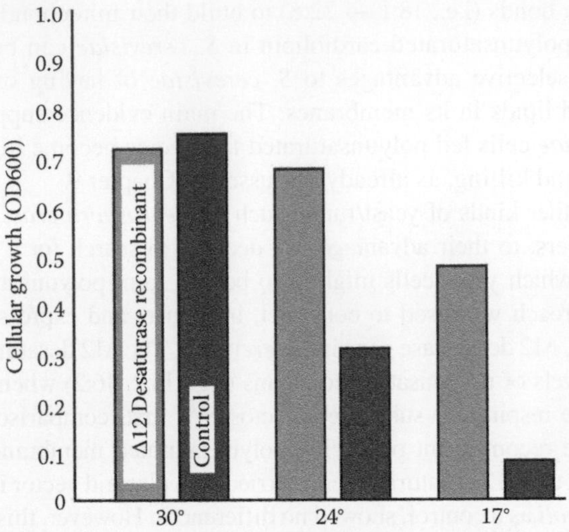


**FIGURE 12.3** Polyunsaturated fatty acids enhance respiratory growth of yeast at cool temperatures. Replacing monounsaturated chains with polyunsaturated chains in membranes strongly stimulates respiratory growth of colonies of *S. cerevisiae* but only at cool temperatures (i.e., 17°C in contrast to 30°C). The mechanism is unknown, but might involve incorporation of 18:2/16:2 chains into cardiolipin working as a mitochondrial energy booster. These data provide another example where mutant cells are able to outperform wild-type cells, simply by increasing the level of membrane unsaturation.

toward a decoupling of unsaturation with temperature. Indeed, the results are dramatic in the sense that cardiolipin retained its relatively high unsaturation levels during the upshift from 22°C to 37°C in contrast to unsaturation levels of the other major classes of phospholipids. For example, the relative abundance of the most highly unsaturated class of cardiolipin (i.e., eight or nine double bonds) dropped only from 35.7% to 33.2% during the shift from 22°C to 37°C. In contrast, the most highly unsaturated class of conventional phospholipid, phosphatidylcholine (PC), plummeted from 30% 18:3 to undetectable levels. A similar dramatic drop occurred with phosphatidylethanolamine (PE; i.e., from 16% 18:3 to 0%). This marked downshift in unsaturation of PC and PE is likely required to maintain membrane function. In contrast, we speculate that the lack of significant shift in unsaturation of cardiolipin is a sign that fluidizing power in the acyl chains of this structure are somehow marginalized or stabilized such that bioactivity is independent of temperature. These data are consistent with a unique role of cardiolipin, perhaps at the level of membrane permeability, and further support the concept that the headgroup of cardiolipin may set the fluidizing properties of this molecule.

The final case history involves fast heartbeats of small mammals. Recall that heart rates across mammals increase from about 30 per minute for the largest animals to more than 1000 for the smallest mammals. Researchers studying the roles of DHA membranes in heart muscles have shown that DHA levels in heart tissue climb in lockstep with heart rate. For example, DHA levels in phospholipids of human heart are very low compared to mice and voles with high heartbeats. Like flight muscle cells, heart cells of hummingbirds (beating at a rate of >1000/minute) are also are likely subject to energy stress. Thus DHA membranes, especially DHA cardiolipin, might be considered as a mechanism to relieve energy stress by increasing electron transport in heart cells as well as flight muscle cells.

## 12.4 NATURAL DOPING WITH DHA/EPA FOR ENDURANCE FLIGHT

It is clear that cells and organisms target or time synthesis or incorporation of DHA/EPA into their membranes to coincide with periods of maximum need and as a means to avoid oxidative damage associated with these molecules. In other words, organisms have evolved mechanisms that maximize the benefits versus risks of DHA/EPA. This principle might help explain an interesting ecological relationship between long-distance flight by a migratory shorebird and its DHA/EPA food source, an amphipod (Maillet and Weber, 2007; Weber, 2009; Figure 12.4A). As first considered in Chapter 9, during their fall migration from the Arctic to South America, sandpipers stop in the Bay of Fundy in Nova Scotia, Canada, to gorge on shrimp-like crustaceans called amphipods. This food source is particularly rich in DHA/EPA, which allows sandpipers not only to bulk up on oils as energy food for their long flight, but also to remodel their membranes. This might be a necessary step to improve energy output of mitochondria before undertaking a nonstop 4500-km flight across the open ocean to South America, lasting close to 72 hours. The birds need to double their weight in 10 to 20 days by feeding on the rich mud flats of the Bay of Fundy, exposed by enormous tides characteristic of this area. As the tide ebbs each hungry sandpiper feasts on 9000 to 20,000 Fundy mud shrimp per day. Approximately three million semipalmated sandpipers, or about 75% to 95% of the world population of this species, refuel in the Bay of Fundy. The use of stored fat as energy source makes this long migratory flight possible because an estimated 85% to 95% of total energy comes from fatty acids. However, DHA/EPA chains are also used to remodel membranes (Figure 12.4B). The authors of these interesting papers hypothesize that DHA/EPA somehow enhances bioenergetic performance. It would be of interest to know whether DHA/EPA from Fundy mud shrimp were incorporated into DHA/EPA cardiolipin in flight muscle, tracing these chains and their putative turbocharging effect to the level of the inner mitochondrial membrane.

There are clues in the literature and from our personal experience that timing of DHA/EPA synthesis or incorporation into membrane phospholipids is of more widespread importance in nature than currently recognized. For example, geese and ducks migrating through southeast Alaska in the fall gorge on DHA/EPA-rich fish eggs deposited by spawning salmon. When cooked, the flesh of these birds reeks of lipoxidized fish oil to the point that it is inedible. Do these migratory birds also harness DHA/EPA in their mitochondria? A modified type of natural doping with DHA/EPA might even be found among human cells. For example, human nutritionists have shown that DHA/EPA levels spike in red blood cells following meals rich in these chains. Scenarios are developed in Chapters 13 and 21 in which human cells benefit from DHA/EPA when available sporadically in the diet.

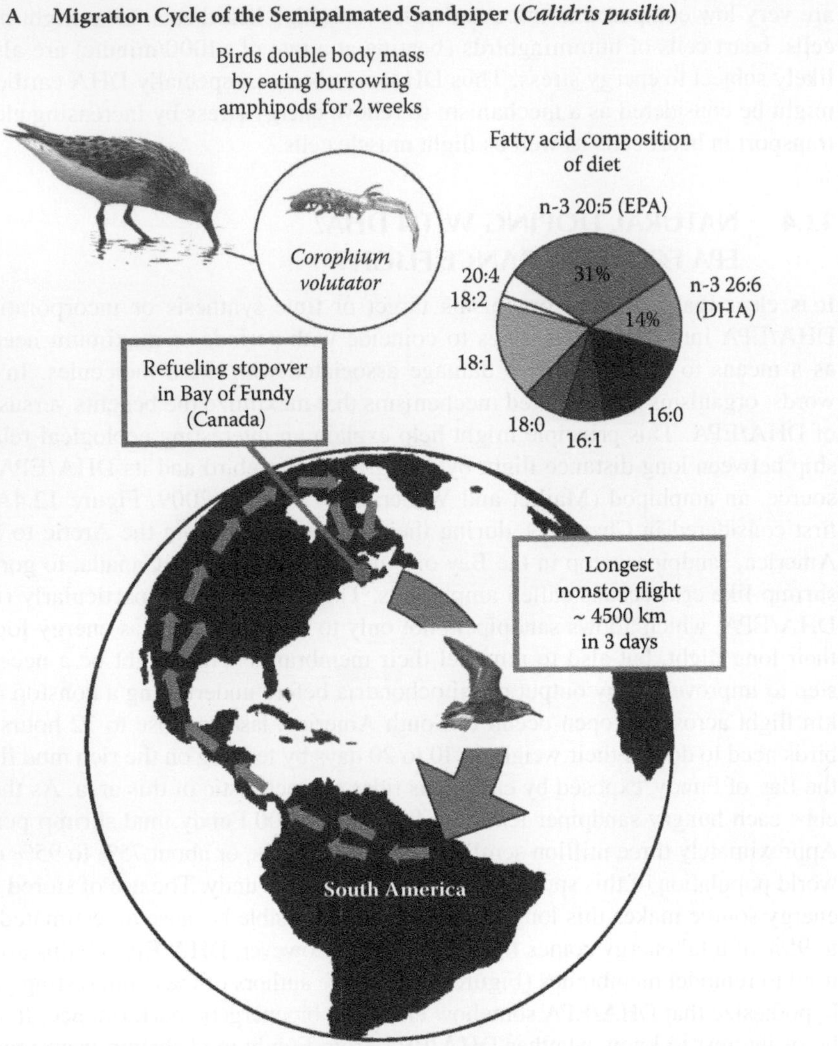


**FIGURE 12.4A** Transient synthesis of DHA/EPA phospholipids as an adaptation for endurance flying. (A) The diagram shows how migrating sandpipers gorge on DHA/EPA-rich food prior to starting a 3-day overwater flight of about 4500 km. (Reproduced from Weber, J.-M., *J. Exp. Biol.*, 212:593–597, 2009. With permission.)

## 12.5 SUMMARY

DHA-cardiolipin, generally targeted to the inner membranes of mitochondria, is a remarkable structure with up to 24 double bonds. A widely accepted molecular glue theory of the roles of cardiolipin in bioenergetics does not adequately explain the need for such dramatic acyl structures. We suggest that cardiolipin plays multiple

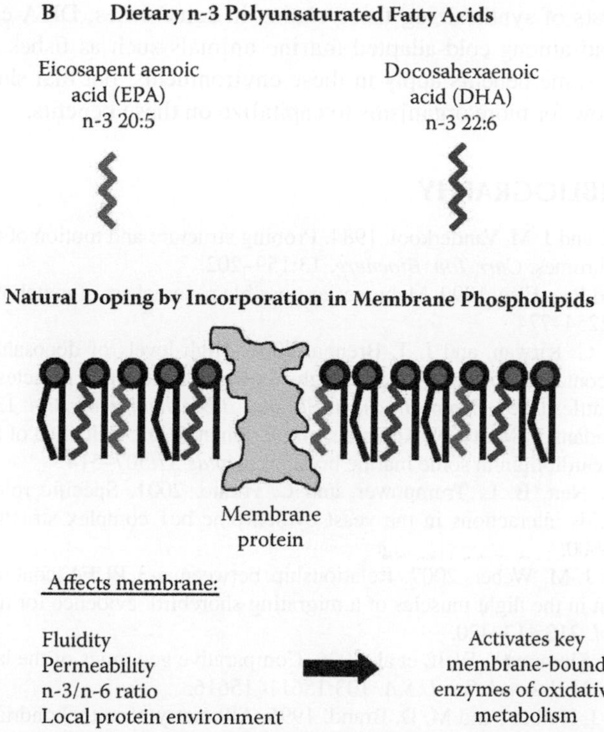


**FIGURE 12.4B** Transient synthesis of DHA/EPA phospholipids as an adaptation for endurance flying. (B) Much of the stored DHA/EPA is consumed directly as energy food during the fall migration flight but these chains are also incorporated into phospholipids of flight muscles, presumably including mitochondrial membranes. This raises the interesting possibility that DHA/EPA phospholipids, perhaps in the form of DHA/EPA cardiolipin, effectively turbo-charge electron transport chains and help sustain energy production during the grueling 72-hour migratory flight.

roles. In contrast to standard membrane lipids, the conjoined anionic headgroup of cardiolipin may provide the bulk of the fluidizing power for this molecule, with the unsaturated tails serving to fill the gaps between adjacent lipids. This model seemingly explains the prevalence of cardiolipin in specialized energy-generating organelles, the general association of cardiolipin with unsaturated chains, the lack of temperature regulation on cardiolipin unsaturation level, and the apparent benefit of cardiolipin to a strain of lactic acid bacteria. Thus, DHA-cardiolipin structure is envisioned to provide two important benefits to mitochondria: lateral membrane motion and a suitable permeability barrier against cations, collectively leading to enhanced net energy availability. We envision that the mitochondria of fast muscles represent the extreme case where natural selection has led to a membrane structure optimized for energy output. Still there appears to be such a fine line between benefits and risks that only a few endothermic animals have evolved DHA-cardiolipin. We suggest this is primarily due to the oxidative instability of DHA, but also to the bioenergetic costs of synthesizing these chains in some cases. DHA-cardiolipin is more widespread among cold-adapted marine animals such as fishes and oysters. We suggest the same benefits apply in these environments, but that slower rates of lipoxidation allow for more organisms to capitalize on these benefits.

## SELECTED BIBLIOGRAPHY


Dixit, B. P. S. N., and J. M. Vanderkooi. 1984. Probing structure and motion of the mitochondrial cytochromes. *Curr. Top. Bioenerg.* 13:159–202.

Hulbert, A. J., and P. L. Else. 1999. Mebranes as possible pacemakers of metabolism. *J. Theor. Biol.* 1999:254–274.

Infante, J. P., R. C. Kirwan, and J. T. Brenna. 2001. High levels of docosahexaenoic acid (22:6n-3)-containing phospholipids in high-frequency contraction muscles of hummingbirds and rattlesnakes. *Comp. Biochem. Physiol. B Biochem. Mol. Biol.* 130:291–298.

Kraffe, E., P. Soudant, Y. Marty, N. Kervarec, and P. Jehan. 2002. Evidence of a tetradocosahexaenoic cardiolipin in some marine bivalves. *Lipids* 37:507–514.

Lange, C., J. H. Nett, B. L. Trumpower, and C. Hunte. 2001. Specific roles of protein-phospholipids interactions in the yeast cytochrome bc1 complex structure. *EMBO J.* 20:6591–6600.

Maillet, D., and J.-M. Weber. 2007. Relationship between n-3 PUFA content and energy metabolism in the flight muscles of a migrating shorebird: evidence for natural doping. *J. Exp. Biol.* 210:413–420.

Makarova, K., A. Slesarev, Y. Wolf, et al. 2006. Comparative genomics of the lactic acid bacteria. *Proc. Natl. Acad. Sci. U.S.A.* 103:15611–15616.

Porter, R. K., A. J. Hulbert, and M. D. Brand. 1996. Allometry of mitochondrial proton leak: Influence of membrane surface area and fatty acid composition. *Am. J. Physiol. Regul. Integr. Comp. Physiol.* 271:R1550–R1560.

Rome, L. C., D. A. Syme, S. Hollingworth, S. L. Lindstedt, and S. M. Baylor. 1996. The whistle and the rattle: The design of sound producing muscles. *Proc. Natl. Acad. Sci. U.S.A.* 93:8095–8100.

Schlame, M., S. Brody, and K. Y. Hostetler. 1993. Mitochondrial cardiolipin in diverse eukaryotes. *Eur. J. Biochem.* 212:727–733.

Schlame, M., D. Rua, and M. L. Greenberg. 2000. The biosynthesis and functional role of cardiolipin. *Prog. Lipid Res.* 39:257–288.

Stuart, J. A., T. E. Gillis, and J. S. Ballantyne. 1998. Remodeling of phospholipid fatty acids in mitochondrial membranes of estivating snails. *Lipids* 33:787–793.

Suarez, R. K. 1998. Oxygen and the upper limits to animal design and performance. *J. Exp. Biol.* 201:1065–1072.

Suarez, R. K., J. R. Lighton, G. S. Brown, and O. Mathieu-Costello. 1991. Mitochondrial respiration in hummingbird flight muscles. *Proc. Natl. Acad. Sci. U.S.A.* 88:4870–4873.

Walenga, R. W., and W. E. M. Lands. 1975. Effectiveness of various unsaturated fatty acids in supporting growth and respiration in *Saccharomyces cerevisiae*. *J. Biol. Chem.* 250:9121–9129.

Weber, J.-M. 2009. The physiology of long-distance migration: Extending the limits of endurance metabolism. *J. Exp. Biol.* 212:593–597.

# 13 Sperm: *Essential Roles of DHA Lead to Development of a Mechanical Stress Hypothesis*

Hypothesis: DHA targeted to sperm tail membranes is an important adaptation for life in the female reproductive tract, improving survival and motility.

Sperm cells are unusual in spending a crucial part of their life cycle traveling as "free-living" cells in the female reproductive tract, where conditions are fundamentally different from those inside the body. For example, sperm face $O_2$ deprivation and fluctuations of pH that would kill many cells. In addition, during the long transit to the egg, sperm face uncertain levels of energy substrates and variable levels of cations, essential for their excitatory nature. It is now clear that sperm have adapted to their free-living environment both biochemically and structurally. Specifically, sperm appear to have evolved DHA membranes as an adaptation to the free-living state, allowing them to successfully traverse the hostile environment of the female reproductive tract. Interestingly, considering the roles of DHA membranes of sperm from an ecological perspective has helped illuminate in molecular terms why sperm tail membranes are enriched with DHA and why this adaptation is essential in the human reproductive cycle.

As background, the journey of sperm in fertilizing the egg is more dependent on energy-rich substrates, nutrients, and other metabolites provided by the female than previously suspected. For example, a portion and perhaps even a majority of the energy for powering sperm movement has been traced back to sugars present in the female reproductive tract, not exclusively internal stores as once suspected. The current picture is that sperm moving in the relatively low $O_2$ environment of the female reproductive tract require glucose as a source of ATP. Glycolytic enzymes are believed to be strategically located throughout the length of the tail region and in proximity to the machinery driving flagellar beating. In contrast, sperm such as that of sea urchin functioning in a highly aerated environment depend on a single mitochondrion (note that human sperm have many) in combination with a unique creatine phosphate shuttle system for energizing motion along the full length of the long flagellum. The definitive finding on the essential role of glycolysis as a source of ATP in mammalian sperm involves knockout mutations of key enzymes resulting in defective sperm. How is bioenergetics of sperm related to DHA structure-function?

Mammalian sperm are enriched in DHA, primarily in tail membranes that house the core machinery for driving motion—a highly energy intensive process. Tail membranes have been shown to be excitatory in nature. The biochemistry of electrophysiology is itself an energy intensive process involving energizing of pumps needed to maintain cation homeostasis—high $Na^+$ outside and high $K^+$ inside. Thus, the view of bioenergetics of sperm has changed dramatically, from ample ATP produced by mitochondria to a paucity of energy available from glycolysis. This is consistent with a concept of sperm in which available energy might become rate limiting for motion, thus elevating the importance of membrane-based mechanisms of energy conservation, which effectively makes more energy available. We suggest that membrane architecture and the roles of DHA as a permeability barrier against energy wastage caused by futile cycling of $Na^+/K^+$ need to be considered in the ecology of sperm.

Sperm tail membranes are excitatory in nature, but there is a major difference from other membranes discussed in previous chapters. That is, tail membrane lipids are likely subject to mechanical stress caused by violent cycles of tail whipping. We hypothesize that the molecular architecture of tail membranes has evolved to compensate for defects in permeability generated by mechanical stress. We further hypothesize that tightening the permeability barrier against $Na^+/K^+$ is an energy-saving mechanism and represents an adaptation necessary for the free-living state. The conformational dynamics of DHA, both flexible and space filling, seems ideally suited for this function. Recent advances are presented as a series of surprises, from the perspective of the authors, that provide a new way of thinking about DHA conformational dynamics in relation to sperm function.

## 13.1 SURPRISE NUMBER ONE: DHA IS LOCALIZED IN TAIL MEMBRANES

The shape of sperm cells, a head region packed with DNA connected to a long whiplike tail, is familiar to most readers. Sperm contains three separate membrane domains, including distinct top and bottom head regions, in addition to the tail. Among the three domains, only the membrane of the tail is enriched with DHA (Figure 13.1). That is, DHA is localized primarily in the tail membrane domain, which surrounds the enzymatic and structural machinery needed to develop thrust or motility. Surprisingly, destruction of the integrity of the tail membrane using detergents does not completely inhibit motion, which can be restored by adding "energizing" metabolites, including ATP and $Mg^{2+}$. However, these membrane-defective sperm wander aimlessly and have obviously lost critical functions attributable to the membrane itself.

During spermatogenesis in the testes, DHA needed for membrane development originates from dietary or biosynthetic pools. Interestingly, dietary or genetic deficiencies can dramatically lower DHA levels in sperm, resulting in decreased motility and abnormal morphologies. For example, patients with retinitis pigmentosa, a major genetic cause of blindness, have low sperm counts, and their sperm contain low DHA content, exhibit reduced motility, and display abnormal structure. Thus, a strong correlation has been established between DHA content and function of sperm. As already mentioned, the production of sperm in the testes requires that this organ

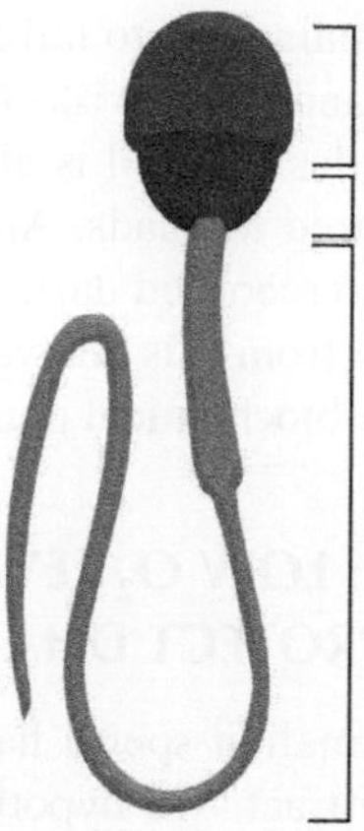


**FIGURE 13.1** DHA in sperm cells is targeted to the tail membrane domain, which is one of the most mechanically active bilayers in the body. A mammalian sperm is a single cell covered by a continuous plasma membrane that is divided into three distinct domains. DHA phospholipids are found primarily in the tail region where we propose these chains play important biochemical roles. (Copyright Alberts et al., *Molecular Biology of the Cell,* 2008. By permission of Garland Science/Taylor & Francis, LLC.)

must obtain a continuous supply of DHA by synthesis or from the diet. This raises the possibility of an unusual requirement in adult males for dietary contributions of DHA to bolster the biosynthetic pool needed for delivery and synthesis of sperm membranes in the testes. Thus, a prudent strategy for maintaining healthy sperm should include a diet rich in DHA.

Studies on spermatogenesis show that sperm production is an imperfect process under the best of conditions, subject to numerous forms of environmental stress. In normal semen, about 20 million sperm are present per cubic centimeter, of which about 5 million are defective. The sensitivity of sperm biogenesis to environmental stresses helps explain why sperm counts may plummet following sessions in saunas, hot tubs, or after wearing tight fitting cycling shorts, each of which cause heat to build up around the scrotum. Indeed, the external location of the scrotum is thought to be a cooling mechanism against heat stress, leading to higher levels of viable sperm. The fact that sperm tail membranes are highly enriched with DHA plays into this scenario, raising the likelihood that protection of DHA might be important in both male fertility as well as female fertility, the latter point discussed in detail later.

The general importance of DHA as a building block for healthy sperm is well established, but surprise number one concerns the apparent strategic localization or targeting of DHA. Indeed, one of the most unusual chemical features of sperm membranes is the differential enrichment of the tail membrane versus head membrane, with 19.6% DHA in the tail, compared to 1.1% in the head based on analysis of monkey sperm. These values are expected to apply to human sperm. Arachidonic acid (20:4) is also preferentially targeted to the tail of monkey sperm, 6.4% versus 1.6%. The overall differences in total unsaturated fatty acids are 34.1% in the tail versus 12.1% in the head membrane. These data support the view that mechanisms exist to target DHA away from head domains and to tail membranes presumably to maximize benefits in this critical region and reduce risks to the head region. In addition to DHA, the cholesterol derivative desmosterol is also targeted to tails, being ~100-fold more abundant in tails compared to heads. An approximate doubling of cholesterol in the tail membrane was also recorded during lipid analysis of monkey sperm. The main conclusion we surmise from this analysis is that DHA is targeted to the tail membrane domain for specific biochemical purposes.

## 13.2 SURPRISE NUMBER TWO: LOW $O_2$ LEVELS IN THE FEMALE REPRODUCTIVE TRACT PROTECT DHA FROM OXIDATION

Increasing evidence suggests that mammalian sperm have adapted to the low $O_2$ environment of the female reproductive tract. We hypothesize that this adaptation confers both major advantages and disadvantages. As background, early chemical studies on animal sperm, chiefly the ram, established that DHA-enriched phospholipids are highly susceptible to oxidation and this process is associated with progressive and irreversible loss of structural integrity, motility, viability, and metabolic activity of the sperm cells. Indeed, exogenously applied lipid peroxides, the first product of lipoxidation, are powerfully spermicidal against washed human sperm treated with as little as 30 nM of lipid peroxide per milliliter. Treated sperm become irreversibly immobile within a few minutes. Classic chemical oxidation protocols using iron or ascorbate catalysis of oxidation of sperm phospholipids show that DHA is preferentially targeted and rapidly lost from the membrane in contrast to the 16:0, which remains relatively constant as expected.

Also, $18{:}1_{n9}$, $18{:}2_{n6}$ and cardiolipin, the latter diagnostic of sperm mitochondrial phospholipids, are relatively stable toward this lipid oxidation protocol. However, $20{:}4_{n6}$ present at levels of approximately 10% of DHA is subject to oxidation, though at somewhat lower rates than DHA. The stability of cardiolipin is consistent with low levels of DHA being incorporated into ram sperm mitochondrial membranes in contrast to tail membranes.

The targeting of DHA to tail membranes keeps this potentially damaging source of reactive oxygen species (ROS) away from the stored DNA. DHA might also be excluded or targeted away from membranes of multiple mitochondria that fill a space lying just underneath the DNA packaged in the headspace. According to this model, both space and time work to prevent damage to germ-line DNA caused by the accumulation of toxic ROS radicals, which are readily derived from DHA.

As described in Chapter 9, avoidance of $O_2$, which readily drives lipoxidation of DHA, has been shown to be a simple but effective mechanism to protect DHA-enriched cells against oxidative stress. The avoidance approach maximizes benefits of DHA and minimizes oxidative risk. However, it comes as a surprise that $O_2$ avoidance by virtue of low oxygen levels in the female reproductive tract might be a mechanism for protecting sperm and its DNA against DHA-mediated oxidative stress. The apparent lack of most conventional ROS defenses in sperm is also consistent with a diminished need for such antioxidation mechanisms in the low $O_2$ world of the female reproductive tract.

## 13.3 SURPRISE NUMBER THREE: MOTION IS ENERGIZED BY SUGAR, A WEAK ENERGY SOURCE, RATHER THAN MITOCHONDRIA

Sperm have evolved a novel adaptation to supply ATP in a low $O_2$ environment. Previously, mitochondria located in the midpiece just above the tail were believed to be the major energizer of sperm motility. It now appears that glycolytic enzymes located along the length of the tail are likely the more important source of ATP, at least in the free-living state. That sperm produce lactic acid from glucose was established more than 30 years ago, and appropriate biochemical activity for various glycolytic enzymes, including glucokinase, lactate dehydrogenase, and glyceraldehyde 3-phosphate dehydrogenase, has since been found in the tail region of a variety of mammalian sperm. It has also been established that "hyperactivated motility" is dependent on ATP from glycolysis and that inhibitors against mitochondrial ATP production do not block fertilization. These observations raise the specter of energy generation being rate limiting for sperm motion.

Formal proof as to the importance of glycolysis in the tail region has been provided by studies of knockout mutations in mice. Specifically, targeted deletion of glyceraldehyde 3-phosphate dehydrogenase results in male sterility and severe aberrations in sperm motility. Thus, it appears that mammals have evolved a unique ATP-producing pathway necessary for energizing sperm motion and fertilization to proceed in an environment low in $O_2$. However, glycolysis is not a perfect solution for energy production because of the need for a continuous supply of glucose presumably contributed by the female; the energy yield from glycolysis is also low compared to respiration. Interestingly, this may subject sperm to energy stress, which, in turn, focuses attention on the membrane itself as an important mechanism to conserve energy.

## 13.4 SURPRISE NUMBER FOUR: SPERM TAIL MEMBRANES ARE EXCITATORY IN NATURE AND MUST EXPEND CONSIDERABLE ENERGY FOR MAINTAINING $Na^+/K^+$ BALANCE

Recent breakthroughs in understanding the molecular biology of sperm function also contribute to understanding the role of DHA. One critical advance is the finding that DHA-enriched membranes of sperm are excitatory in nature, housing critical cation pumps/gates that in turn consume a considerable amount of energy. As background, mammalian spermatozoa are quiescent on first entering the female reproductive tract. During their transit through the female reproductive tract, sperm undergo dramatic changes that enable them to fertilize the egg. During this period, sperm acquire progressive motility and finally develop hyperactivated motility. These states of motility are regulated by intracellular pH of sperm whose cytoplasm becomes increasingly alkaline from the vagina (pH 5.0) to the cervical mucous (pH 8.0). Alkalinization has a dramatic effect on membrane potential primarily mediated by $K^+$ currents originating from the principal piece of the sperm flagellum. Efflux of $K^+$ sets the membrane potential to negative potentials, opening $Ca^{2+}$ gates and driving $Ca^{2+}$ into the cell.

Thus, both $Ca^{2+}$ and $K^+$ ion currents modulate flagella thrust, and molecular details of signaling systems for activation/hyperactivation of sperm are now coming to light.

Recall that the sperm tail is filled almost entirely by the protein machinery of motility, the axoneme. This structure, which converts ATP energy to thrust for propulsion, sprouts from the basal body and extends to the tip. As already mentioned, disruption of the tail membrane does not immediately stop flagellar beating provided $Mg^{2+}$-ATP is added as substrate. This fact, along with advances mentioned above, raises the likelihood that the DHA-enhanced tail membrane plays a central role in regulating the electrophysiology and directionality of motility.

The present working model is that DHA-enriched bilayers of sperm have evolved as a permeability defense against energy wastage caused by futile cycling of electrically active cations: $Na^+$, $K^+$, and $Ca^{2+}$, but not $H^+$. One prediction of this model is that sperm membranes, especially in the tail region, are relatively permeable to $H^+$, thus exposing the interior of the lumen of the tail and perhaps the head as well to the variations in pH encountered in different regions of the reproductive tract. According to this model, the molecular architecture of DHA-enriched bilayers as a permeability barrier is an important parameter in regulating the electrophysiology of sperm. However, this process requires a great deal of energy, which is difficult to generate in a low $O_2$ world.

Thus, the tail membrane is believed to play at least two important electrochemical functions, working as permeability barrier against futile cycling of cations and as a matrix for housing gates/pumps/receptors involved in electrophysiology. The list of membrane-bound enzymes present in the tail is not complete but is likely to include critical cation pumps, voltage-gated channels, and signaling proteins necessary for the electrophysiology of motility. For example, the possibility that the mode of action of $Ca^{2+}$ pumps present in excitatory membranes might require a highly fluid lipid environment has already been discussed in Chapter 3. In other words, it is difficult to imagine how an electrical spike could be propagated along the sperm membrane should critical gates/pumps become trapped within lipid rafts. In essence, DHA-enriched phospholipids of the sperm tail can be considered as an antifreeze defense to ensure the freedom of motion of certain membrane enzymes important in excitatory biochemistry. The role of DHA in disruption of lipid rafts has already been covered in Chapter 7 and is proposed to apply to sperm membranes as well.

## 13.5 SURPRISE NUMBER FIVE: DYNAMIC SPACE-FILLING CONFORMATION OF DHA

The springlike or helical shape model once envisioned for DHA chains has now given way to a dynamic conformational model. A DHA chain is now viewed as a highly flexible molecule with rapid transitions between large numbers of conformers on the time scale from picoseconds to hundreds of nanoseconds. This unusual flexibility has been explained on the basis of low barriers to torsional rotation about C-C bonds linking *cis*-double bonds with the methylene carbons between them. In this section we explore how the flexibility of DHA chains might be harnessed to enhance the bioenergetic properties of sperm tails. The compression or compactness of the DHA chain is also of interest from the standpoint of sperm tail function as developed in more detail below. Indeed, the intimate relationship between conformation-function of DHA chains seems to stand out in the case of sperm tail membranes operating in an energy-starved environment.

One of the most interesting and least understood aspects of DHA function in sperm tails involves molecular roles of these chains in cation permeability. This is important for both electrophysiology and energy conservation. From first principles, permeability properties of sperm tail membranes must involve conformations of the DHA chain itself. We hypothesize that DHA chains in sperm tail membranes are best suited among fatty acids in containing sufficient bulk in the form of their dynamic conformations to rapidly fill any molecular voids, pores, or defects occurring within their intramembrane molecular domains (Figure 13.2). In essence, what we are proposing is that the dynamic conformation of DHA chains allows these molecules to rapidly readjust their conformations when the membrane surface of the tail region is mechanically stressed, as occurs during cycles of flagellar whipping (Figure 13.2). This molecular model assumes the enhanced formation of water wires at "stretch points" along the tail caused by cyclical stretching and compression of the bilayer—DHA is envisioned to rapidly fill these defects so as to inhibit the formation of water wires.

Another aspect of DHA structure-function worth considering involves the nature of mechano-sensitive excitatory channels and enzymes. It is well known that ion channels displaying various types of mechano-sensitive gating are ubiquitous in natural membranes. Recent studies focus on various channels whose molecular structures are known, where it has been found that the membrane environment exerts profound effects on voltage-dependent $K^+$ channels. Interestingly, the mechanical state of the membrane was also discovered to exert a prominent effect on interactions between channels and toxins that bind to these proteins. If one assumes that tail membranes of

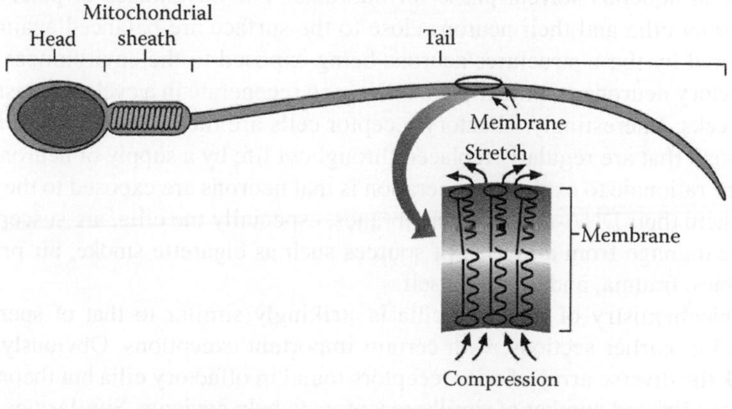


**FIGURE 13.2** DHA chains are hypothesized to be harnessed to seal membrane defects generated during violent tail whipping (i.e., mechanical stress). DHA chains are able to rapidly shift between distinct conformations, perhaps acting as a self-sealing mechanism against defects generated by repetitive cycles of mechanical stress along the flagellum.

sperm are subject to mechanical stress, then this raises the possibility that the activity of ion channels and excitatory enzymes critical to sperm function are simultaneously altered. That is, in addition to functions involving permeability and motion, DHA phospholipids in intimate contact with ion channel proteins might help reduce excessive lipid-protein forces otherwise strong enough to disfigure channel conformation.

## 13.6 SURPRISE NUMBER SIX: LESSONS FROM SPERM APPLIED TO NEUROSENSORY CILIA AND OTHER MECHANO-SENSITIVE MEMBRANES

In this section lessons learned from studies of DHA in tails of sperm are applied to other kinds of mammalian cells whose membranes might be subject to mechanical stress. For example, when the 10,000 or so smells sensed by humans enter the nasal passage, these molecules pass over more than one billion olfactory cilia whose membranes are enriched in DHA (Figure 13.3). These whiplike hairs, 30 to 200 μm in length, extend from the ends of ~40 million olfactory receptor neurons distributed over the surface of the nasal cavity. As in the case of sperm tails, which they resemble, olfactory cilia are located in the outside world with respect to other cells of the body. The membranes of these specialized cilia provide a vast surface area needed for embedding many odor-sensing proteins, which trigger the olfactory cascade. These odor receptors belong to the large superfamily of G-protein–mediated signaling proteins such as rhodopsin. However, unlike the detection of light, in which about 1000 DHA-enriched membrane disks are stacked one on top of the other, in the outer segment of rod cells, membranes for detecting odors take the form of hairlike cilia attached to the olfactory neurons. About eight to twenty cilia from each olfactory receptor neuron extend not into the air phase but rather into a blanket of mucus forming the surface of the nasal passage. This mucus traps moisture and provides an aqueous solvent phase for odorants. The advantages of placement of odor-sensing cilia and their neurons close to the surface are balanced against risks encountered by these structures/neurons being exposed to the environment. Recall that olfactory neurons continually grow, die, and regenerate in a cycle that lasts about 4 to 8 weeks. Interestingly, olfactory receptor cells are the only neurons in the nervous system that are regularly replaced throughout life by a supply of neuronal stem cells. One rationale to explain regeneration is that neurons are exposed to the outside world where their DHA-enriched membranes, especially the cilia, are susceptible to oxidative damage from a myriad of sources such as cigarette smoke, air pollution, toxic fumes, trauma, and even $O_2$ itself.

The biochemistry of olfactory cilia is strikingly similar to that of sperm tails described in earlier sections, with certain important exceptions. Obviously, sperm tails lack the diverse array of odor receptors found in olfactory cilia but theoretically might use a limited number of similar receptors to help navigate. Similarities include their excitatory nature, DHA-enriched membrane composition, internal machinery for creating motion, likely sensitivity to $O_2$ and other oxidants, and a relatively short life cycle driven by environmental exposure. Olfactory cilia might also depend on certain "external" nutrients and minerals as described for sperm. There are insufficient

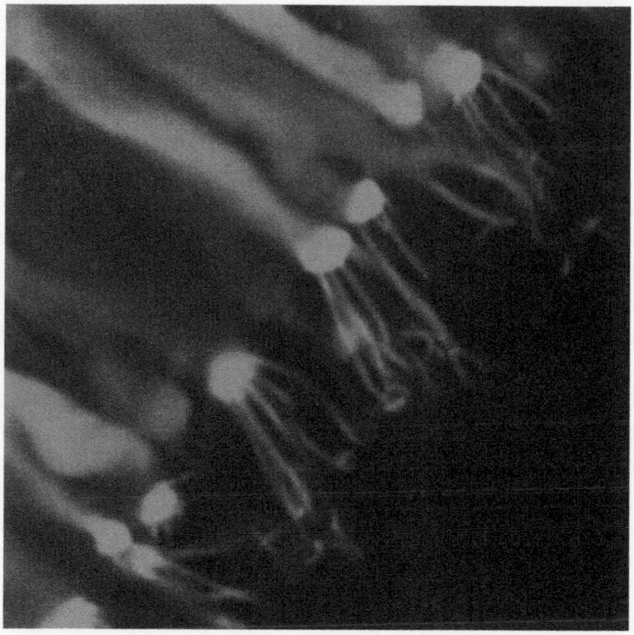

**FIGURE 13.3** (A color version of this figure follows p. 140.) Membranes of olfactory cilia attached to the tips of olfactory neurons share properties with sperm tails. Membranes of olfactory cilia are excitatory in nature and house olfactory receptors, whose mode of action depends on the physical state of these bilayers. Olfactory membranes along with rhodopsin disks serve as striking examples in showing the essential biochemical roles contributed by membranes in neurosensory transduction systems. Recall that hundreds of different olfactory receptors present in these membranes combine to create the sense of smell. (Photos of olfactory cilia of larval *Xenopus* courtesy of Ivan Mangini.)

data to evaluate whether the bioenergetics of sperm and olfactory cilia are similar. However, harnessing DHA as a mechanism for conserving energy at the membrane level as hypothesized for sperm might also apply to olfactory cilia.

One of the major benefits of DHA in olfactory transduction is hypothesized to be motion, which allows numerous G-protein–mediated odor receptors to collide rapidly with their G-proteins. This raises a kinetic issue, the need for rapid detection and signaling in response to transient odorants, before they are swept out of the nasal cavity. For example, speed is of the essence in order to react quickly to air-borne toxic molecules and fumes. In addition to response time, temperature also must be considered as an important environmental variable because of the variability in temperature of air entering the nasal cavity. In the case of olfactory cilia recall that their operational temperature is not constant and depends to some extent on air temperatures, especially cold air, which can begin to lower the temperature around cilia because of their location in or near the surface of the mucus layer. The point is that DHA, working as powerful antifreeze, seems a good choice to maintain rapid odor reception across a range of temperatures.

As suggested for sperm and olfactory cilia, the role of DHA as defense against mechanical stress may be more important and widely distributed in nature than currently appreciated. Another example is cardiac muscle cells and other fast muscles such as flight muscle of hummingbirds that seem to be subject to mechanical stress as a natural part of their fast pumping or muscular action. Recall that heart rates across mammals vary more than twentyfold (see Chapter 12) and, importantly, that DHA levels in heart muscles increase in a roughly linear fashion with increasing heartbeats. Based on this interesting set of data, Hulbert and colleagues have proposed that DHA membranes modulate respiratory rates in fast muscles (Hulbert and Else, 1999; see Chapter 12 references). Specific biochemical roles of DHA in fast muscle cells are discussed in Chapter 12 and the case of hummingbird flight muscle is especially interesting. Approximately one third of the total volume of flight muscle cells is filled by mitochondria, many of which seem to be compressed between powerful muscle sheaths. This raises the possibility that the extremely rapid firing of these strong muscles might subject their adjacent mitochondrial membranes to continuous cycles of mechanical stress during flight. Movies taken of individual mammalian cardiac cells show what appear to be continuous cycles of stretching, shrinking, and contortion of cell volume. We suggest that the pumping action characteristic of individual cardiac cells subjects the plasma membrane to mechanical stress similar to violent whipping of sperm tails. Red blood cells dramatically downsize and change shape as a prerequisite to entering the narrow passageways of capillaries. As discussed in Chapter 12, red cells are opportunistic with respect to acquisition of omega-3 fatty acid from the diet for membrane synthesis, and this mechanism might be an adaptation against mechanical stress. A final example relating to the potential roles of DHA in accommodating mechanical stress comes from bacteria. Interestingly, the initial discovery of omega-3 fatty acids (i.e., EPA, not DHA) in bacteria involved a marine bacterium exhibiting an unusual mechanism of flagellar-less propulsion that is mediated by continuous bending actions of individual cells. Are the plasma membranes of these cells also subjected to mechanical stress? In addition to EPA, the membranes of these whip-like bacterial cells also contain high levels of methyl-branched fatty acids (see Chapter 10), which may replace the function of cholesterol-like molecules found in sperm for the purpose of further stabilization of membrane functions during mechanical stress.

## 13.7 SURPRISE NUMBER SEVEN: GENDER DETERMINATION INFLUENCED BY DIET?

The final surprise concerns one of the most interesting and controversial reports in the fertility field (Mathews, Johnson, and Neil 2008). This involves data suggesting that what a woman eats before pregnancy may play a role in whether the baby is a boy or a girl. The research involved 700 first-time pregnant women, and is the first to show a link between a woman's diet and the gender of her offspring. The women were asked about their eating habits the year before getting pregnant. In general, women who have a hearty appetite, eat a lot of potassium-rich food like bananas, and don't skip breakfast are more likely to have a boy. In addition, women who routinely ate breakfast involving cereals were more likely to have boys compared to women who seldom ate cereals for breakfast, a possible sign that the latter group was skipping breakfast. Among women with the highest calorie intake before pregnancy (but still within the normal, healthy range) 56% had boys versus 45% of the women with the lowest calorie intake. Women who had boys also ate an additional 300 mg of potassium daily on average and about 400 more net calories compared to women who had girls.

Various notions have been proposed to explain how the female diet might influence gender determination. Indeed, the validity of the central conclusions has been called into question. Other ideas focus on the role of glucose on the differential development of the fertilized male versus female embryo. However, we are interested in the revolutionary prospect that female diet potentially benefits the male sperm not only after fertilization but also during the journey of the sperm to the egg. This idea is based on recent advances in understanding how environmental signals characteristic of different regions of the female reproductive tract can dictate profound changes in behavior of sperm. pH was the first to be explained at the molecular level and now glucose has emerged as a potential dietary modulator provided by the female.

The current picture is that the female might have more influence on sperm biochemistry and regulation and hence gender determination than previously expected. Major signals and nutrients dictated by the female might include temperature, pH, $K^+$, $Na^+$, $Ca^{2+}$, $O_2$, and glucose. It is not difficult to envision a scenario in which sperm carrying one gender determinant or the other are nutritionally or biochemically favored by conditions encountered in the female reproductive tract. Indeed, there is a surprising match between the molecular model discussed here, which we developed over the past several years, and the recent dietary study in which starchy and $K^+$-rich foods skewed the ratio of males to females in favor of males. This may turn out to be a molecular version of the "tortoise and hare story" in which male sperm are biochemically predetermined to be faster or more accurate than female sperm provided they are saturated with glucose and $K^+$ during the race to the egg. A role for DHA remains possible.

## 13.8 SUMMARY

The human reproductive cycle is dependent on the functionality of tail membranes of sperm, which are highly enriched with DHA. Studies on the biochemical roles of DHA in sperm have lagged behind research on their electrophysiology. However, combining data from these two fields help define the roles of sperm membranes as excitatory bilayers similar to neurons. Mechanical stress is introduced as a new form of energy stress at the membrane level, which might physically disrupt the molecular architecture of sperm tail membranes as well as olfactory cilia, perhaps benefiting from the conformational dynamics of DHA for rapidly filling defects. The low $O_2$ world of the female reproductive tract is proposed to reduce risks associated with DHA membranes, and thereby maximize net benefit. In contrast, during biogenesis in the testes, sperm appear to be highly sensitive to environmental stresses, resulting in large numbers of defective sperm.

## SELECTED BIBLIOGRAPHY

Aitken, R. J., and J. S. Clarkson. 1987. Cellular basis of defective sperm function and its association with the genesis of reactive oxygen species by human sperm. *J. Reprod. Fertil.* 81:459–469.

Alvarez, J. G., and B. T. Storey. 1995. Differential incorporation of fatty acids into and peroxidative loss of fatty acids from the phospholipids of human spermatozoa. *Mol. Reprod. Dev*. 42:334–346.

Connor, W. E., D. S. Lin, D. P. Wolfe, and M. Alexander. 1998. Uneven distribution of desmosterol and docosahexaenoic acid in the heads and tails of monkey sperm. *J. Lipid Res.* 39:1404–1411.

Connor, W. E., R. G. Weleber, C. DeFrancesco, D. S. Lin, and D. P. Wolfe. 1997. Sperm abnormalities in retinitis pigmentosa. *Invest. Ophthalmol. Vis. Sci.* 38:2619–2628.

Dratz, E. A., and L. L. Holte. 1992. The molecular spring model for the function of docosahexaenoic acid ($22{:}6_{n3}$) in biological membranes. In *Essential fatty acids and eicosanoids*, ed. A. Sinclair and R. Gibson, 122–127. Champaign, IL: American Oil Chemists' Society.

Feller, S. E. 2008. Acyl chain conformations in phospholipid bilayers: A comparative study of docosahexaenoic acid and saturated fatty acids. *Chem. Phys. Lipids* 153:76–80.

Malnic, B., P. A. Godfrey, and L. B. Buck. 2004. The human olfactory receptor gene family. *Proc. Natl. Acad. Sci. U.S.A.* 101:2584–2589.

Mathews, F., P. J. Johnson, and A. Neil. 2008. You are what your mother eats: Evidence for maternal preconception diet influencing foetal sex in humans. *Proc. R. Soc. B Biol. Sci.* 275:1661–1668.

Miki, K., W. Qu, E. H. Goulding, et al. 2004. Glyceraldehyde 3-phosphate dehydrogenase-S, a sperm-specific glycolytic enzyme, is required for sperm motility and male fertility. *Proc. Natl. Acad. Sci. U.S.A.* 101:16501–16506.

Mukai, C., and M. Okuno. 2004. Glycolysis plays a major role for adenosine triphosphate supplementation in mouse sperm flagellar movement. *Biol. Reprod.* 71:540–547.

Navarro, B., Y. Kirichok, and D. E. Clapham. 2007. KSper, a pH-sensitive $K^+$ current that controls sperm membrane potential. *Proc. Natl. Acad. Sci. U.S.A.* 104:7688–7692.

Schmidt, D., and R. MacKinnon. 2008. Voltage-dependent $K^+$ channel gating and voltage sensor toxin sensitivity depend on the mechanical state of the lipid membrane. *Proc. Natl. Acad. Sci. U.S.A.* 105:19276–19281.

Soubias, O., and K. Gawrisch. 2007. Docosahexaenoyl chains isomerize on the sub-nanosecond time scale. *J. Am. Chem. Soc.* 129:6678–6679.

Tabarean, I. V., P. Juranka, and C. E. Morris. 1999. Membrane stretch affects gating modes of a skeletal muscle sodium channel. *Biophys. J.* 77:758–774.

Turner, R. M. 2006. Moving to the beat: A review of mammalian sperm motility regulation. *Reprod. Fertil. Dev.* 18:25–38.

Wang, H.-Y. J., S. N. Jackson, and A. S. Woods. 2007. Direct MALDI-MS analysis of cardiolipin from rat organs sections. *J. Am. Soc. Mass Spectrom.* 18:567–577.

# Section 5

## Lessons and Applications

In this section the concepts developed in the earlier chapters are used as a lens to describe cellular phenomena ranging from nutritional mutualism to oil biodegradation, winemaking, ocean warming, molecular farming, and human disease. In each instance membrane fatty acids play critical roles. Lessons are drawn from cases of oil biodegradation and winemaking, and generalized to membrane function, but do not involve DHA or EPA. The remaining chapters apply previous lessons to understand mutualism between animals and DHA bacteria, the mortality of coral endosymbionts in the warming ocean, the challenges facing genetic engineering of omega-3 crop plants, and the roles of DHA/EPA in human conditions such as aging, Alzheimer's disease, and colon cancer. The common theme that runs through these chapters is the fine balancing of simultaneous benefits and risks associated with DHA/EPA-based membranes—the DHA principle.

# 14 DHA/EPA Mutualism between Bacteria and Marine Animals

Hypothesis: DHA/EPA-producing bacteria have evolved to thrive in the gastrointestinal tracts of marine animals and to supply these oils to their hosts.

Mutualism defines a symbiotic relationship in which organisms belonging to different species provide benefit to each other. The importance of mutualism between animals and the microbial communities living in their digestive systems has long been established (i.e., for ruminants) but is now being recognized more broadly, including for human health. In the marine realm, genomic sequencing of marine bacteria has revealed many genes contributed by these bacteria that might benefit their hosts. For example, many marine animals eat organisms having exoskeletons composed of chitin, whose degradation can benefit from chitinoclastic bacteria. Chitin is the second most abundant polymer formed in nature, with annual production in the aquatic biosphere estimated to be several billion metric tons. Chitin degrading genes have been identified in many marine bacteria, a fact that is consistent with a mutualistic role. In another example, pioneering research using germ-free zebrafish (Figure 14.1) shows that gut bacteria favored by the host animal quickly reestablish themselves when sterile conditions are removed. This finding is consistent with the view that at least some of these bacteria-host relationships have evolved to be mutually beneficial.

These existing relationships provide a conceptual basis for exploring potential mechanisms and applications of bacterial DHA/EPA mutualism in marine fishes and animals. DHA/EPA-producing bacteria are known to inhabit the gastrointestinal tract and potentially provide marine animals with an auxiliary set of genes for de novo synthesis of omega-3 chains not represented in the host. Ultimately, the evolution of DHA/EPA mutualism depends on contributions of these chains toward a fitness advantage in the host animal. It is well documented that DHA/EPA are essential nutrients for growth and development of marine fish as well as other marine animals. Given that the levels of these chains in the diet are often variable and potentially rate limiting for growth, DHA/EPA mutualism might provide host animals with an alternative source of these chains to augment dietary supplies.

## 14.1 GASTROINTESTINAL TRACT OF FISH IS A SUITABLE HABITAT FOR DHA/EPA MUTUALISM

A generalized picture of the gut habitat of a marine fish is shown in Figure 14.2, which highlights conditions favoring growth, specifically DHA/EPA-enriched

**FIGURE 14.1** Germ-free zebrafish have been used in pioneering research on defining the importance of the bacterial population of the gastrointestinal tract of fish. Interestingly, lessons learned about the dynamics of bacterial populations in fish apply to humans. (Courtesy of Charles Badland, Program in Neuroscience, Florida State University.)

bacterial biomass. Few marine bacteria store excess DHA/EPA as triacylglycerols, in contrast to animals and plants. Thus, any mechanism transferring DHA/EPA from bacteria to host requires lysis of bacterial cells or efflux, a point discussed in some detail below. While the advantage of this mutualism to the host is the DHA/EPA, the advantages to the bacteria include the availability and ample supply of energy- and nitrogen-rich substrates in the diet. Essential micronutrients such as iron are also expected to be readily available compared to seawater because of greater stability of ferrous ions ($Fe^{2+}$) under anaerobic conditions, greater solubility of ferric iron ($Fe^{3+}$) at slightly acidic pH characteristic of the gut, as well as the more abundant or concentrated dietary supply of iron present in food sources. The advantages to the bacteria are balanced against disadvantages.

For the bacteria, one major disadvantage to the mutualism involves cold temperatures characteristic of ectothermic sea life, which is expected to slow metabolism. However, DHA/EPA bacteria are characteristically psychrophilic, grow rapidly, and are accustomed to temperatures approaching 0°C. Indeed, there are no reports of bacteria producing DHA above about 20°C, which means that these cells not only tolerate but may require cool temperatures characteristic of fish habitats for growth. Seemingly in support of this assertion, Japanese researchers have observed that DHA levels in a transgenic strain of *E. coli* plummet approximately fivefold as temperature increases to 20°C, with little detectable DHA at 25°C. Thus, bacterial DHA production appears to be most active over the range of about 0°C to 15°C, dropping sharply as temperature approaches 20°C. Recall that marine fishes from which DHA/EPA bacteria have been isolated inhabit niches roughly mirroring this same range of temperatures.

One of the most limiting metabolites for DHA/EPA production in the fish gut environment is likely $O_2$ itself. To counter the low oxygen, DHA/EPA-producing bacteria have evolved powerful terminal oxidases designed for growth under these

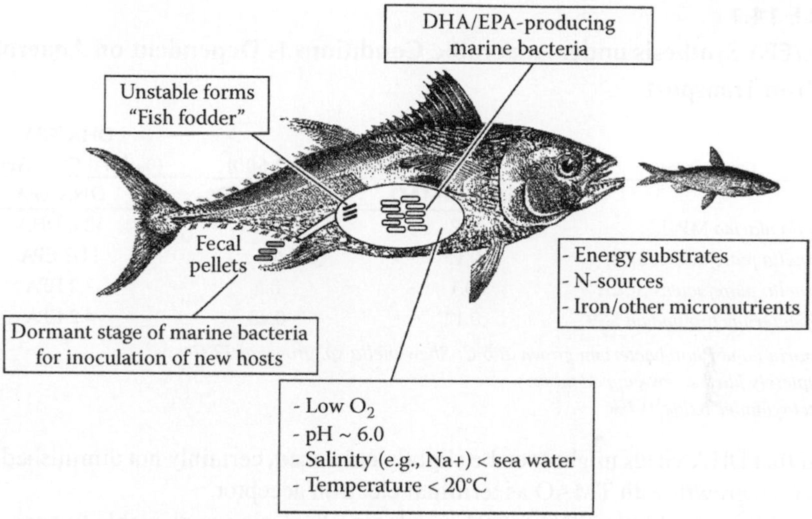


**FIGURE 14.2** DHA/EPA-producing bacteria inhabit the gastrointestinal tracts of certain marine fish and mollusks, with the highest levels being reported for certain cold-adapted mollusks. Physiological and genomic analyses show that the gut of a marine fish is a suitable habitat for supporting robust growth of potentially beneficial bacteria, including DHA/EPA-producing strains. Numerous possible mutualistic genes besides DHA/EPA have been identified in marine bacteria consistent with the view that nutritional mutualism is important for growth and survival of marine animals.

conditions, typical of the fish gut. Nevertheless, rates of terminal electron transport to $O_2$ are expected to be limiting in the gut environment. The alternative of filling this low-oxygen niche with strictly fermentative organisms, such as have evolved to degrade cellulosic polymers in the rumen of animals, does not seem logical for several reasons. These include an often-low dietary intake of fermentable carbohydrate, with the exception of chitin for growth of fermentative cells, and a relatively low energy yield derived from fermentation. Genomic analysis shows that marine bacteria have evolved highly sophisticated, branched pathways of electron transport. This means that more energy can be derived from anaerobic electron transport compared to fermentation. It is thought that a variety of alternative electron acceptors, such as trimethylamine oxide (TMAO), are available in the digestive tract. For example, TMAO is common in the marine environment and is characteristically used by marine bacteria driving relatively robust biomass production and DHA/EPA production in the absence of $O_2$. The point is that, whereas the energy yield with alternative acceptors is lower compared to $O_2$, these yields still far exceed energy production via anaerobic fermentation. Thus, fish gut bacteria seem to specialize in using alternative terminal electron carriers such as TMAO, which yield far more energy and translate to more rapid cellular growth. Conditions expected in the fish gut are readily recreated in the laboratory, as summarized in Table 14.1, and show that DHA/EPA are produced during anaerobic respiration. In various experiments, we have

**TABLE 14.1**
**DHA/EPA Synthesis under Anaerobic Conditions Is Dependent on Anaerobic Electron Transport**

| Organism[a] | Anaerobic Growth[b] (OD 600) | | DHA/EPA (% Total Fatty Acids) |
|---|---|---|---|
| | Without TMAO | With TMAO[c] | DHA/EPA |
| *Moritella marina* MP-1 | 0.2 | 1.5 | 12.8 DHA |
| *Shewanella putrefaciens* 2738 | 0.15 | 1.1 | 11.2 EPA |
| *Shewanella putrefaciens* SC2A | 0.1 | 0.5 | 7.2 EPA |
| *Photobacterium profundum* SS9 | 0.12 | 0.28 | 4.2 EPA |

[a] *M. marina and Photobacterium grown at 6°C; Shewanella sp. grown at 12°C.*
[b] *Completely filled screw-capped tubes.*
[c] *Trimethylamine oxide, 0.4%.*

noted that DHA yields might even be slightly enhanced, certainly not diminished, by anaerobic growth with TMAO as terminal electron acceptor.

In addition to TMAO, DHA/EPA-producing bacteria are also able to conserve energy by shuttling electrons to other acceptors likely present in the gut. For example, marine bacteria have evolved a terminal oxidase that maintains electron flow and energy production by passing electrons to fumarate, a common by-product of dietary protein degradation. Recall that fumarate is formed as an end product of anaerobic catabolism of amino acids by gut microbes and is reduced to succinate as an end product providing a valuable metabolite needed by the host for energizing its mitochondria. Indeed, succinate produced in the gut is predicted to be one of the most important nutrients contributed by mutualistic marine bacteria to their hosts, which continuously supply amino acids in the form of their food, possibly in return for succinate. We suggest a number of anaerobic respiratory processes potentially active in the gut of the marine fish, catalyzed by mutualistic microbial communities. These include respiration with alternative electron acceptors such as $NO_3^-$, $NO_2^-$, TMAO, dimethyl sulfoxide (DSM), fumarate and $Fe^{3+}$. Also included are Stickland fermentations, which are a class of reaction in which different amino acids act as electron donor and receptor. This interesting redox couple is energy yielding for the bacterium while providing useful end products for the host.

Genomic analysis shows that marine bacteria, such as the DHA-producing isolate *Moritella* sp. PE36, obtain energy by a variety of mechanisms, including fermentation of available carbohydrates (e.g., chitin) and anaerobic production of ATP from arginine and nitrogenous bases such as purines. The importance of amino acids for the bioenergetics of *Moritella* is illustrated by the presence of over 50 peptidase-like genes allowing these cells to process proteins as sources of energy, carbon skeletons, and amino acids needed for protein synthesis. Carbamyl phosphate, a direct precursor for anaerobic synthesis of ATP, appears to be generated from multiple ureido-compounds including arginine, and perhaps purines and pyrimidines as well. Interestingly urea, a by-product of host metabolism, readily condenses with glyoxylic acid, such as produced in the fish gut, to yield ureidoglycolic acid, thus creating a potential new source of carbamyl phosphate → ATP. Energy-yielding pathways for fermenting valine, isoleucine, leucine and alanine via their α-ketoacid derivatives also appear to be operational in *Moritella*. These pathways appear to depend on ferredoxin-linked hydrogenases, which work as dumps for excessive reducing power generated from oxidative decarboxylation of these acids. All of the above pathways can be considered as adaptations to the gastronintestinal tract of marine animals, and are consistent with mutualism between the gut bacteria and the animal host.

The relatively large genome sizes of bacteria such as *Shewanella* and *Moritella* allow these marine bacteria to effectively "mine" energy from a diversity of energy-yielding substrates found in the diet of the fish they inhabit. For example, the low $O_2$ environment of the gut elevates the need for multiple anaerobic electron transport chains as well as fermentative pathways for energy generation. These bacteria certainly benefit from the close association with their animal host, and the potential for supply of chemicals such as EPA/DHA to the host likely define the mutualism.

## 14.2 DHA/EPA-PRODUCING BACTERIA INHABIT THE INTESTINAL TRACTS OF CERTAIN MARINE FISH AND MOLLUSKS

The convergent evolution of an anaerobic, polyketide pathway for DHA/EPA synthesis, as already discussed in Chapters 4 and 5, is presumed to be at the heart of any omega-3 mutualism in the gut environment. This pathway also provides a target set of genes present in bacteria that might benefit the host animal. About 20 years ago, a massive screening program was undertaken by Japanese researchers to determine the distribution of DHA/EPA bacteria among fish gut isolates with an eye toward harnessing these bacteria in aquaculture and for possible industrial production of DHA/EPA (Yazawa et al., 1988). This survey involved isolation and testing of tens of thousands of isolates of fish gut bacteria mainly from temperate-water fishes such as tuna and mackerel. This study led to important generalizations regarding the ecology and distribution of DHA/EPA bacteria in marine fishes inhabiting temperate waters.

DHA-producing bacteria, in contrast to EPA-producing strains, are largely missing from temperate isolates, in agreement with earlier studies indicating that DHA-producing bacteria seem confined to the deeper ocean or otherwise extremely cold environments. EPA-producing bacteria were readily detected in certain temperate fishes but often represented only a small percentage of the total isolates, with an interesting bias toward "blue-back fishes" such as mackerel. These data shows that any natural EPA mutualism in temperate fishes seems restricted to species such as tuna. However, DHA/EPA-producing bacteria may be more important for mutualism in cold environments such as deep-sea fishes, where an increasing number of DHA bacteria are being recovered and studied. DHA/EPA-producing bacteria represent a high percentage of the total gut population in certain mollusks, including ice mollusks.

Marine microbiologists have recovered DHA/EPA bacteria from sediment samples collected from the deepest regions of the ocean (e.g., >10,000 m; see Chapter 5) though it is not yet established whether these bacteria grow in the free-living state under these conditions or are "inoculated" from the excrement of deep-water fishes. Recall that DHA-producing bacteria have evolved resistant states allowing them to retain viability in seawater for periods of several years. Newly hatched deep-water fish might become "inoculated" from the reservoir of dormant cells in the surrounding seawater; after entering the stomach, these dormant cells are envisioned to become activated for vegetative growth by favorable conditions encountered in the animal gut. In some preliminary experiments in this area, we have identified possible trigger molecules, including iron and ammonium. Our present view is that DHA-producing bacteria are often present in large numbers in the guts of deep-water fishes, perhaps accounting for the discovery of these bacteria deep in the water column and sediment samples taken at depth. We suggest that mutualism seems likely under these conditions.

## 14.3 MECHANISMS FOR RELEASE OF DHA/EPA TO BENEFIT THE HOST

In order for the host animal to benefit from mutualism based on DHA/EPA, these molecules must be effectively delivered. Two primary questions arise: how and in what form might DHA/EPA chains be delivered to the host animal? This important step has received little attention and has implications for understanding how nutrients in general are passed from bacterial cells to host. As background, recall that spiny, marine fish constantly lose water by osmosis and must replenish their water supply by drinking seawater. Powerful osmoregulation systems make this possible by effluxing $Na^+$ via the gills and excessive $Mg^{2+}/Ca^{2+}$ in urine. This creates a gradient of osmotic strength that might be sensed by bacteria inhabiting the gastrointestinal tract. As already discussed in Chapter 5, marine bacteria have evolved sophisticated osmotic sensing systems for modulating DHA/EPA levels, and this circuitry might be taken over by the host as a trigger mechanism for release of DHA/EPA for use by the host. Alternatively, bacteria producing DHA/EPA tend to lyse readily, especially during environmental changes. We suggest that these cells could be subject to a bacterial form of programmed cellular death, perhaps under the control of the host. Spontaneous or host-directed lysis is expected to disrupt the membrane, directly exposing phospholipids as substrates for phospholipases likely present in the fish gut. Thus, the simplest model for delivery of DHA/EPA as well as other nutrients held inside the cell involves releasing cellular components by triggering cellular lysis.

An alternative mechanism involves efflux rather than lysis. For example, it has been shown that even mild temperature up-shock of a concentrated cell paste of *Shewanella* sp (Yazawa et al., 1988). 2738, a fish-gut isolate, results in release of a vast majority of membrane EPA as the free acid. This method was used by Japanese scientists to essentially purify EPA, normally a tedious, multistep process, in a single step from phospholipids of *Shewanella* sp (Yazawa et al, 1988). The release of EPA in such a short time from whole cells might be triggered by lysis, but these data also raise the possibility of specific recognition and excision of EPA chains from membrane phospholipids followed by efflux rather than via lysis. However, biochemical pathways remain obscure.

An efflux mechanism is especially interesting from a biotechnological standpoint since this might open the door to a new kind of continuous industrial fermentation for DHA/EPA. According to this design a fixed bed of cells engineered to efflux

DHA/EPA could be fed nutrients leading to continuous synthesis of DHA/EPA that is essentially already purified and appears in the effluent stream. This circumvents a major drawback of DHA/EPA bacteria and marine algae, which store DHA/EPA chains in their membranes or in triacylglycerol reserves inside the cell. Because of the high product specificity of anaerobic DHA/EPA synthetases, the effluxed stream of DHA or EPA would be expected to be highly purified by the bacterium itself, essentially free of other contaminating polyunsaturated chains.

From a mutualistic perspective, an efflux mechanism is attractive because it allows DHA/EPA bacteria living in the animal gut to remain viable, in essence behaving as miniature DHA/EPA chemostats for continuous production of a chemical form readily used by the host animal. Recall that low $O_2$ levels likely present in the fish gut environment both protect DHA/EPA chains from oxidation and discourage hydrogenation, such as occurs in the strictly anaerobic environment of the rumen. This allows DHA/EPA chains safe passage from the gut into the host. Genomic analysis shows that marine bacteria have evolved a variety of metabolite efflux pumps, many of whose natural substrates have not yet been identified, pumps that might accommodate hydrophobic substrates such as DHA/EPA. There are several important topics that need to be addressed in pursuing any putative efflux mechanism, such as identifying the presence of enzymes for excision of DHA/EPA from the membrane, the nature of exit portals across the outer membrane, as well as the mechanism of putative fatty acid efflux pumps across the inner membrane. Research on mechanisms for transport of lipophilic products from the cytoplasm to the medium deserves more attention because of potential important applications for microbial-based, industrial fermentations of products ranging from high value compounds such as DHA/EPA to the next generation of biofuels with greater energy content than ethanol.

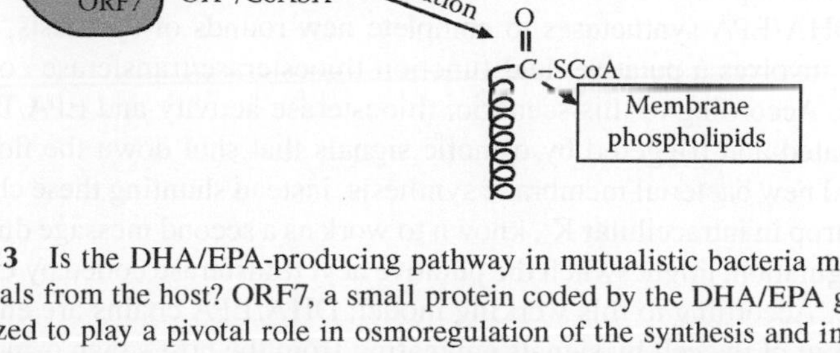

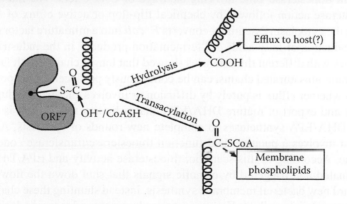


**FIGURE 14.3** Is the DHA/EPA-producing pathway in mutualistic bacteria modulated by osmotic signals from the host? ORF7, a small protein coded by the DHA/EPA gene cluster, is hypothesized to play a pivotal role in osmoregulation of the synthesis and incorporation of newly synthesized chains. According to this model osmotic signals from the host switch ORF7 back and forth between hydrolytic and transacylation modes of action and thus direct these chains toward efflux benefiting the host or new membrane synthesis needed for bacterial growth.

Based on studies of osmoregulation of DHA/EPA synthesis described in Chapter 5, we are considering the possibility of an osmotic switch that shunts the flow of DHA/EPA chains away from bacterial membrane synthesis toward efflux for the benefit of the host, avoiding cellular lysis (Figure 14.3). Recall that DHA/EPA synthetases are large enzymes organized like an assembly line with a striking array of five or six covalently bound acyl binding sites. It seems likely that the purpose of multiple pantethenate sites is to maintain growing DHA/EPA chains tightly bound to the enzyme surface. This stands in contrast to other fatty acid elongases, such as operating in *E. coli*, in which growing chains leave the complex as acyl–acyl carrier protein (ACP), which enters the cytoplasmic acyl-ACP pool on the way to a new round of synthesis. In other words, it is envisioned that pantethenate sites are occupied by immature chains, with mature chains of EPA/DHA remaining covalently bound to their final pantethenate-binding site on the enzyme until removed. One obvious feature of this model is that if EPA/DHA chains are not constantly removed by mechanisms such as thioesterases or transacylases, then the enzyme would be clogged eventually by immature chains and catalysis would stop, essentially shutting down the assembly line.

Previous studies with *E. coli* show that the assembly line for synthesis of fatty acids is shut down when growth ceases or when new membrane surface is no longer needed. However, fatty acid synthesis is reactivated in the presence of foreign thioesterases that hydrolytically cleave acyl-ACP bonds yielding free fatty acids and regenerating CoASH for a new round of acylation. Free fatty acids must be shunted or effluxed out of the cell to avoid toxicity caused by energy uncoupling. That fatty acid chains can be removed from cells has been demonstrated using a lauric acid recombinant of *E. coli*. In this case fatty acid crystals composed of lauric acid have been shown to accumulate around recombinant colonies of cells genetically engineered with a plant thioesterase gene favoring cleavage of C-12 esters. The main point is that thioesterase action followed by chemical flip-flop or active efflux of free fatty acids from the cytoplasm essentially converts *E. coli* into a miniature factory for continuous production of fatty acids as "fermentation product," in the industrial sense. Later studies with different thioesterases showed that longer chain length fatty acids, including monounsaturated chains, can be continuously produced and exported. It is not known whether efflux is purely by diffusion or involves energized efflux.

Release and export of mature DHA/EPA chains by any mechanism is expected to permit DHA/EPA synthetases to complete new rounds of synthesis. A possible mechanism involves a putative dual function thioesterase/transferase coded by the ORF7 gene. According to this scenario, thioesterase activity and EPA/DHA efflux are coordinated and triggered by osmotic signals that shut down the flow of EPA/DHA toward new bacterial membrane synthesis, instead shunting these chains out of the cell. A drop in intracellular $K^+$, known to work as a second message during bacterial osmoregulation, might switch the putative acyl transferase coded by ORF7 into a thioesterase. According to this working model, DHA/EPA chains are envisioned to be shunted out of the cell by signals emanating from the host's own osmotic regulatory system and sensed by gut bacteria.

## 14.4 SUMMARY

Mutualism involving gastrointestinal bacteria likely plays an important role in the growth and development of marine animals, such as fishes. For example, there is strong evidence supporting the important roles of chitinoclastic bacteria involved in mutualistic associations with fish. Fish may produce their own chitinases, but chitin-degrading bacteria offer a variety of important nutrients such as succinic acid, as a by-product of anaerobic respiration. The list of mutualistic associations is likely to grow. Understanding the molecular biology of omega-3 mutualism between DHA/EPA-producing bacteria and their marine animal hosts is in its infancy. This area deserves significant attention, with implications for aquaculture, industrial production of omega-3 fatty acids, and marine ecology. A spin-off of this research involves the transfer and expression of DHA/EPA genes in land plants, leading to commercialization.

## SELECTED BIBLIOGRAPHY

Cahill, M. M. 1990. Bacterial flora of fishes: A review. *Microb. Ecol.* 19:21–41.

Goodrich, T. D., and R. Y. Morita. 1977. Bacterial chitinase in the stomachs of marine fishes from Yaquina Bay, Oregon, USA. *Mar. Biol.* 41:355–360.

Heidelberg, J. F., I. T. Paulsen, K. E. Nelson et al. 2002. Genome sequence of the dissimilatory metal ion-reducing bacterium *Shewanella oneidensis*. *Nat. Biotechnol.* 20:1118–1123.

Hunt, D. E., D. Gevers, N. M. Vahora, and M. F. Polz. 2008. Conservation of the chitin utilization pathway in the Vibrionaceae. *Appl. Environ. Microbiol.* 74:44–51.

Jøstensen, J.-P., and B. Landfald. 1997. High prevalence of polyunsaturated-fatty-acid producing bacteria in arctic invertebrates. *FEMS Microbiol. Lett.* 151:95–101.

Keyhani, N. O., and S. Roseman. 1999. Physiological aspects of chitin catabolism in marine bacteria. *Biochim. Biophys. Acta* 1473:108–122.

Leys, D., A. S. Tsapin, K. H. Nealson, T. E. Meyer, M. A. Cusanovich, and J. J. Van Beeumen. 1999. Structure and mechanism of the flavocytochrome C fumarate reductase of *Shewanella putrefaciens* MR-1. *Nat. Struct. Biol.* 6:1113–1117.

Meyer, T. E., A. I. Tsapin, I. Vandenberghe, et al. 2004. Identification of 42 possible cytochrome C genes in the *Shewanella oneidensis* genome and characterization of six soluble cytochromes. *OMICS* 8:57–77.

Nichols, D. S. 2003. Prokaryotes and the input of polyunsaturated fatty acids to the marine food web. *FEMS Microbiol. Lett.* 219:1–7.

Rawls, J. F., M. A. Mahowald, R. E. Ley, and J. I. Gordan. 2006. Reciprocal gut microbiota transplants from zebrafish and mice to germ-free recipients reveal host habitat selection. *Cell* 127:423–433.

Yazawa, K., K. Araki, N. Okazaki, et al. 1988. Production of eicosapentaenoic acid by marine bacteria. *J. Biochem.* 103:5–7.

# 15 Membrane Adaptations for an Oily Environment: *Lessons from a Petroleum-Degrading Bacterium*

Hypothesis: An oil-degrading bacterium subject to proton uncoupling of its membrane has remodeled its respiratory chain to accept $Na^+$, a far less permeable cation.

Contamination of marine waters by petroleum from accidental spills, terrestrial runoff, and natural seeps is a major global ecological problem (Figure 15.1). This has spurred interest in the molecular biology and applications of oil-degrading bacteria such as the ubiquitous marine organism *Alcanivorax borkumensis*. Genomic, proteomic, biochemical, and microbiological studies show the unique properties of these cells that have evolved as first responders following oil contamination of marine waters. Major adaptations have occurred at the membrane level because these cells specialize in living in close contact with oil droplets, a normally toxic environment. Oil is the primary energy source for *A. borkumensis*, but many compounds found in oil are also known to act as solvents to destroy the membrane, with toluene given as an example in Chapter 10.

We chose to consider obligate oil-consuming bacteria here because the adaptations help define the importance of membrane permeability in a natural ecosystem. Note that *A. borkumensis*, though a marine organism, does not produce DHA or EPA. These cells do not use a lipid-based strategy for adapting to an oily environment but instead modify their electron transport chain to accommodate $Na^+$. The necessity for this radical alteration of typical proton bioenergetics points toward a potent energy stress that must be managed before life is possible on an oil droplet. We hypothesize that petroleum hydrocarbons working as a potent membrane solvent effectively uncouple the proton permeability barrier of the cell without fully destroying the barrier against $Na^+$. Organic acids present in petroleum likely serve as proton uncouplers. According to this model *A. borkumensis* is able to exploit a fundamental property of membrane permeability, which favors $Na^+$ fidelity over $H^+$ fidelity. By using $Na^+$ as primary cation for bioenergetics, *A. borkumensis* is able to move rapidly into oil-contaminated environments without losing membrane fidelity. However, this is not a simple change and depends on a series of mutational events required to remodel several powerful $H^+/Na^+$ pumps, events highlighted during genomic analysis by experts in this field.

**FIGURE 15.1** The chronic oil slicks at Coal Oil Point, California, originate from the natural seepage at the sea floor. An estimated 20 to 25 tons of crude oil are emitted daily at this location. (Photo by D. L. Valentine.)

## 15.1 GENOMIC ANALYSIS REVEALS THAT $Na^+$ BIOENERGETICS EVOLVED AS A MECHANISM TO MARGINALIZE PROTON LEAKAGE CAUSED BY PETROLEUM

Genomic analysis shows that *A. borkumensis* has evolved a novel electron transport chain that pumps $Na^+$ as well as $H^+$, seemingly remodeling its fundamental system of bioenergetics in order to adapt to an oily world (Figure 15.2). This dramatic change in bioenergetics involves shifting from proton electrochemical gradients to sodium gradients as the primary energy form for driving cellular functions. The apparent chemical basis behind switching to $Na^+$ involves the differential permeability properties of $Na^+$ versus $H^+$ across the lipid permeability barrier of the membrane and has already been discussed in Chapter 8. It is well known that proton/sodium fidelity across lipid portions of the membrane is negatively affected by increasing temperature and can likely be generalized to include any environmental change resulting in membrane hyperfluidity. Lipophilic solvents are known to damage the membrane permeability barrier in this case, apparently resulting in chronic proton leakage (see Chapter 10). Hydrophobic acids common in petroleum also act to diminish proton gradients by passively transporting protons through the membrane in the protonated form, often requiring active transport of the conjugate base out of the cell. Interestingly, the molecular basis of membrane adaptation in *A. borkumensis* does not involve changes of lipid chains but rather remodeling of formerly proton-pumping enzymes to accept $Na^+$. Changes in these pumps occur at or around former proton (i.e., hydronium ion) binding sites of primary proton pumps to accommodate the larger bulk of the $Na^+$ ion and permit the electron transport chain to efficiently pump $Na^+$ as the primary bioenergetic cation.

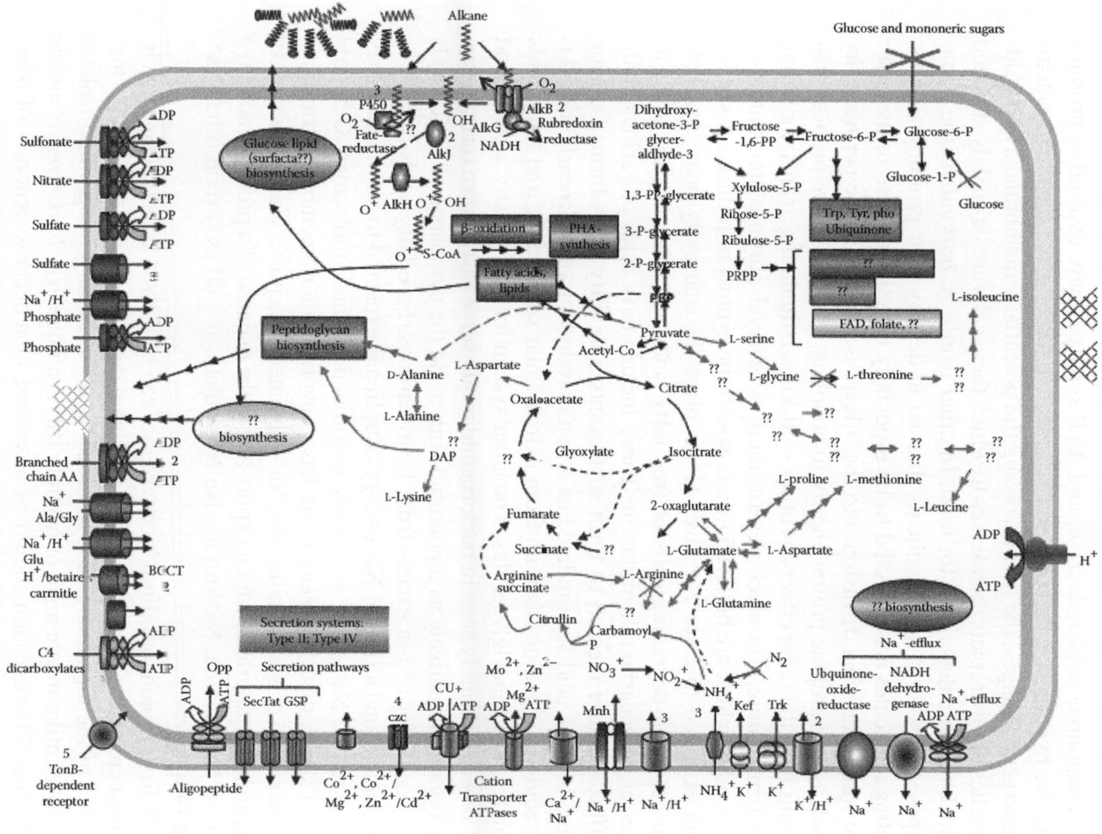


**FIGURE 15.2** Genomic and proteomic analyses of the ubiquitous oil-degrading bacterium *Alcanivorax borkumensis* show that these cells have evolved a system of $Na^+$ bioenergetics. It is hypothesized that fresh petroleum modifies membrane permeability properties leading to chronic uncoupling of proton energy gradients. To capitalize on fresh oil, this organism has evolved to use $Na^+$ to energize cellular processes, as this larger ion moves through membranes at rates dramatically slower than $H^+$, thus providing a competitive advantage. (Courtesy of P. N. Golyshin and K. N. Timmis.)

Systems of sodium bioenergetics have evolved in other bacteria. For example, the roles of sodium motive force in DHA-producing bacteria and in a thermophilic bacterium, as a way around proton leakage across the membrane, have already been discussed in Chapters 5 and 8. These changes are not as extensive as the case of *A. borkumensis*. Analysis of the genome of *A. borkumensis* shows that virtually the complete set of proton pumps of the electron transport chain, including NADH dehydrogenase, ubiquinone oxidoreductase, and ATP synthase, have evolved to primary $Na^+$ pumps. We hypothesize that at least two selective forces drive such major membrane bioenergetic changes, the first involving the potentially membrane hyperfluidizing properties of oil droplets themselves to which these cells are intimately attached and on which they are dependent for food. In other words, *A. borkumensis* are constantly exposed to solvent-like chemicals that act to destroy their proton gradients. The solvating molecules are expected to be especially toxic in the case of the inner membrane where chemical diffusion into the bilayer is likely to cause permeability defects such as catastrophic proton energy uncoupling. In essence, we propose that hydrocarbons have the same effects as other well-known fluidizing agents, including increasing temperature, too much desaturation, or general purpose sterilants (see Chapter 10), all of which are believed to increase futile cycling of protons. A second mechanism expected to favor $Na^+$ over $H^+$ involves the unusually heavy trafficking in and out of oil-degrading cells of free fatty acids and other organic acids as metabolites of hydrocarbon degradation. These include fatty acids of various chain lengths effluxed out of the cell to form surfactants, free fatty acids formed during β-oxidation, benzoic acid accumulated as an end production of hydrocarbon oxidation, fatty acids salvaged or taken up from extracellular surfactants, and fatty acids entering and exiting triacylglycerides. The latter system for storing fatty acids, found in *A. borkumensis*, is rarely seen in bacteria and might serve as a storage form of carbon or energy or alternatively as a potential sink to rid the cytoplasm of toxic levels of free fatty acids formed in excess from hydrocarbons. Free fatty acids behave as proton uncouplers, and a shift to $Na^+$ energy coupling is expected to diminish energy uncoupling because $Na^+$ is not as readily mobilized by these molecules. Genomic analysis also shows that *A. borkumensis* has coevolved many membrane transport systems to handle $Na^+$ as energizing ion. $Na^+$-mediated transporters include a variety of sodium/proton antiporters, which export $Na^+$ at the expense of a proton gradient, a symporter for $Na^+$/alanine, $Na^+$/sulfate, $Na^+$/glutamate, as well as several other $Na^+$-dependent symporters.

The shift from proton to sodium bioenergetics can be viewed as a defensive mechanism against energy stress caused by oil-mediated destruction of the proton permeability barrier of the membrane. It is too early to tell if other oil-degrading isolates have evolved a similar mechanism or whether this is unique to *A. borkumensis* and its nearest relatives. This question should be answered soon because genomes of other isolates are being completed. There is also the question of how readily the genes for the primary $Na^+$ pumps are transferred from one strain to the other. Whereas any one of the genes coding for the $Na^+$-specific enzymes would be expected to be advantageous to a new oil-degrading recipient, transfer of all of these pumps as a cluster offers the greatest selective advantage. Thus, as a slow-growing cell attached to an oil droplet and where toxic levels of free fatty acids might be prevalent, it makes sense that *A. borkumensis* switch to $Na^+$ as a far less permeable energy coupling cation. The switch to $Na^+$ is not absolute since these pumps still accommodate protons, which perhaps benefits the cell during the long stages of its lifecycle when there is a paucity of oil. The energy savings by switching to $Na^+$ are expected to be large, perhaps a doubling of energy efficiency helping explain the dominance of *A. borkumensis* as pioneering organisms in petroleum bioremediation.

## 15.2 OUTER MEMBRANE (OM) LIPID STRUCTURE AS A PHYSICAL BARRIER AGAINST OIL

The above adaptations central for $Na^+$ bioenergetics occur in the cytoplasmic membrane. However, the outer membrane (OM) itself is the first membrane exposed to toxic effects of oil and can also be considered as a defensive layer against oil as a lipophilic toxicant or solvent (Figure 15.3). Recall that the OM of *E. coli*, which has been studied in great detail, contributes an important barrier function especially against lipophilic solutes. These include permeability barrier against hydrophobic antibiotics (e.g., macrolides, novobiocins, rifamycins, and actinomycin D), detergents (e.g., bile salts, SDS, and Triton X-100) and hydrophobic dyes (e.g., methylene

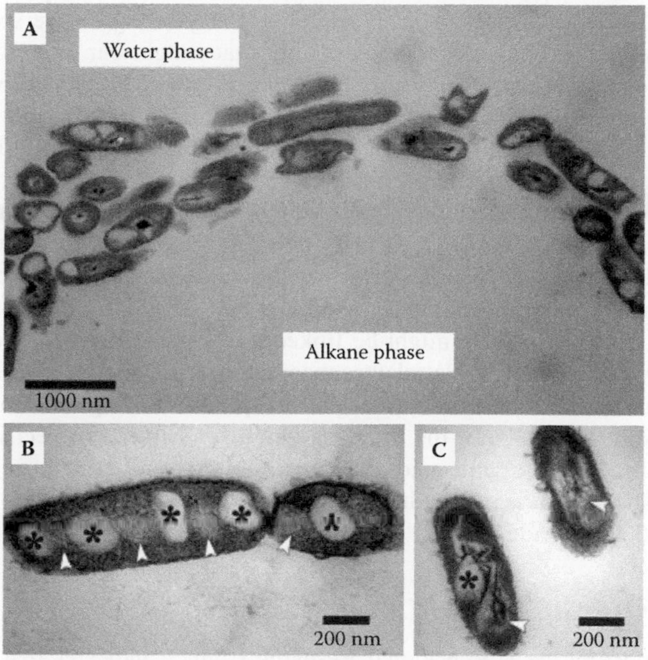


**FIGURE 15.3** Cells of *A. borkumensis*, which live in close contact with oil droplets, also depend on their robust outer membrane for protection. (A) Cells surrounding an oil droplet. (B) and (C) At higher magnifications lipid storage particles comprised of triacylglycerols are revealed and are thought to serve as an energy reserve and as a possible mechanism to modulate levels of free fatty acids, which are toxic. (Reproduced from J.S. Sabirova et al., *J. Bacteriol.*, 188:3763–3773, 2006. With permission.)

blue and acridine dyes). Various assays have been used to show that the OM bilayer is less permeable to lipophilic substrates by a factor of 50 to 100, in comparison with the cytoplasmic membrane. Such low permeability is consistent with the low fluidity and tight architecture of the outer leaflet of the OM. The tight packing of the OM is broken in the presence of chemical chelators that compete for $Mg^{2+}/Ca^{2+}$ or in medium with low levels of these ions. This might help explain the high $Mg^{2+}/Ca^{2+}$ requirement for *A. borkumensis* when growing on oil. Recall that these divalent ions are plentiful in seawater, the natural environment of these cells.

## 15.3 OTHER CHANGES

In addition to membrane changes, various secondary metabolic adaptations might also be important for adaptation to an oily, marine environment. For example, *A. borkumensis* has evolved powerful multilayered osmoregulatory systems that prevent dehydration in full strength seawater as well as at higher salinity. The osmoregulatory systems of *A. borkumensis* are powerful enough to permit growth in environments with severalfold the salinity of seawater. This raises questions of the environments for which such strong osmoregulatory systems are needed. Briny, natural oil seeps come to mind as the possible evolutionary proving ground for these cells, as do evaporation ponds. This lifestyle might have set the stage for adaptation to $Na^+$ bioenergetics. A system that caught our attention involves possible mechanisms that decrease intracellular levels of $K^+$, which, when present in high intracellular levels to balance the osmolality of seawater as seen in bacteria such as *E. coli*, has the effect of diminishing the driving force of $Na^+$ electrochemical gradients. The presence of ectoine/betaine biosynthetic genes in *A. borkumensis* might reduce the need for cytoplasmic $K^+$ as bulk osmoregulant. Interestingly, cation-energized $K^+$ transporters have been identified, but determinants for ATP-dependent $K^+$ transporters are absent, which might indicate that $K^+$ is less important in osmoregulation compared to *E. coli*. Thus, minimizing the levels of $K^+$ in the cytoplasm might be linked to the switch to a $Na^+$ economy. In addition to working against $Na^+$ electrochemical gradients, $K^+$ as inhibitor might compete with $Na^+$ at the level of former proton-specific binding sites now enlarged to include $Na^+$. Alternatively, the low $K^+$ might also enable the selective efflux of certain anions such as organic acids, by exerting tight control on the anion content of the cytoplasm. That is, it may be easier to efflux an organic oxyanion when not bathed in 0.5 M chloride.

## 15.4 ECOLOGICAL SUPPORT

A recent report by the U.S. National Research Council indicates that a majority of oil entering the ocean is from natural sources, particularly hydrocarbon seeps. The chemical composition of crude oil, the natural substrate of *A. borkumensis*, is complex and not yet fully defined. Recent advances with comprehensive two-dimensional gas chromatography and Fourier transform ion cyclotron resonance mass spectrometry have revealed tens of thousands of individual compounds, each potentially serving as an energy source to bacteria. Two important classes of compounds present in crude oil are potentially toxic to bacteria. These include membrane fluidizers such as the low molecular weight aromatics, and organic acids such as the naphthenic acids that can cause futile proton cycling. Both classes of compounds tend to be physically removed from oil by evaporation or dissolution within hours to days of environmental exposure. We hypothesize that the ecology of *A. borkumensis*, characterized as one of rapid response, hinges in part on the ability to cope with high concentrations of these compound classes present in freshly exposed oil. We suggest that the switch in this organism to $Na^+$ bioenergetics may enable it to be the first responder, in that other organisms would be inhibited by the high concentrations of proton uncouplers in the fresh oil. This lifestyle is consistent with observations that frequently show *A. borkumensis* as the first organism to take hold in a fresh spill, followed by a phase-out as other petroleum degrading bacteria become dominant. The transition of the community has been attributed to factors such as substrate quality and availability, and we further suggest that the loss of proton uncouplers from the oil may serve as an additional control.

The selective advantage to *A. borkumensis* provided by $Na^+$ bioenergetics might also be used actively by this organism to inhibit competition. This organism has transient access to an extensive reservoir of hydrocarbons as evidenced by its physical association with oil droplets and its capacity for intracellular carbon storage. This organism has also developed mechanisms to excrete hydrophobic materials, and we suggest that the excretion of proton uncoupling hydrophobic acids may be an ecological deterrent to other petroleum-degrading bacteria. The excretion of toxins into the marine environment is normally costly on account of biosynthesis and dispersion. For *A. borkumensis* there is little bioenergetic cost, since it can excrete metabolites such as fatty acids that have already yielded substantial energy, and these compounds are likely to partition into the oil droplets and not to be dispersed. This mechanism to poison the environment with proton uncouplers could prolong the time that *A. borkumensis* dominates the oil droplets and thus confer a competitive advantage.

## 15.5 SUMMARY

Oil-degrading bacteria face a dilemma at the membrane level and must overcome the toxic effects of petroleum behaving as a potent solvent and uncoupler of proton energy gradients. Consistent with either a water-wire mechanism or hydrophobic acids, $Na^+$ is greatly favored over $H^+$ in this environment. *A. borkumensis* cells appear to leverage this fundamental difference governing cation permeability and transport to their great advantage in adapting to an oily world. These cells are required to remodel their entire electron transport chain, including ATP synthase, in order to shift from a system of proton bioenergetics to $Na^+$.

An understanding of the genes enabling *A. borkumensis* to be the initial agent in bioremediation of marine oil contamination is growing, and in the future, it might be possible to develop improved organisms as inoculant for oil spills. In addition, understanding membrane adaptations for an oily environment has implications for improving industrial fermentations involving products or substrates with properties destructive to the membrane. One of the most promising places to look for applications involves bioproduction of diesel-like fuels. Studies of membrane adaptation by

*A. borkumensis* even have implications in the health arena because human pathogenic bacteria might use similar mechanisms to evade toluene and other sterilants now in widespread use.

## SELECTED BIBLIOGRAPHY

Barquera, B., P. Hellwig, W. Zhow, et al. 2002. Purification and characterization of the recombinant $Na^+$-translocating NADH: Quinone oxidoreductase from *Vibrio cholerae. Biochemistry* 41:3781–3789.

Hara, A., K. Syutsubo, and S. Harayama. 2003. *Alcanivorax* which prevails in oil-contaminated seawater exhibits broad substrate specificity for alkane degradation. *Environ. Microbiol.* 5:746–753.

Mulkidjanian, A. Y., P. Dibrov, and M. Y. Galperin. 2008. The past and present of sodium energetics: May the sodium-motive force be with you. *Biochim. Biophys. Acta* 1777:985–992.

Ruis, N., D. Kahne, and T. J. Silhavy. 2006. Advances in understanding bacterial outer-membrane biogenesis. *Nat. Rev. Microbiol.* 4:57–66.

Sabirova, J. S., M. Ferrer, D. Regenhardt, K. N. Timmis, and P. N. Golyshin. 2006. Proteomic insights into metabolic adaptations in *Alcanivorax borkumensis* induced by alkane utilization. *J. Bacteriol.* 188:3763–3773.

Schneiker, S., V. A. P. Martins dos Santos, D. Bartles, et al. 2006. Genome sequence of the ubiquitous hydrocarbon-degrading marine bacterium *Alcanivorax borkumensis. Nat. Biotechnol.* 24:997–1004.

# 16 Lessons from Yeast: *Phospholipid Conformations Are Important in Winemaking*

Hypothesis: A unique blend of phospholipids in yeast membranes drives wine fermentation to completion.

Winemaking is an ancient art but is yet to be fully understood or mastered. For example, "stuck fermentations," in which the fermentation of sugar in grape juice is incomplete, compromise flavor and are costly to the industry. Recent advances in understanding membrane fatty acid function provide insight into why yeast (i.e., *Saccharomyces cerevisiae*) remodels its membrane phospholipids during fermentation, and why it is import in winemaking (Figure 16.1). Attention here is focused on a phospholipid conformation unique to fermentative yeast cells. We suggest that the paucity of similar structures in nature is based on their harmful and powerful energy uncoupling activity. We also suggest that while this phospholipid conformation is harmful for most cells, yeast harnesses these structures for their benefit and, by happenstance, the benefit of winemakers.

## 16.1 DISCOVERY AND ROLES OF ASYMMETRICAL PHOSPHOLIPIDS IN YEAST

When growing in the wine vat, yeast synthesizes large amounts of membrane phospholipids in which a shortened fatty acid is paired with a much longer chain. This structure was discovered almost 50 years ago by Konrad Bloch and his student Howard Goldfine (Goldfine and Bloch, 1963), but the biochemical roles of these molecules are not fully understood. The current understanding is that *S. cerevisiae* has evolved the short chain pathway as an alternative mechanism for fluidizing its plasma membrane when its $O_2$-dependent desaturase systems needed for forming long-chain unsaturated fatty acids are shut down because of anaerobic growth. This alternative mechanism involves the biosynthesis of chain-shortened fatty acids, $C_8$ through $C_{14}$, occupying the *sn*-2 position of phospholipids, which replace $C_{16}$ and $C_{18}$ monounsaturated chains found in membranes of aerobically grown cells. Relative abundances for $C_8$ are usually low, about 1% of total fatty acids, whereas $C_{10}$ (6%), $C_{12}$ (23%) and $C_{14}$ (14%) are more abundant. This class of short-chain fatty acids disappears when cells are grown aerobically and thus are usually found only in membranes of cells grown fermentatively.

Growth of yeast cells requires that at least a portion of its membrane surface be in a liquid or fluid state. In aerobically grown cells monounsaturated chains satisfy this

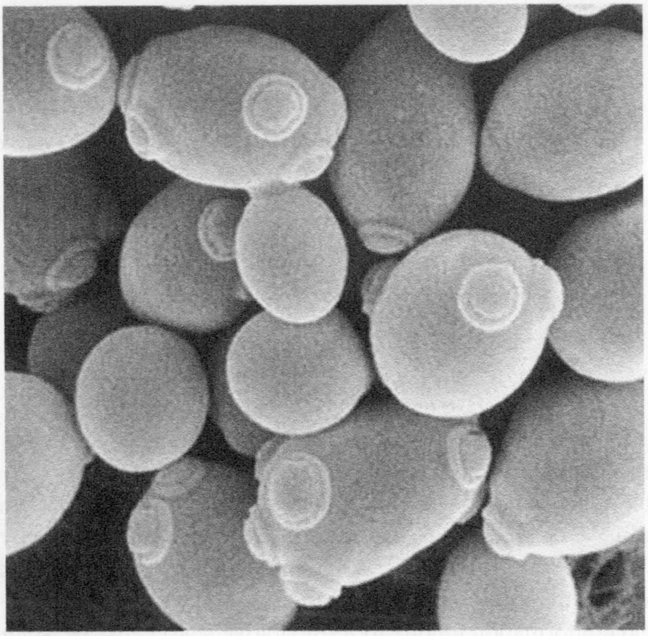

**FIGURE 16.1** Scanning electron micrograph of yeast cells used in winemaking (*Saccharomyces cerevisiae*). (Courtesy of Eric Schabatach.)

requirement, in contrast to fermenting cells for which chain-shortened phospholipids seem to play this role. These results show that chain-shortened phospholipids are fluidizing, in the sense of providing essential motion and keeping the plasma membrane in a fluid state. In essence, fermentative yeast cells have evolved a short-chain mechanism to fluidize their membranes. This means that chain-shortened phospholipids behave as surrogates for unsaturated fatty acids such as oleic acid ($18{:}1_9$) found in aerobic cells. However, the conformation of this structure has serious drawbacks.

The bioactivity of chain-shortened phospholipids in mitochondria of yeast was first studied some years ago. The morphology of mitochondria isolated from yeast tricked into producing these organelles under low $O_2$ conditions was found to be comparable to cells grown aerobically. Moreover, the electron transport chain, measured as rates of $O_2$ consumed, was fully functional in mitochondria containing the following levels of short chains: 2.3% $C_8$, 15.3% $C_{10}$, 13% $C_{12}$ and 13.6% $C_{14}$. However, biochemical studies show that proton bioenergetics, or some other critical bioenergetic function, appears to be completely uncoupled when mitochondrial membranes are enriched in short-chain fatty acids. Potassium circulation is also disrupted. From these experiments, it is concluded that phospholipids in which one fatty acid chain is much shorter than the other likely behave as potent proton uncouplers when present in the mitochondrial membrane. We suggest this is likely in any membrane, as discussed below. A possible mechanism involves the destruction of the permeability barrier for protons caused by excessive formation of water wires in mitochondrial membranes enriched in chain-shortened phospholipids.

## 16.2 MEMBRANE ALTERATIONS ACCOMPANYING FERMENTATION

The 7- to 10-day life cycle of wine yeast is arbitrarily divided into three stages, as follows: (I) energy is coupled to growth, (II) energy becomes uncoupled from growth, and (III) death and dormancy due to chronic energy stress. Because the cell population in a fermentation vat is not synchronized, this simple division is obviously oversimplified. Ethanol production is about 50% completed by the end of the first stage and is completed during the next stages. We suggest that numerous chemical changes occurring during the natural fermentation cycle work as signals to dramatically change the molecular architecture and functions of yeast membranes. These changes include decreased levels of $O_2$, increased levels of ethanol, decreased levels of the cholesterol-like molecule ergosterol with increased levels of squalene, and increased levels of $C_8$ to $C_{14}$ saturated fatty acids at the expense of long-chain monounsaturated fatty acids. Increased ethanol, a lipophilic molecule that readily enters the membrane, is an obvious change, as is anaerobicity that occurs at the beginning of the fermentation. Recall that decreasing $O_2$ levels simultaneously switches the sterol pathway from ergosterol (requiring $O_2$) to squalene (anaerobic), but, perhaps more importantly, decreasing $O_2$ levels also switches the membrane lipids away from 16:1-18:1 monounsaturates to the short saturated chains described above. Ergosterol is important among yeast sterols as a membrane stabilizer. During the approximately four to five generations of growth in the wine vat, much of the original content of this $O_2$-dependent membrane structural components is "diluted out" with the new membrane surface, which is enriched with chain-shortened building blocks. We suggest that these changes in membrane composition equate to significant changes in function.

Microbiologists studying the effects of ethanol on yeast physiology have shown clearly that accumulation of ethanol causes proton energy uncoupling at the level of the plasma membrane. Recall that actively fermenting yeast cells must generate and maintain proton electrochemical gradients for the purpose of energizing various metabolite uptake pumps. ATP produced by glycolysis of (grape) sugar is used to drive powerful ATP-$H^+$ efflux pumps, essentially the reverse of ATP synthase. Some researchers have proposed that modifications of $H^+$-ATPase are responsible for energy uncoupling, though we favor an alternative mechanism involving water wires as discussed below.

## 16.3 ENERGY UNCOUPLING IN THE PLASMA MEMBRANE OF YEAST IS ADVANTAGEOUS IN NATURE

What is the competitive advantage for yeast to uncouple catabolism from energy conservation? The mechanism used by fermentative yeast cells to uncouple their energy metabolism while continuing to produce heat and ethanol as end products is reminiscent of the process of thermogenesis (i.e., purposeful uncoupling of energy metabolism) used by animals to warm vital organs. Even plants use thermogenesis, an example being skunk cabbage, which uses the heat to melt through snowdrifts in search of sunlight. In its natural environment, such as inside a ripe fig, yeast must compete against other fermentative microbes such as bacteria, which seek to inhabit this same niche. It has been proposed that yeast have evolved a powerful and generalized chemical warfare agent that is used against their competitors. This agent is ethanol, a natural fermentation product that becomes toxic to many cells at the levels produced by yeast. *S. cerevisiae* itself is highly tolerant, with many strains surviving exposure to levels of ethanol as high as 17% to 19% (v/v), whereas levels from 4% to 7% are inhibitory to most other organisms. The majority of strains of wine yeast readily produce between 11% and 13% ethanol during the fermentation of grape juice. In the case of yeast, we propose that it is not the heat but rather the ethanol that seems to be beneficial as a growth retardant against microbial competition in natural environments. The general mechanism of chemical warfare among microbes is similar to gramicidin A (Chapter 8) and to the excretion of lipophilic acids (Chapter 15), in that it targets the permeability barrier against protons to supply a selective advantage to the cell. In this case the selective advantage stems from ethanol tolerance. This property also benefits the winemaker by allowing yeast to complete the alcoholic fermentation.

## 16.4 WATER WIRE THEORY CAN EXPLAIN UNCOUPLING IN YEAST

Mixed-chain phospholipids such as C:18/C:10 phosphatidylcholine (PC) perhaps initially stabilized by residual ergosterol present during the early part of the fermentation cycle are believed to form tongue-in-groove structures (Figure 16.2) late in the fermentation cycle. This process might be retarded initially due to some cell-to-cell carryover of 18:1/16:1 chains from inoculant cells. This means that several cycles of growth are likely required in the wine vat before mixed-chain phospholipids strongly influence membrane function. Another point is that, if any additional $O_2$ enters the vat early in the fermentation, then both 18:1/16:1 and ergosterol levels will spike, in essence postponing the transition to shortened chains. There is convincing chemical evidence that the thick membrane structure shown on the left of Figure 16.2 is stabilized by ergosterol and that this structure is of average membrane thickness.

Note the dramatic thinning of the membrane shown on the right side of the diagram. It is interesting that ethanol, which works as a lipophilic membrane solvent and is produced in increasingly high concentrations by *S. cerevisiae* as the fermentation proceeds, favors thin membranes. Thus, the average thickness of yeast membranes is hypothesized to drop gradually, being replaced by an unusually thin structure. We suggest that this shift in membrane structure is the basis of proton energy uncoupling in the plasma membrane.

The suggestion that ethanol itself, along with perhaps a lowering of ergosterol levels, triggers a thick-to-thin transition is consistent with two well-established facts about the fermentation cycle of yeast. The first is that the proton uncoupling attributed to thin membranes occurs at about the same rate that vegetative growth slows. Note that during the vegetative stage energy production is coupled to production of new cellular biomass. This is in contrast to the uncoupled stage in which biomass production diminishes and an increasing amount of metabolic energy is wasted as heat. Heat production in the fermentation vat is not trivial, especially during large-

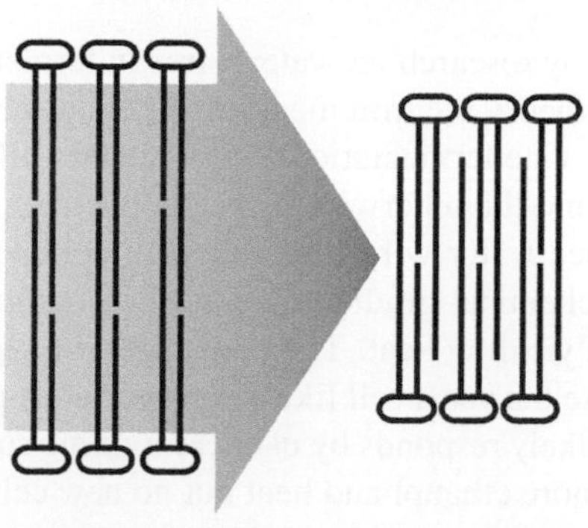


**FIGURE 16.2** Mechanism of membrane interdigitation. Chemists have shown that fatty acid chains of asymmetrical phospholipids, such as present in yeast membranes, can slide past one another in the presence of ethanol. Cholesterol repels the interdigitation process. Water wires are known to form in such drastically thinned membranes, resulting in uncoupling of proton energy gradients. We propose that yeast have evolved to capitalize on membrane thinning as a beneficial mechanism for energy uncoupling, which allows yeast cells in nature to elevate ethanol to levels toxic for competing cells. This mechanism benefits winemakers by driving breakdown of the last residues of grape sugars and creating a dry wine.

scale fermentation, and care must be taken to maintain temperatures within a physiological range. It is likely that, as more and more of the yeast population enters a prolonged uncoupled phase, the chronic energy stress eventually causes the death of many yeast cells. Thus, energy uncoupling may be a winemaker's friend, but this mechanism is not friendly to the yeast cells themselves.

Thus, three changes in membrane chemistry seem to be required during winemaking, the first two dictated by the environment and the third by accumulation of ethanol itself. First, lack of $O_2$ as substrate for critical fatty acid desaturases leaves yeast with no option but to synthesize short, saturated chains to fluidize their bilayers. This results in phospholipids with one chain dramatically shorter than the other (e.g., 18:0/10:0). In the presence of ethanol, acyl chains in this class of phospholipids interdigitate across the lumen, essentially sliding past similar structures in the opposing leaflet and compressing the bilayer as much as 30% to 40% (that is, the bilayer is dramatically thinned). Membrane biophysicists studying the effects of membrane thickness on proton permeability have shown that model membranes similar to those found naturally in yeast are catastrophically leaky for protons. A second important change in membrane chemistry involves the gradual dilution of ergosterol. This sterol plays an important role in initially stabilizing chain-shortened phospholipids against water wire formation in the plasma membrane, a property that facilitates growth of yeast during the first stage of winemaking. The third chemical change comes from the presence of abundant ethanol and, as mentioned above, facilitates the gradual thinning of the membrane. These data, along with studies demonstrating that yeast membranes become porous for protons, lead us to predict that water wires form in plasma membranes of actively fermenting yeast cells in levels that can convert these cells into sugar-consuming heat machines that simultaneously produce abundant ethanol.

We are not aware of any research on water wires in fermentative yeast. However, according to water wire theory the thin membrane structure surrounding yeast cells in the later stages of the wine fermentation is predicted to allow unacceptable seepage of water molecules into the otherwise hydrophobic membrane interior. The rest is chemically driven since water wires are excellent conductors of protons, which short-circuit the electrochemical gradient needed by growing yeast cells, draining their energy supply, and yielding heat. The conversion of sugar to ethanol does not stop immediately because the yeast cell likely senses that its proton fidelity has been compromised. The cell likely responds by using remaining sugar for energy to pump out protons, producing more ethanol and heat but no new cells.

## 16.5 SUMMARY

From a practical standpoint, during the fermentation of grape juice to wine, all of the sugar remaining in the juice must be converted to ethanol by yeast in order to produce the most valuable dry wines. In an ideal situation, yeast uses all of the sugar while producing about 13% ethanol or higher, but many cells die in the process. Unfortunately, problem fermentations are common, poorly predictable, and hard to restart, resulting in loss of time and money. Thus, winemaking economics has stimulated interest in knowing more about the molecular biology of yeast membranes during the relatively short lifetime of these cells in the fermentation vat. The proved method of purging the wine vat with a burst of air appears to rejuvenate membrane architecture, but creates unacceptable off-flavors in wine. However, this disadvantage does not apply to beer or biofuels fermentations, where the membrane can be maintained in a healthy state by bursts of air. Thus, an increasing knowledge base of yeast membrane dynamics might lead to more predictable, restartable or even improved fermentations with possible spin-off toward improving biofuels production technology. Currently, ethanol production for biofuels in the United States suffers under the weight of poor energy balance, which would be improved by increasing ethanol yields.

Many interesting challenges lie ahead in applying membrane water wire theory toward winemaking and bioethanol fuel production. The wine yeast *S. cerevisiae*, a workhorse in fermentations, is now under the scrutiny of scientists interested in optimal production of fine wines and fuels. The membrane is now considered an important target toward improving the alcoholic fermentation with applications to biofuels production in general.

## SELECTED BIBLIOGRAPHY

Ambesi, A., M. Miranda, V. V. Petrov, and C. W. Slayman. 2000. Biogenesis and function of yeast plasma-membrane H(+) ATPase. *J. Exp. Biol.* 203:155–160.

Boulton, R. B., V. L. Singleton, L. F. Bisson, and R. E. Kunkee. 1996. *Principles and Practices of Winemaking*. New York: Chapman & Hall.

Cartwright, C. P., J.-R. Juroszek, M. J. Beavan, F. M. Ruby, M. F. de Morais, and A. H. Rose. 1986. Ethanol dissipates the proton-motive force across the plasma membrane of *Saccharomyces cerevisiae. J. Gen. Microbiol.* 132:369–377.

Goldfine, H., and K. Bloch. 1963. Oxygen and biosynthetic reactions. In *Control mechanisms in respiration and fermentation,* ed. B. Wright, 81. New York: Ronald Press Co.

Haslam, J. M., T. W. Spithill, A. W. Linnane, and J. B. Chappell. 1973. Biogenesis of mitochondria. The effects of altered membrane lipid composition on cation transport by mitochondria of *Saccharomyces cerevisiae. Biochem. J.* 134:949–957.

Huang, C. 1990. Heinrich Wieland—Prize Lecture: Mixed-chain phospholipids and interdigitated bilayer systems. *Klin. Wochenschr.* 68:149–165.

Tierney, K. J., D. E. Block, and M. L. Longo. 2005. Elasticity and phase behavior of DPPC membrane modulated by cholesterol, ergosterol, and ethanol. *Biophys. J.* 89:2481–2493.

Umebayashi, K., and A. Nakana. 2003. Ergosterol is required for targeting of tryptophan permease to the yeast plasma membrane. *J. Cell Biol.* 161:1117–1131.

# 17 DHA Principle Applied to Global Warming

Hypothesis: A small rise in ocean temperature enables destabilizing conformations of DHA/EPA that result in uncoupling of photosynthesis.

With global climate change now a reality, the twenty-first century may turn out to be the golden era of ecological science. Already there is great public awareness of the immense harm to ecosystems that might occur if global climate change continues unabated. The intricate and vital relationship among mankind and other life forms is now widely appreciated. The case of DHA/EPA provides an opportunity to apply the DHA principle toward understanding the impact of global climate change on oceanic productivity. We start with the effect of warming ocean temperatures on the phytoplankton that are the major producers of DHA/EPA in the biosphere and follow the consequences as these chains move through the marine food web (Figure 17.1).

Marine algae or phytoplankton, which form the base of the food chain, produce most of the DHA/EPA in the biosphere, whereas land plants do not synthesize these molecules. However DHA/EPA-producing phytoplankton play important roles in the fertility of freshwater lakes and rivers. At the second trophic level, small animals called zooplankton graze on phytoplankton to obtain energy and important nutrients including DHA/EPA. While eukaryotic phytoplankton density is low over vast areas in the ocean gyres, upwelling of mineral-rich deep ocean waters feeds vast blooms (Figure 17.2). Because of their central role in the marine food chain sophisticated satellites have been deployed to monitor the status of phytoplankton on a global scale. Data acquired through satellite imaging has already provided some major surprises. For example, even though phytoplankton are often present in relatively small numbers in seawater and tend to die off rapidly, primary production in the sea versus the land has been calculated to be roughly equivalent. These data elevate the importance of marine photosynthesis and once again show how the future of man is tied to the sea.

## 17.1 MARINE PRODUCTIVITY IS THREATENED BY EVEN MODEST THERMAL UPSHOCKS

Unfortunately, recent data support a "worst case scenario" regarding the health of phytoplankton and ultimately the human food chain, resulting from ocean warming. It has long been recognized by marine biologists that phytoplankton (i.e., total mass and distribution) are a measure of the health of the oceanic world. In the face of possible ocean warming, NASA has developed and carried out a 10-year satellite monitoring program. Whereas to human eyes the ocean appears as shades of blue, sometimes blue-green, from earth orbit satellite sensors can be tuned to distinguish

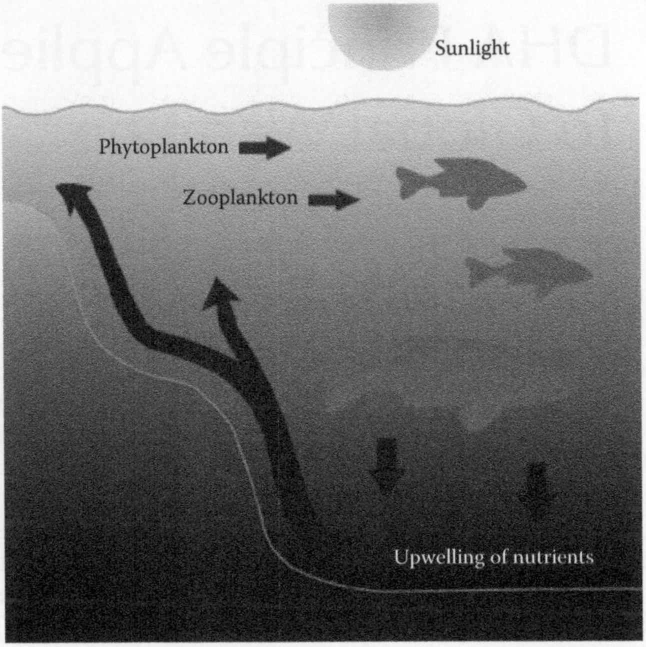


**FIGURE 17.1** Simplified marine food web. Note that phytoplankton are the primary producers of DHA/EPA that flow to upper levels of the food chain and eventually to man. (Courtesy of Bob Worrest.)

even slight variations in color. Owing to their photopigment, chlorophyll, phytoplankton preferentially absorb the red and blue portions of the visible electromagnetic spectrum (for photosynthesis) and reflect greenish light. Oceanic regions with "blooms" or high concentrations of phytoplankton will appear as certain shades, depending on the type and density of the phytoplankton population there. A phytoplankton bloom off the California coast captured by satellite imagery is shown for example in Figure 17.2. Time-series satellite coverage over the entire ocean surface has recently been summarized in an article in *Nature* and shows for the first time that the growth of phytoplankton is tightly linked to climate change and would be significantly altered by a warming ocean (Behrenfeld et al., 2006). As temperatures increased from 1999 to 2004, phytoplanktonic productivity dropped by about 200 million tons a year; some regions saw as much as a 50% drop. Also, major shifts toward cooler regions of the ocean were noted.

It is also interesting that satellites monitoring the oceanic phytoplankton crop relay this information on a daily basis to commercial fishing vessels seeking the most fertile fishing grounds such as waters off the Chilean coast. Fishermen have long recognized the close linkage between phytoplankton "fertility" and fish populations further up the food chain. Satellite data has also been useful to ecologists for comparing the total fertility of marine versus terrestrial ecosystems and has led to some unexpected conclusions. It is now recognized that net primary production of biomass in the marine world, though scattered over a wide area, is similar to productivity by

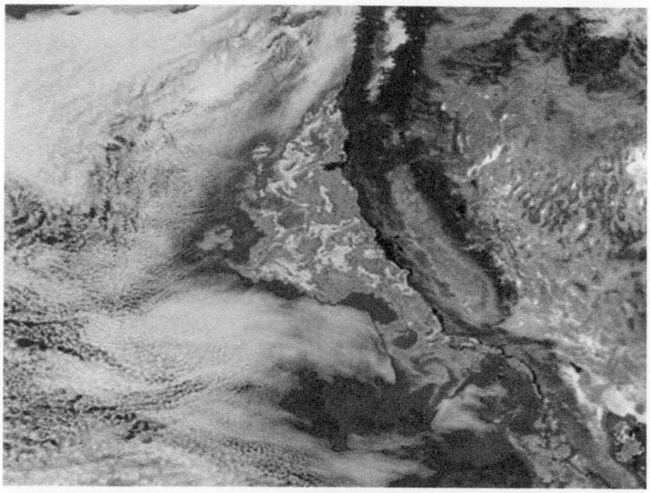

**FIGURE 17.2** (A color version of this figure follows p. 140.) A NASA satellite detects a phytoplankton bloom off the California coast. Periodic upwellings of mineral-rich waters, characteristic of this region of the coast, result in large blooms of algae. One of the greatest salmon fisheries of all time once flourished here, with enormous runs navigating up the Sacramento River and its tributaries each year to spawn. (Image courtesy of the *SeaWiFs* Project, NASA/Goddard Space Flight Center, and ORBIMAGE.)

land plants. This means that any effects of climate change on life in the sea will have a major impact on the productivity of the biosphere as a whole.

## 17.2 HIGHLY UNSATURATED MEMBRANES OF SYMBIOTIC ALGAE OF CORALS HAVE ALREADY BEEN IMPLICATED AS "REPORTERS" OF GLOBAL WARMING

Over the past 30 years, bleaching and death of ocean corals dependent on photosynthetic symbionts have been documented across the range of global distribution (Figure 17.3). Though not the sole cause, small increases in sea surface temperatures (e.g., 2°C) and light intensity have been correlated with bleaching and death. Because losses of symbiotic algae occur as a result of a modest up-shock in temperature, the coral-algae system represents a sensitive reporter or sensor of events expected during oceanic warming (Figure 17.4). It now seems clear that unsaturated fatty acids of chloroplast membranes are a key determinant of thermal-stress sensitivity, not in the coral itself, but in membranes of the algal symbionts. Analyses of photosynthetic membranes (i.e., thylakoid bilayers) reveal that the critical threshold temperature separating thermally tolerant from sensitive species of algae is determined by the levels of the omega-3 fatty acid 18:4 enriched in membranes of the sensitive strain of coral. The thermal killing cascade of the algal symbiont likely involves multiple steps, opening up the possibility of a systematic approach for understanding individual steps as discussed below.

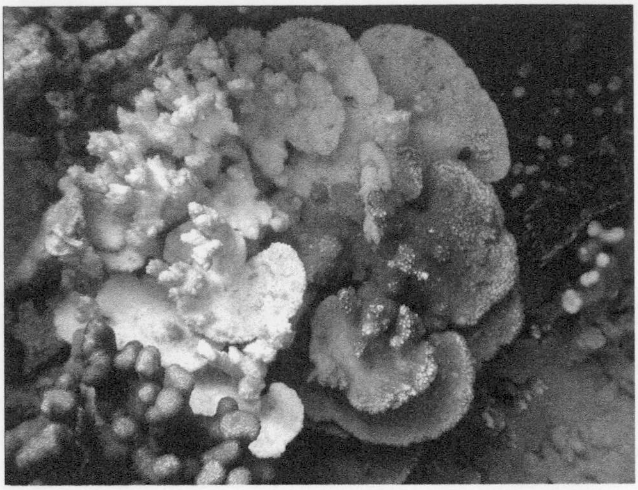

**FIGURE 17.3** The health of coral reefs is in worldwide decline in the face of anthropogenic pressures and global climate change. The coral shown here has undergone extensive bleaching, such as occurs with a modest up-shock in temprature. (Courtesy of Forest Rohwer.)

Research on corals supports the operation of a lethal oxidative cascade essentially as follows: thermal trigger reaction → block in ATP production that restricts carbon assimilation → increase in reactive oxygen species (ROS) caused by reaction between photosynthetically derived $O_2$ and an excess of photochemically generated "high energy electrons" → ROS chain reaction results in a burst of oxidative damage → terminal stages of ROS death cascade killing first the algae and spilling over to damage the host coral → finally the symbiotic algae is literally bleached of chlorophyll and/or expelled from their hosts → the host dies of energy/oxidative stresses. It is interesting to recall that parts of this scheme have already received a great deal of attention during studies of the importance of oxidative stress and the roles of ROS in critical human diseases and conditions, such as cancer and aging. For example, DHA and membrane unsaturation levels as possible triggers of aging in mammals has recently been proposed and is discussed in Chapter 19. Also, recall that some of the most potent broad-spectrum herbicides such as paraquat are designed to attack the chloroplast, effectively accelerating many of the above steps into a time frame of minutes or hours, leading to plant death (interestingly, this destroys the myth that plants die slowly). Among the steps of the death cascade for corals as outlined above, the so-called trigger reaction(s) that initiates the death cascade is one of the least understood.

## 17.3 THERMAL KILLING OF DHA-PRODUCING BACTERIA AS A SURROGATE FOR MARINE CHLOROPLASTS

The mechanism of the trigger reaction of the thermal death cascade in marine chloroplasts is unknown, and we suggest this process might be studied in simpler

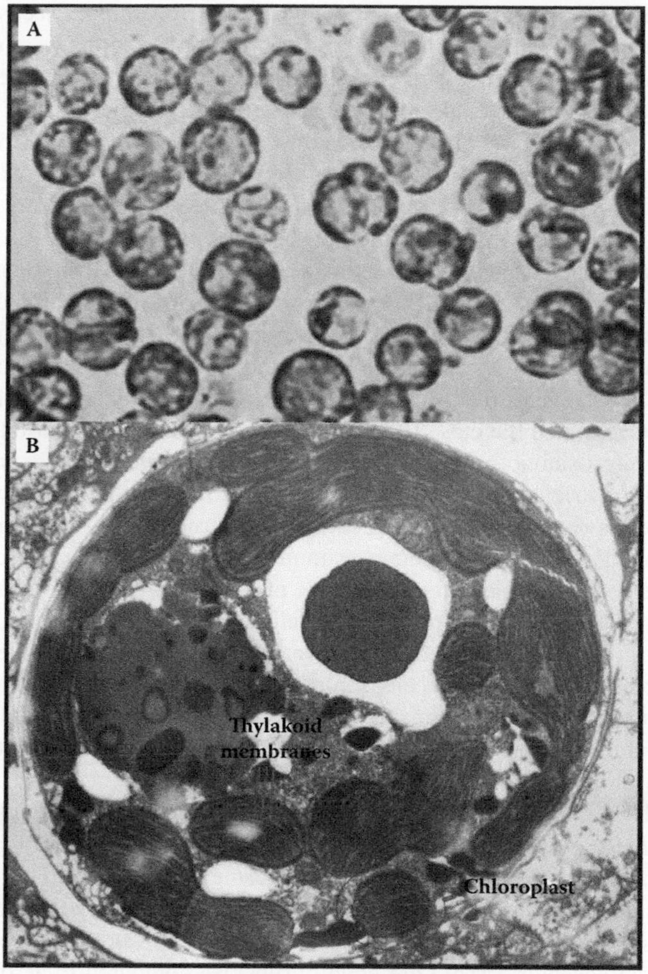


**FIGURE 17.4** Omega-3-producing algae called Xooxanthellae are a weak link in symbiotic corals during periods of ocean warming. (A) Symbiotic cells belonging to the genus *Symbiodinium* provide photosynthetically derived nutrients essential for growth of their host. (Courtesy of Misaki Takabayashi.) (B) Photosynthetic membranes of Xooxanthellai are often enriched with Omega-3 fatty acids, such as 18:4. This makes these cells sensitive to thermally triggered oxidative stress thta kills the algae and that spills over and destroys its host. Corals harboring symbiotic algae with more saturated membranes survive under similar conditions. (Courtesy of Misaki Takabayashi.)

cells such as DHA-producing bacteria. The rationale is that since this cascade has evolved in marine chloroplasts, which are thought to be bacterial in origin, perhaps DHA-producing bacteria might be subject to thermal death triggered by a similar mechanism. Recall that *Moritella* are marine bacteria with DHA-enriched membranes and are highly sensitive to thermal changes. Indeed, *M. marina* MP1 cells are sensitive to even moderate up-shocks in temperature. Unfortunately, early studies on physiological analysis of the temperature-triggered death cascade in *M. marina* were carried out more than a decade before DHA chains were discovered in membranes of these cells. Thus, mechanisms other than membranes received most of the attention for explaining the exceptionally rapid thermal killing of these cells (e.g., protein denaturation). With the benefit of hindsight, we suggest that DHA might be involved in thermally triggered cellular death characteristic of these cells.

In addition to being sensitive to thermal changes, the growth of *M. marina*, as well as other marine isolates, is also known to be sensitive to other stresses such as osmotic, ionic, and pressure shifts. For example, transfer of certain deep-sea bacteria adapted to intense hydrostatic pressure, equivalent to the greatest depths of the ocean, to the laboratory causes cells to undergo lethal morphological changes and cellular death, events that might also be triggered at the membrane level. Early researchers recognized the close linkage between temperature-hydrostatic pressure and temperature-osmotic pressure. For example, *M. marina* becomes increasingly psychrophilic, restricted to a maximum temperature range below 10°C, when growing in diluted seawater. When seawater levels are returned to full strength these cells regain the ability to divide at their maximum temperature up to 21°C. A similar principle has been established for hydrostatic pressure showing that barophilic cells growing near their maximum growth temperature become dependent on hydrostatic pressure for growth and die at this temperature when pressure is relieved. Thus, there is increasing evidence that environmental signals trigger change not only in the composition of membrane phospholipids but also perhaps in the conformational dynamics and interactions of lipids, altering critical biochemical properties such as proton permeability, which in turn creates energy stress.

In our labs we have confirmed a number of the original observations consistent with a thermally mediated cellular death cascade in *M. marina* MP1 and further found that molecular $O_2$ is not required for triggering this response. This is based on experiments done by Laura Gillies and the senior author in Bruce German's laboratory demonstrating that thermally treated cells growing anaerobically on the alternative electron acceptor TMAO are equally sensitive compared to aerobic cells in terms of their thermal stress response. These data seem to rule out a mechanism based on initial oxidative damage as a primary trigger for cellular death. Instead we propose that thermally triggered changes to the dynamics of DHA phospholipids initiate the cellular death cascade of these marine bacteria. Once again, this brings energy stress possibly caused by water wires front and center as a possible key trigger mechanism of cellular death in these cells.

## 17.4 CONFORMATIONAL MODEL

We propose a molecular model in which DHA/EPA phospholipids in chloroplasts are both the antenna and amplifiers of thermal change. This model is essentially an extension of the waterwire hypothesis developed in Chapters 8 and 18 and involves thermally triggered changes to DHA/EPA chains enabling an unacceptable amount of water to enter the membrane. The chemical changes to the DHA/EPA might include a further reduction in intermolecular stabilization with adjacent lipid tails, a greater probability of contorting to a destabilizing conformation, or transitioning between conformations at such a rate so as to enable void spaces and weaken membrane architecture; in each case the result is likely to be increased formation of water wires. Convincing data indicate that thermal energy drives conformational changes in phospholipid structure, which in turn causes proton leaks. That is, thermally triggered changes in DHA/EPA phospholipids in photosynthetic membranes are believed to directly affect the molecular architecture and consequently proton permeability properties of the bilayer. This leads to the concept that changes in DHA/EPA conformational dynamics caused by thermal changes become inhibitory in terms of physiological functions important in chloroplasts.

Interestingly, it is known that the putative states of DHA/EPA membranes are reversible and do not appear to depend on covalent modification of phospholipid structure. Thus, changes in conformational dynamics seem a reasonable explanation. Perhaps the best biochemical information about the nature of these putative harmful fatty acid states comes from studies of thermal effects on proton permeability. Studies with whole cells show that environmental signals seem to trigger similar changes (e.g., pressure has the opposite effect of heat and so forth). Based on information in the literature we surmise that acceptable conformational dynamics of DHA reflect structures, intermolecular interactions, and rates of transition that help form molecular plugs or shields against proton leakage via water wires. One can envision a DHA-enriched membrane subdivided into elementary units of permeability represented by individual phospholipids contributing to proton fidelity. The "weaving" of the chloroplast membrane into a permeability barrier tight enough for protons involves suitable conformations of DHA chains themselves. According to this simple idea, a DHA phospholipid representing a fundamental permeability unit of the bilayer is reversibly converted from a state of proton fidelity to a proton leaky state by environmental signals such as temperature and pressure. Changes in permeability states can also be achieved by remodeling phospholipids with different fatty acid tails that counter environmental effects. For example, the environmental signals modulating the quantity of DHA entering the bilayer of *M. marina* cells (see Chapters 5, 10, and 14) are the same signals suggested to drive major changes in conformational dynamics.

The chemistry underlying DHA's conformational dynamics is not entirely clear, but evidence does support two or more membrane states represented by the extremes: high proton fidelity versus leaky for protons. Temperature and pressure are known to thin or thicken the bilayer. This means that at a finer level DHA phospholipids themselves must adjust to this change or expose the bilayer to an increased probability of formation of lethal levels of water wires. Adding membrane bulk such as the case of extra methyl groups as discussed in previous chapters can also be viewed as a defensive mechanism based on fatty acid conformations that modify the permeability barrier. Interestingly, the high-EPA–producing bacterium *Shewanella* 2738 also produces high levels of methyl-branched lipids that might be needed to counterbalance any negative effects of EPA. Individual DHA phospholipids might gain or lose bulkiness or chain length due to physical forces in the environment. It is reasonable that among fatty acids these changes reach an extreme with DHA.

## 17.5 DHA/EPA ARE NEEDED TO BUILD EFFICIENT NEUROSENSORY MEMBRANES IN ZOOPLANKTON AND OTHER MARINE ANIMALS

Zooplankton acquire DHA/EPA for their membranes by feeding on phytoplankton (Figure 17.5). Zooplankton clearly do not need DHA/EPA for synthesizing light harvesting membranes, as in the case of phytoplankton. Instead, we hypothesize that zooplankton require DHA/EPA as building blocks for a class of membranes unique to animals, neurosensory membranes of neurons, and sensory cells. The concept that DHA/EPA enhance the efficiency of neurons of a simple animal, *C. elegans*, is introduced in Chapter 6 and applied here to the marine food chain. Unfortunately, we are not aware of any study quantifying the global stock or production of DHA/EPA, or even at the level of a marine ecosystem. Our own studies (Jones et al., 2008) show that in the surface waters of the upwelling region off the coast of southern California, DHA constitutes almost one quarter of all fatty acids present, and is approximately

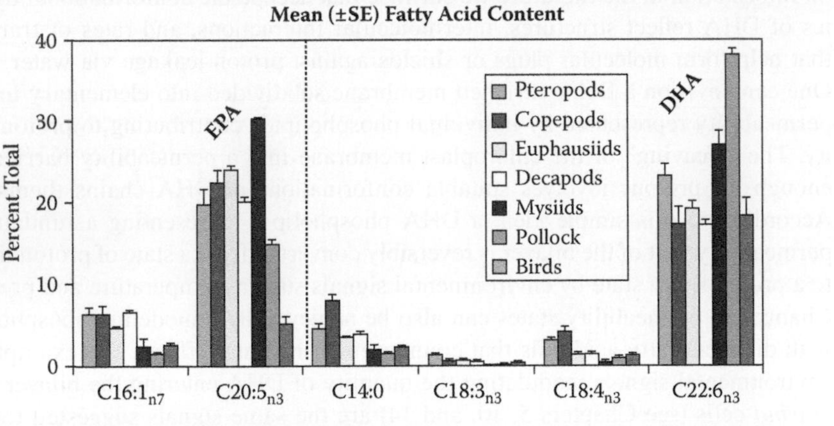


**FIGURE 17.5** (A color version of this figure follows p. 140.) Distribution of DHA/EPA in a hierarchy of animals of the eastern Bering Sea. Data are arranged with the diatom-indicating FA on the left of the vertical line and the small cell-indicating FA to the right of the line. Pteropods = *Limacina heliciana;* Copepods = *C. marshallae, N. cristatus, M. pacifica, E. bungii;* Euphausiids = *T. raschii, T. inermis, T. spinifera, T. longipes*, furcilia; Decapods = three unidentified taxa, two zoea, one megalopes; Mysiids = Unidentified; Birds = two common murres (*Uria aalge*), three short-tailed shearwaters (*Puffinus tenuirostris*). Summary of data from all taxa averaged over all stations. This survey of animal life around the Pribilof Islands shows the importance of DHA/EPA in marine animals starting with zooplankton and ending with pollock and birds. Note that whereas DHA/EPA are roughly equally distributed among zooplankton, DHA predominates further up the chain. Also note the high total levels of DHA plus EPA in these animals. Recall that DHA/EPA are an important energy food for animals but also play important structural roles in membranes of mitochondria and neurons. We hypothesize that DHA/EPA are needed to build efficient neurosensory membranes in both endothermic and ectothermic marine animals. (Courtesy of J. M. Napp, L. E. Schaufler, G. L. Hunt, and K. L. Mier, unpublished data.)

2- to 10-fold greater in concentration than EPA. A strong correlation to temperature is assumed throughout the ocean.

Aquatic ecologists at the University of California Davis have shown that both growth and egg-laying capacity of freshwater zooplankton, close relatives of marine zooplankton, depend on the levels of EPA chains present in their phytoplankton food sources, which vary seasonally with water temperatures (Müller-Navarra, Brett, Liston, and Goldman, 2000). In these studies EPA levels in phytoplankton at different seasons (i.e., high in cool weather, low in warm seasons) were found to be a suitable gauge for monitoring food quality for zooplankton. This effect for dietary EPA provided by phytoplankton can now be explained based on synthesis of a more efficient nervous system in zooplankton grazing on EPA-enriched phytoplankton. This mechanism is based on later studies using the tiny worm *C. elegans* as a model for explaining why EPA/DHA are important for sensing food, motion, eating, egg laying, growth, and, ultimately, competition and survival. The importance of DHA/EPA for synthesis of efficient neural membranes points toward a tight ecological linkage between phytoplankton and zooplankton that we suggest persists for animals occupying upper tiers of the marine food chain.

Studies of DHA/EPA in the complete marine food web suggest important roles and dynamic cycling. For example, when herring fry are fed a DHA-deficient diet, their vision in dim light is impaired. This is important in nature because these fish spend much of their life cycle in dim light and predators often attack from the depth. In another study fish larvae were fed field-collected zooplankton, mainly copepods. The DHA content of the predominant species of copepod (*Acartia*) was found to fluctuate dramatically within a period as short as 12 days, in the same open ocean area. These latter data are consistent with the view that severe fluctuations in available DHA might modulate fish growth and development. The dramatic variability in DHA content within the same group of copepods is also surprising and deserves further attention. Other data suggest that within more confined marine waters, such as certain bays with restricted flows, a more stable source of DHA/EPA may persist. Is it possible that herring have evolved to spawn in areas where DHA-enriched zooplankton are abundant? Interestingly, too much DHA might also be harmful. For example, a chronic reproductive defect in salmon in the Baltic Sea might be linked to excessive levels of DHA in the diet.

Judging from their distributions in nature, DHA and EPA appear to serve similar roles, though strong preferences in different organisms suggest important distinctions too. Starting from top to bottom of the food chain, membranes of human neurons contain far more DHA than EPA. In contrast the ratio of DHA to EPA in herring oil is roughly unity as is the case for membranes of neurons from lobster. The greatest variability in terms of levels of unsaturation and classes of fatty acids occurs among phytoplankton. For example, some marine algae do not synthesize DHA or EPA at all, while others may produce mainly EPA but not DHA. Environments such as the upwelling zone off southern California clearly favor phytoplankton with DHA. Great variability is also observed for chains with four double bonds, mainly 20:4 and 18:4. As these unsaturated fatty acids move up in trophic level from phytoplankton to zooplankton it is reasonable to expect that these chains might be subject to remodeling to meet the needs of the animals at the next trophic levels. Indeed, one of the simplest scenarios to explain shifts in ratios of different classes of highly polyunsaturated species is that membrane specialization (e.g., in neurons) drives partitioning of these chains through different anabolic routes, to storage lipids, or to catabolism.

## 17.6 SUMMARY

Increasing evidence suggests that the balance between benefits and risks of highly unsaturated membranes in marine phytoplankton is dramatically shifted into the risk category by even slight oceanic warming. The current working model is that the fatty acid tails of phospholipids such as DHA/EPA serve as the chemical transducers of thermal change, working via a water wire mechanism of proton uncoupling leading to energy stress/oxidative stress. It is difficult to grasp why oceanic productivity could evolve to be so dependent on such a narrow range of temperatures, particularly considering the massive temperature swings revealed by the geologic record. One explanation is that marine chloroplast membranes have evolved to operate near their maximum threshold efficiency for balancing critical biochemical processes such as long-distance electron transport versus proton permeability. This specialization leaves little tolerance for any further environmental changes, such as temperature up-shocks of even a few degrees. If this scenario is correct, then it is easy to predict that ocean warming will generate a domino effect rippling through the marine food chain, likely resulting in dramatic changes in productivity and trophic interactions. Perhaps the second greatest concern involves the health of zooplankton that require photosynthetically derived omega-3 chains for building an efficient nervous system. In a worst case scenario, photosynthetic organisms of one kind or another will likely continue their role as primary producers in the marine food chain even after significant ocean warming has occurred. However, the life cycle of zooplankton might be eroded if these organisms are forced to graze on omega-3–deficient phytoplankton, resulting in a significantly weakened nervous system. It remains unclear how far this "omega-3 effect" will propagate up the food chain. Could severe ocean warming broadly stunt neural development of marine mammals, including whales and dolphins?

We suggest that the power of DHA/EPA membranes and membranes in general to directly sense and amplify effects of oceanic warming to the detriment of mankind is genuine. Perhaps, the most surprising thing about omega-3–mediated cellular death cascades, as already documented for corals, is that relatively small changes in the thermal environment are able to initiate these cascades. Chemical modeling of thermal effects on DHA conformations offers an interesting new tool for explaining the threat of oceanic warming on marine life. One of the most illuminating future experiments being discussed involves tracking changes in DHA conformations driven by only a few degrees up-shock in temperature using physical-chemical techniques. However, before this experiment is carried out, a healthy debate is needed to establish what temperature should be chosen as baseline. Biological evidence predicts significant conformation changes for DHA in the range of about 15°C to 21°C, essentially marine conditions.

## SELECTED BIBLIOGRAPHY


Ahlgren, G., L. Van Nieuwerburgh, I. Wanstrand, M. Pedersen, M. Boberg, and P. Snoeijs. 2005. Imbalance of fatty acids in the base of the Baltic Sea food web: A mesocosm study. *Can. J. Fish. Aquat. Sci.* 62:2240–2253.

Arts, M.T., R. G. Ackman, and B. J. Holub. 2001. Essential fatty acids in aquatic ecosystems: A crucial link between diet and human health and evolution. *Can. J. Fish. Aquat. Sci.* 58:122–137.

Baker, A. C. 2001. Reef corals bleach to survive change. *Nature* 411:765–766.

Behrenfeld, M. J., R. T. O'Malley, D. A. Siegel, et al. 2006. Climate-driven trends in contemporary ocean productivity. *Nature* 444:752–755.

Bell, M.V., R. S. Batty, J. R. Dick, K. Fretwell, J. C. Navarro, and J. R. Sargent. 1995. Dietary deficiency of docosahexaenoic acid impairs vision at low light intensities in juvenile herring (*Clupea harengus* L.). *Lipids* 30:443–449.

Berkelmans, R., and M. J. H. van Oppen. 2006. The role of zooxanthellae in the thermal tolerance of corals: A "nugget of hope" for coral reefs in an era of climate change. *Proc. Biol. Sci.* 273:2305–2312.

Budge, S.M., C.C. Parrish, and C. H. Mckenzie. 2001. Fatty acid composition of phytoplankton, settling particulate matter and sediments at a sheltered bivalve aquaculture site. *Mar. Chem.* 76:285–303.

Davis, M.W., and B. L. Olla. 1992. Comparison of growth, behavior and lipid concentrations of walleye pollock *Theragra chalcogramma* larvae fed lipid-enriched, lipid-deficient and field-collected prey. *Mar. Ecol. Prog. Ser.* 90:23–30.

Falkowski, P. G., and M. J. Oliver. 2007. Mix and match: How climate selects phytoplankton. *Nature Rev. Microbiol.* 5:813–819.

Imbs, A. B., D. A. Demidkova, Y. Y. Latypov, and L. Q. Pham. 2007. Application of fatty acids for chemotaxonomy of reef-building corals. *Lipids* 42:1035–1046.

Irigoien, X., R. P. Harris, H. M. Verheye, et al. 2002. Copepod hatching success in marine ecosystems with high diatom concentrations. *Nature* 419:387–389.

Jónasdóttir, S.H. 1994. Effects of food quality on the reproductive success of *Acartia tonsa* and *Acartia hudsonica*: Laboratory observations. *Mar. Biol.* 121:67–81.

Jónasdóttir, S. H., H. G. Gudfinnsson, A. Gislason, and O. S. Astthorsson. 2002. Diet composition and quality for *Calanus finmarchicus* egg production and hatching success off south-west Iceland. *Mar. Biol.* 140:1195–1206.

Jones, A., A. Sessions, B. Campbell, C. Li, and D. L. Valentine. 2008. D/H ratios of fatty acids from marine particulate organic matter in the California borderlands. *Organic Geochemistry* 39(5):485–500.

Mayzaud, P., H. Claustre, and P. Augier. 1990. Effect of variable nutrient supply on fatty acid composition of phytoplankton in an enclosed experimental ecosystem. *Mar. Ecol. Prog. Ser.* 60:123–140.

Morita, R. Y. 1975. Psychrophilic bacteria. *Bacteriol. Rev.* 39:144–167.

Müller-Navarra, D. C., M. T. Brett, A. M. Liston, and C. R. Goldman. 2000. A highly unsaturated fatty acid predicts carbon transfer between primary producers and consumers. *Nature* 403:74–77.

Tchernov, D., M. Y. Gorbunov, C. de Vargas, et al. 2004. Membrane lipids of symbiotic algae are diagnostic of sensitivity to thermal bleaching in corals. *Proc. Natl. Acad. Sci. U.S.A.* 101:13531–13535.

# 18 DHA Principle Applied to Molecular Farming

Hypothesis: Terrestrial plants lack DHA/EPA because the risks of lipoxidation and of destabilizing conformational dynamics exceed benefits. Designing crop plants to produce DHA/EPA will only be achieved through careful consideration of their ecological roles.

Human nutritionists warn that the supply of DHA is not keeping up with demand and predict a global health crisis. This dire forecast was previously blunted because of uncertainty concerning the dietary roles played by DHA in the body. That the global supply of DHA, especially from the marine fishery, is plateauing in the face of increasing population growth is a sound argument. Safety issues related to accumulation of heavy metals in the flesh of long-lived fish such as tuna are also of concern. Recent advances concerning the mechanism of action of DHA reinforce the concerns of human nutritionists and focus attention toward new commercial sources of DHA. The commercial area that has received the most attention involves the need for highly purified and safe sources of DHA targeted to supplementation of baby formula. This is a large market justified by convincing research showing that dietary DHA is needed and incorporated during the period of rapid brain development in infants, and is also preferentially shunted into mothers milk.

To meet the needs for DHA in the twenty-first century, plant geneticists are attempting to design land plants that produce marine oils in bulk and have met with some success. However, the primary industrial target in this field is DHA, which is proving elusive. We suggest that plant geneticists may be confronted with a powerful ecological challenge involving the DHA principle in attempting to engineer DHA into land plants. Seldom is a macroecological pattern clearer than the case of global distribution of DHA in primary producers, which is confined to marine plants and is absent in land plants. As already summarized in Chapters 9 and 17 there is a strong biochemical and chemical basis for this distribution pattern, which works against plant genetic engineers. Key questions are whether technology is currently available for circumventing this principle or whether more fundamental research is required. We tell the story of the quest for DHA crops from an historical perspective, including some personal experience.

## 18.1 ASILOMAR CONFERENCE IN 1975 ON RECOMBINANT DNA USHERED IN THE ERA OF GENETICALLY ENGINEERED CROP PLANTS

In the spring of 1975, the Nobelist Paul Berg helped organize an international meeting for the purpose of alerting the public for the first time concerning the benefits and risks of using recombinant DNA technology to harness the gene. In his introductory remarks, Berg suggested that the advancement of DNA technology would be better served by a more transparent discussion of this powerful new tool and its future impact on public health, agriculture, and the environment. He also pointed out a hard lesson learned by nuclear scientists who failed to alert the public regarding the benefit and risks of harnessing of the atom, and in retrospect helped create a long-term distrust of scientists by the public.

The senior author was asked to speak on the role of applied molecular genetics in agriculture in the age of genetic engineering. In order to draw attention to the scientific marriage between agriculture and molecular genetics that was just beginning, the title "Molecular Biology of Farming," shortened to "Molecular Farming," was chosen and the prospects of transferring nitrogen fixation genes (*nif*), which we had named in 1971, to corn and other crop plants was presented using a benefit/risk approach. At the time of this meeting, which came to be known as the Asilomar (California) Conference on Recombinant DNA, it was already apparent that the yields of potential transgenic, $N_2$ fixing crops such as corn would likely be negatively impacted because of the high cost of adding another energy-intensive process to compete for the plants' limited resources. Interestingly, the exorbitant energy cost of ~16 ATP per $N_2$ fixed provided the best rationale against "run away" *nif* genes destroying the biosphere, a question asked by Berg at the end of the author's talk. The complexity of handling so many *nif* genes in symbiotic bacteria exceeded the limits of available technology, and this scientific hurdle soon discouraged commercial interest in developing new $N_2$ fixing crops.

Recognizing the need for a simpler "gene bullet" compatible with available genetic vector systems for plants, the conceptual groundwork for the idea of using round-up (glyphosate)-resistance genes from bacteria for introduction into crop plants was laid down in the senior author's laboratory in the Plant Growth Laboratory of University of Calfornia Davis around 1976. This idea circulated for several years in the academic community before its timeliness and potential application were first embraced by the venture capitalist Norm Goldfarb and the senior author in 1980. This led to the formation of the plant biotechnology company Calgene (now a campus of Monsanto). It soon became apparent that the choice of round-up resistance genes was a good one since this herbicide was widely used and had already been determined to be safe (i.e., readily biodegraded and without an enzyme target in humans). Also, the technology was greatly simplified because only a single core gene was involved. These considerations turned out to be critical from an applied standpoint, but nevertheless it took scientists at Calgene and later Monsanto more than a decade of extensive research to commercialize round-up ready crops, currently planted on >100,000,000 acres. It is safe to say that round-up resistance genes helped launch the modern plant biotechnology industry.

By 1985, there was a movement toward improving not only plant protection traits such as herbicide tolerance but also harnessing genes with impact on human health. This led to a new generation of plant geneticists turning their attention to increasingly complex traits, such as producing heart-healthy oils. Discussions on modifying oil crops such as canola and soybean for production of beneficial fish oils, specifically DHA, began at Calgene in the mid-1980s with specific research programs being initiated by teams formed in 1993 (see Chapter 1). Some important basic insights on membranes gained as a spin-off from this work are highlighted in this book. Today, several industrial laboratories, including Monsanto, are competing in a race to develop omega-3 crops, and some successes are being reported. This research is driven by the need to maintain an adequate supply of DHA for human health and development in the face of increasing global populations and a dwindling supply of marine food products. The importance of DHA in human health and preventive medicine is discussed in Chapters 6, 19, 20, and 21.

## 18.2 DHA IS CURRENTLY PRODUCED FROM MARINE ALGAE, BUT CROP PLANTS ARE BEING CONSIDERED

Production of DHA by crop plants is attractive for several reasons. For example, chemical synthesis of DHA on a large scale has not been successful because of the difficulty surrounding the generation of six precisely located double bonds along the chain. The fractionation of widely available crude sources of DHA, such as fish oil, has proved difficult because such oils are a mixture of fatty acids with similar chemical properties, such as EPA, which are costly to separate; also there are questions regarding the long-term viability and safety of the marine fishery as a source of these fatty acids. The first commercially successful production system now supplying DHA (needed for brain development in infants) for supplementation of baby food involves industrial fermentation using phytoplankton selected from the marine environment and grown in industrial-scale fermentors (Figure 18.1). These naturally occurring strains are selected for high DHA production but are not genetically engineered. Still, this remains a costly approach. Certain crop plants, such as flax, naturally produce precursors of DHA, but these starting chains contain only three double bonds and require several additional steps, including a possible rate-limiting reaction, needed to produce DHA in the body. In other words, flax seeds do not contain DHA as such, and there is some controversy regarding how fast flax oil as precursor is converted to DHA.

It now seems clear from reports in the literature that DHA can also be produced—so far only in small amounts—by modifying conventional oil crops such as soybean and corn. However, there is one aspect of DHA biotechnology in plants that we suggest is understudied and represents a potentially serious impediment in developing DHA crops. This involves protecting the transgenic plant itself from damage caused by the presence of DHA, which is predicted to destabilize critical plant membranes under field conditions. This warning is based on the well-established observation that DHA is absent in all land plants, presumably because of its harmful properties in this environment. We suggest that the evolution of membranes and oils in land

**FIGURE 18.1** Naturally occurring marine algae are used for commercial production of DHA. (A) Marine phytoplankton selected for high DHA yields are maintained on agar slants in the laboratory. (B) Large fermentation facility used for commercial production of DHA-rich algae. Chemically pure DHA needed for brain growth and development is extracted from algae and used to fortify baby food. (C) Starter cultures are grown in liquid medium. (Courtesy of Martek Biosciences Corporation.)

plants has excluded DHA for a reason(s) that first must be understood in order to successfully harness DHA genes in field crops, such as corn, soybeans, and canola.

## 18.3 DHA PRODUCED BY LAND PLANTS IS PREDICTED TO BE ITS OWN WORST ENEMY

Studies of the mechanism of heat stress in terrestrial plants provide important clues concerning potential harmful properties of DHA in field crops (Figure 18.2). We suggest that DHA will have the same effects as thermal stress in disrupting chloroplast membrane architecture. In essence we envision that DHA will behave as an uncoupler of photosynthetic energy production in chloroplasts. In other words, DHA enrichment of chloroplast membranes, which seems inevitable during the growth cycle of DHA oil crops, is predicted to have the same effect as scorching daytime temperatures (i.e., about 35°C is typical in full sunlight) in inhibiting plant growth. Several possible scenarios for explaining the mechanism of heat stress in plants have emerged over the years, but none has been widely accepted until recently. Interestingly, the current picture is that heat stress is caused by proton energy uncoupling at the level of chloroplast membranes, triggering first a drop in ATP production and later inhibition of carbon dioxide fixation and plant growth. According to this scheme, as temperatures

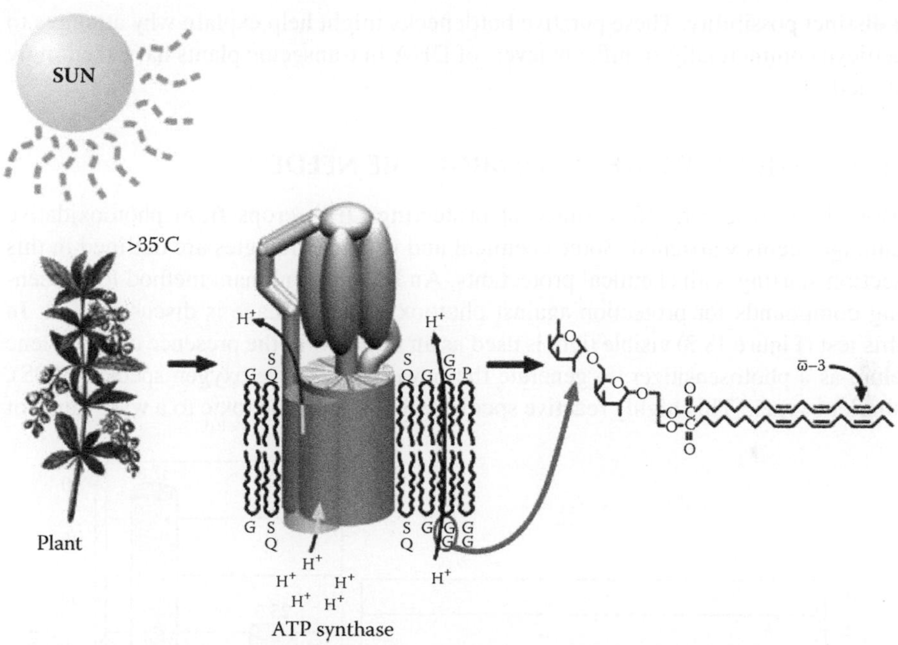


**FIGURE 18.2** DHA as an Achilles' heel in genetic engineering of crop plants to produce marine oils. Studies of heat stress demonstrate that temperatures near the maximum tolerated by crop plants tend to destroy the proton permeability barrier of their polyunsaturated chloroplast membranes, effectively starving the plant of energy. DHA incorporated into chloroplast membranes of crop plants is hypothesized to exacerbate this problem and severely disrupt energy supply for plant growth.

approach 35°C, chloroplast membranes reach a critical state of enhanced permeability and become excessively leaky for protons, shutting down photosynthesis and growth. Depending on the severity of heat stress, crop yields can decline.

Studies by German scientists (Bukhov et al., 1999) show that thermally induced changes in membrane lipid structure trigger the heat stress cascade. Using genetic engineering techniques Japanese researchers (Murakami et al., 2000) showed that lipid conformations seem to be involved. Crop plants, such as corn, often enrich their chloroplast membranes with long chain fatty acid building blocks with up to three double bonds per chain (e.g., 18:2 or 18:3). Thermally induced changes occurring in 18:3-enriched membranes were found to be responsible for triggering heat stress based on the finding that replacing 18:3 chains with 18:2 resulted in significantly greater tolerance to heat. Data from these two experiments lead to the current working model in which heat stress is considered as a specific form of energy stress involving uncoupling of proton gradients across chloroplast membranes. Imagine that DHA (22:6) is incorporated into chloroplast membranes during growth of genetically engineered plants. We suggest this would result in a more heat-sensitive crop plant, unable to grow under normal field conditions. A combination of proton uncoupling along with 22:6 triggering the photooxidative stress cascade in chloroplasts is a distinct possibility. These putative bottlenecks might help explain why attempts to achieve commercially significant levels of DHA in transgenic plants have seemingly stalled.

## 18.4 PHOTO-PROTECTION MIGHT BE NEEDED

More basic research effort aimed at protecting DHA crops from photooxidative damage seems warranted. Some chemical and genetic strategies are outlined in this section starting with chemical protectants. An EPA recombinant method for screening compounds for protection against photooxidative damage is discussed first. In this test (Figure 18.3) visible light is used as an inhibitor in the presence of methylene blue, as a photosensitizer, to generate the powerful reactive oxygen species (ROS), singlet oxygen. This highly reactive species is believed to be toxic to a wide range of

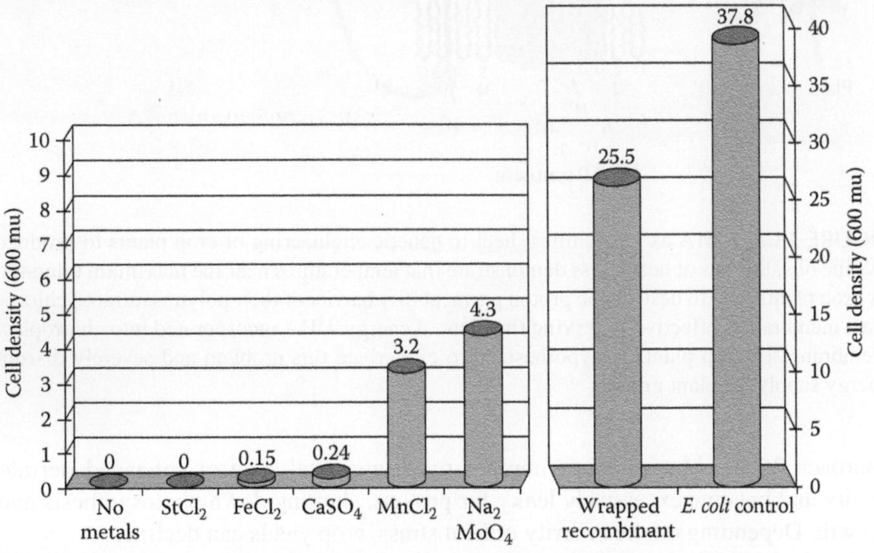


**FIGURE 18.3** An EPA recombinant system as a screen for photoprotective molecules that reduce lipoxidation. EPA recombinants have prove to be useful models for testing compounds that protect cells with highly unsaturated membranes against photokilling. For this assay, several thousand cells of the EPA recombinant were spread on ECLB plates containing 500 μg/L of methylene blue as photosensitizer. Plates were incubated for 4 days in a plant growth chamber maintained at 18°C and illuminated with a light intensity of 190 ft candles (fluorescent bulbs). A recombinant control plate wrapped in aluminum foil served as a positive control and displayed confluent growth. A second control plate was spread with a wild-type strain of *E. coli* as inoculant and exposed to visible light; following an initial lag period lasting several hours, confluent growth was observed. During the course of this experiment, various metals placed on disks in the center of the inoculated plates were tested and some found to protect against photokilling. In order to quantify this effect, cells were washed from the plates with a fixed volume and the resulting optical density quantified. Among the metal ions tested, molybdate offered the most protection, being consistently more potent than manganese.

organisms, from *E. coli* to plants. This test shows dramatic photoinhibition of EPA recombinant cells compared to wild-type cells used as controls. Exposure of several thousand EPA recombinant cells spread on the surface of Petri dishes to visible light essentially sterilizes this population of cells in contrast to foil-wrapped controls (compare "no metals" with "wrapped control"). These results indicate that visible light is the lethal agent. The photosensitizer methylene blue is also required. Also notice that *E. coli* fad $E^-$ used as the wild-type control grows robustly. In contrast, plates inoculated with EPA recombinant cells show no visible colonies. These data, while interesting, require further analysis to show unequivocally that EPA phospholipids are responsible.

Recently, a novel protective system based on catalytic detoxification of reactive oxygen species by manganese in dark-grown cells has been described in bacteria such as *E. coli*. These data show that $Mn^{2+}$-organic acid complexes in bacteria provide a basal level of protection against ROS. This "primitive" chemical mechanism is envisioned to work alongside well-established enzymatic defenses to provide a strong shield against oxidative damage caused by ROS generated in the dark. In order to test the photoprotective effects of metals on EPA recombinant cells, various metals were impregnated in sterilized paper disks placed in the center of pre-inoculated dishes; protection is scored as a function of cellular growth surrounding the disk. As expected, $Mn^{2+}$ displays some protective effects, as summarized in Figure 18.3. However, further testing shows that molybdate is the most effective among all compounds tested so far. This is a surprise since this trace metal has not been reported as being active in protection assays against ROS carried out using a bacterial bioassay. Protection by molybdate is roughly linear as a function of concentration, with levels as low as 100 μg/disk (i.e., diffusing across the plate surface) showing strong protection. The mechanism might involve photoquenching against formation of singlet oxygen.

Genetic strategies might also be needed to protect DHA crops. Since marine plants produce vast quantities of DHA using visible light as energy source, these cells have obviously evolved any necessary protection genes or systems against damage to protect their chloroplast membranes. Thus, DHA-producing marine plants are a logical source of information and genes for protection of DHA crops. Recall that a striking difference between land and marine environments involves the high salinity or osmotic strength surrounding marine plants. Phytoplankton grow in seawater, whereas crop plants are killed when irrigated with even a fraction of seawater. How do marine plants survive when subjected simultaneously to osmotic stress as well as oxidative stresses caused by their highly unsaturated membranes? One possibility is that phytoplankton have evolved a salinity defense that doubles as a mechanism against photooxidative stress. Marine phytoplankton face perpetual osmotic stress. To avoid almost certain death caused by the chemical forces of osmotic pressure drying out their cytoplasm and destroying their chloroplasts, phytoplankton and other marine microorganisms seem to have no choice but to counterbalance the osmotic strength of their cytoplasm to a level at least equivalent to seawater (i.e., roughly 0.44 M NaCl). Biochemical strategies to cope with osmotic stress often involve the synthesis or accumulation of osmoprotective molecules such as trehalose, glycerol and $K^+$. All of these molecules are referred to as compatible solutes because their buildup often in high concentrations is tolerated by cellular processes and enzymatic machinery. We suggest that since phytoplankton must build a strong molecular defense against osmotic stress, they may have evolved a dual-purpose molecule to simultaneously counter photooxidative stress. Owing to the high levels of such osmolytes expected in phytoplankton, even modest photooxidative protection/quenching on the part of a compatible solute such as trehalose might lower ROS levels below a critical threshold enabling conventional defenses to further protect the cell. Thus, there is reason for optimism that, if needed, novel protection systems for DHA crops can be engineered. However, this optimism is tempered by the fact that marine phytoplankton are protected in the water column against full sunlight. That is, phytoplankton living in the uppermost layer of seawater where light energy is most intense are inhibited by the excess of light and must retreat to deeper water to avoid photoinhibition.

Genes present in existing plant germplasm might be exploited in protecting DHA crops. Recall that crop plants have evolved an umbrella of genes essential for protecting their polyunsaturated phospholipids, with up to three double bonds, against photooxidative damage. Thus protective genes might already be operating in some cultivars, allowing plant breeders to make conventional crosses as a strategy to protect DHA crops. In other words, a basal level of protection might already be operational in certain cultivars, allowing these genes to be transferred to germplasm used for developing DHA crops. Conventional protective genes known to be present in the germplasm pool include lipophilic antioxidants such as vitamin E and carotenoids, fatty acid peroxidases, as well as a battery of conventional ROS detoxifying enzymes acting in the cytoplasm. Also, "partitioning genes," which shunt or time DHA away from chloroplast membranes during the growth phase, and into storage lipids, can also be added to the list of possible conventional protection genes as might genes that target DHA phospholipids into cardiolipin. However, even the most efficient mechanism of partitioning, such as operating during filling of the maturing oil seed, might break down during seedling development when lipid stores from the DHA-enriched seed are recruited for new membrane synthesis. The point is that protecting young plants might demand special conditions. Also since chemical oxidation of seed lipids continues in the dark during long-term storage, any membranes previously enriched with DHA would likely require added protection.

## 18.5 SUMMARY

After first being discussed by plant geneticists at Calgene and other organizations as early as the mid-1980s, the concept of moving fish-oil genes into field crops has now reached a stage where this new generation of omega-3–enriched oil seeds is entering the pipeline of new plant products. Forward progress on DHA crops seems to be slowing. Is protection of DHA-enriched membranes a bottleneck to progress? Even though multiple systems have evolved to protect membranes in commercial crops, these mechanisms might not be powerful enough to protect DHA-enriched chloroplast bilayers. Studies of the mechanism of thermal stress in plants provide important clues pointing toward the need for extra protection of DHA crops. This idea has been tested in a bacterial bioassay system involving an EPA recombinant, and the results are consistent with the above scenario. Also, data obtained in studies of thermal killing of a DHA-producing bacterium (see Chapter 17) points toward possible risks of DHA chains during thermal stress. Thus, photoinhibition seems to be a serious impediment in development of DHA crops. Nevertheless, we assert that there are ways around this. Finally, the production of DHA oil seeds on a large scale remains an exciting target for modern agriculture, but more basic research is likely needed to achieve these goals.

## SELECTED BIBLIOGRAPHY

Bukhov, N. G., C. Wiese, S. Neimanis, and U. Heber. 1999. Heat sensitivity of chloroplasts and leaves: Leakage of protons from thylakoids and reversible activation of cyclic electron transport. *Photosyn. Res.* 59:81–93.

Damude, H. G., and A. J. Kinney. 2007. Engineering oilseed plants for a sustainable, land-based source of long chain polyunsaturated fatty acids. *Lipids* 42:179–185.

Graham, I. A., T. Larson, and J. Napier. 2007. Rational metabolic engineering of transgenic plants for biosynthesis of omega-3 polyunsaturates. *Curr. Opin. Biotechnol.* 18:142–147.

Kinney, A. J. 2006. Metabolic engineering in plants for human health and nutrition. *Curr. Opin. Biotechnol.* 17:130–138.

Murakami, Y., M. Tsuyama, Y. Kobayashi, H. Kodama, and K. Iba. 2000. Trienoic fatty acids and plant tolerance of high temperature. *Science* 287:476–479.

Robert, S. S. 2006. Production of eicosapentaenoic acid and docosahexaenoic acid-containing oils in transgenic land plants for human and aquaculture nutrition. *Mar. Biotechnol.* 8:103–109.

Wu, G., M. Truksa, N. Datla, et al. 2005. Stepwise engineering to produce high yields of very long-chain polyunsaturated fatty acids in plants. *Nat. Biotechnol.* 23:1013–1017.

# 19 DHA/Unsaturation Theory of Aging

Hypothesis: Low membrane unsaturation in mammals increases life span and vice versa.

A famous story tells of the person who craved the good life now, and bartered his soul with the devil for instant gratification. A molecular version of this tale is now circulating among scientists in the field of aging. According to this version, mammals such as mice that fully benefit from the presence of DHA in most of their membranes pay with a shortened life span lasting only about 3 to 4 years. In contrast, humans, who target DHA away from most cells and thus shun many of the benefits, can live more than a century. However, since our indispensable and complex neurosensory system requires DHA we may be selectively pushing brain longevity beyond its evolved limit (as described in Chapter 20).

The DHA theory of aging, recently broadened to encompass membrane unsaturation in general, states that the level of unsaturated fatty acid chains in membranes is a key determinant in longevity, an inverse relationship. In simple language, high levels of membrane unsaturation, especially DHA content, predict a short life span; low levels of unsaturation predict a long life span. This fits with the long life span of humans where the overall DHA content is relatively low in most cellular membranes, save sperm and neurons. In contrast, small mammals such as mice have higher levels of both DHA and unsaturation, consistent with their much shorter life span. Recently, the unsaturation theory of aging received a boost from studies of a mouse-sized mammal that seems to break the rules of aging by living almost ten times longer than mice.

## 19.1 DHA CONTENT PREDICTS LONG LIFE SPAN OF NAKED MOLE RATS AND SHORT LIFE OF MICE

The two tiny mammals shown in Figure 19.1, naked mole rats versus mice, are at the center of a critical test of the DHA theory of aging in small rodents, with implications for humans. Naked mole rats are small mammals that exhibit exceptional longevity. This mouse-sized rodent (~35 g) has a captive maximum life span of >28 years, which is the longest known for any rodent species and is eight or nine times greater than the 3- to 4-year life span reported for similar-sized mice. Naked mole rats not only exhibit extraordinary life spans, but these subterranean northeast African rodents also show vitality and continue to breed well into their third decade of life. These tiny rodents are receiving increasing attention as a new animal model to evaluate mechanisms that impact longevity and the aging process in mammals.

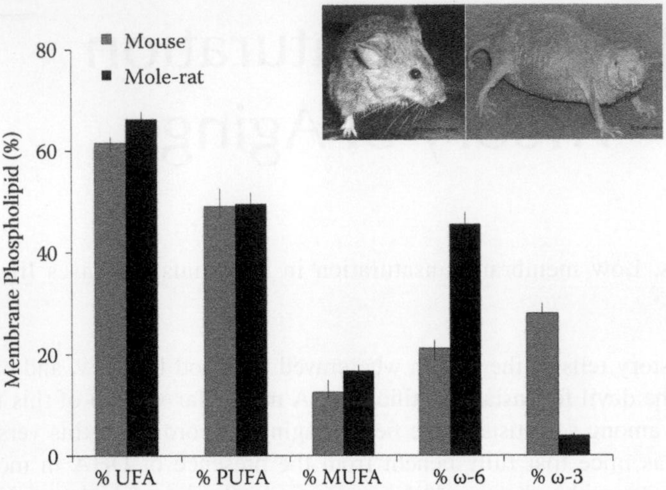


**FIGURE 19.1** DHA/unsaturation theory of aging has been tested in naked mole rats versus mice. The extraordinarily long life span of the naked mole rat (pictured at right), a mouse-sized mammal, may be explained by a dramatic decrease in DHA content compared to mice (pictured at left), displayed graphically as percentage ω-3. That is, naked mole rats live eight to nine times longer than mice and have only about one-ninth the level of DHA in their membranes. The high membrane DHA content and high overall membrane unsaturation levels of mice are consistent with their relatively short life span. (Courtesy of Yael Edrey and Rochelle Buffenstein.)

The "oxidative stress" theory of aging is proposed as a mechanism to explain the striking difference in longevity between these two rodents. At the heart of this concept is reactive oxygen species (ROS), generated as a byproduct of mitochondrial respiration, continuously damaging critical cellular polymers: DNA, proteins and unsaturated membrane lipids. DHA has been singled out because this structure is the most vulnerable to oxidative attack among natural membrane fatty acids (Figure 19.2; also see Chapter 9). Thus, the extreme oxidative instability of these chains has given rise to the DHA theory of aging in which DHA-enriched membranes are considered an important determinant of aging and life span. However, the DHA theory of aging seems less applicable to humans because most of our cells contain only traces of DHA. That is, the DHA theory of aging is treated by some as a special case more applicable to small mammals with relatively high DHA levels. A more general membrane unsaturation theory of aging applicable to humans has been developed during the past several years by Spanish scientists. This concept takes into account the total number of double bonds and includes a wider spectrum of polyunsaturated fatty acids common in most human membranes (Pamplona et al., 2002). Nevertheless, experimental substantiation of the DHA theory of aging in small mammals would go a long way toward establishing the significance of the general membrane unsaturation hypothesis of aging. As already mentioned, some of the most convincing data linking membrane unsaturation to longevity are coming from studies of rodents.

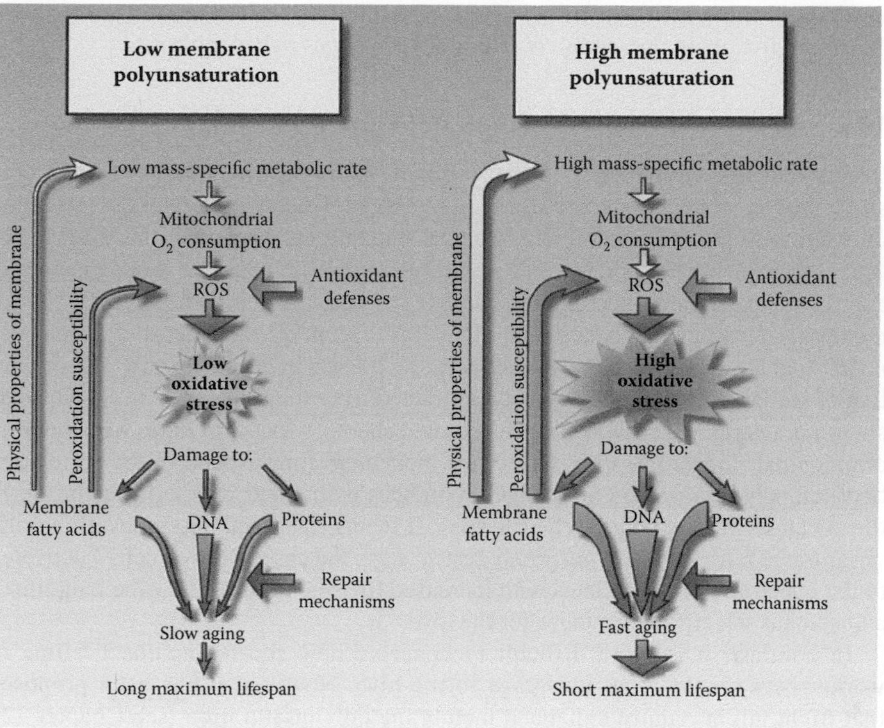


**FIGURE 19.2** Lipoxidation-mediated oxidative stress as a mechanism to explain the membrane unsaturation theory of aging. This theory states that increased levels of unsaturated chains in membranes, especially DHA or other polyunsaturated molecules, elevate rates of lipoxidation, which in turn boost the pool of reactive oxygen species believed to be responsible for oxidative death of aging cells. As applied to humans the presence and levels of polyunsaturated chains distributed in membranes of all classes of cells is proposed to modulate aging. This concept links unsaturation levels in membranes to the widely accepted mitochondrial theory of aging. (Reproduced from Hulbert, A.J. et al. *Physiol. Rev.*, 87:1175–1213, 2007. With permission.)

Recently, a comprehensive analysis of the DHA content of a variety of tissues of naked mole rats has been reported and contrasted with similar studies of mice. The startling finding is that naked mole rats have only about one-ninth the level of DHA compared to mice. For example, phospholipids from skeletal muscle, heart, kidney and liver of naked mole rats contained an average 2.2% DHA (range 0.6% to 6.5%) versus 19.3% (range 11.7% to 26.2%) in mice. Even brain DHA content, which is strikingly similar among most mammalian species, is about 25% lower in naked mole rats compared to mice. This drop in brain DHA might be linked to the fact that naked mole rats are virtually blind and have evolved a fold of skin to prevent direct sunlight from reaching the eye, which retains only the ability to detect day versus night. Interestingly, the percentage of total unsaturated chains in membranes between naked mole rats and mice is similar, which means that a major variable is

DHA itself. Once again, the most interesting result is that mice have nine times more DHA and live about one-eighth to one-ninth as long as naked mole rats.

## 19.2 UNSATURATION THEORY APPLIED TO INSECTS

Recently, it has been suggested that the honeybee is also a suitable organism to study the role of membrane unsaturation in aging (Haddad et al., 2007). Depending on what they are fed, genetically identical female eggs become either workers or "queens" with worker honeybees having maximum life spans of a few months in contrast to queen honeybees, which can live for years. A comprehensive analysis of membrane fatty acids of heads shows that 18:2 content of membranes of queen bees is ~10-fold lower compared to worker bees; 18:3 levels in queens is only one-quarter that of workers. Thus, 18:2/18:3 levels extracted from the heads of honeybee queens show a dramatic reduction in polyunsaturated chains, which correlates with an order of magnitude longer life span. One of the most interesting aspects of the correlation between polyunsaturation and longevity in bees is that nutrients fed specifically to the queen seem to greatly extend life span. It seems clear that dietary levels of polyunsaturated fatty acids are purposely restricted in the case of "royal jelly" delivered to the queen and this correlates with increased life span. Presumably the long life of a queen has selective advantages for the hive.

In contrast, it is more difficult to conceive how shortening the lifetime of worker bees would offer advantage to the hive. However, it has been proposed that flight might require enhanced membrane polyunsaturation (see Chapter 12). Flight is not an advantage in the cases of queens of honeybees and larval insects, and fatty acid analysis shows that polyunsaturation of membranes as expected is significantly depressed. Another interesting finding is that during the first week of activity, membrane polyunsaturation in worker bees rises sharply. A spike in polyunsaturation is also observed during the transition from larva to butterfly. Thus, flight can be considered significant enough for survival of the species to compensate for a dramatically reduced life span. Once again, the concept of benefits versus risks seems to apply to polyunsaturated membranes of insects. Recall that the degree of unsaturation for membrane lipids in insects ranges from one to three double bonds. That is, DHA/EPA are virtually absent in insects compared to mammals. One explanation is that plant-eating insects never encounter such highly unsaturated chains in their diet because of the absence of these molecules from terrestrial plant life. Insects might be an interesting tool for testing the unsaturation theory of aging. Tests might involve force-feeding insects with DHA/EPA to determine if one can decrease longevity of insects and allow them to fly for longer periods and at cooler temperatures. It seems likely that the list of organisms whose longevity correlates with membrane unsaturation levels will be expanded in the future.

## 19.3 LIPOXIDATION MECHANISMS

According to the membrane unsaturation theory of aging, the acyl composition of membrane bilayers is a missing link in understanding the aging cascade. Membranes that have different fatty acid composition are known to differ dramatically in their rates of chemical oxidation as discussed in Chapter 9, and this is proposed to account for much of the observed variation in life span. A working model of the role of membrane unsaturation in aging is described in Figure 19.2. The major new addition to the popular oxidative stress theory of aging involves the feedback loop from membrane lipid oxidation to the ROS pool. The intensity of the contribution of the membrane to the ROS pool is envisioned to be variable and dependent on the level of unsaturation. For example, membranes highly enriched with DHA are predicted to have a large influence on the ROS pool in contrast to membranes composed of more saturated chains.

It is well known that ROS, once produced, will chemically attack cellular macromolecules including DNA, proteins, and lipids. Recall that lipoxidation in the dark is autocatalytic and thus a self-propagating chain reaction that, once triggered, will continue unless stopped by antioxidant defenses. Also recall that once lipoxidation is initiated, lipid peroxides and other products of lipoxidation often remain in the membrane and pose an oxidative threat to neighboring lipid molecules. These properties bring unsaturated phospholipids into the mainstream of aging theory because of their contribution to both the lipid- and aqueous-phase ROS pools.

## 19.4 LIPOXIDATION PRODUCTS MIGHT DIRECTLY UNCOUPLE CATION GRADIENTS IN MITOCHONDRIA AND CREATE ENERGY STRESS

The purpose of mitochondria as ATP generators for the cell is defeated in aging mitochondria whose inner membranes become increasingly porous for protons, effectively converting mitochondria into heat machines. Water wire theory developed with bacteria provides a new mechanism to help explain aging-related proton uncoupling in mitochondria. It is hypothesized that oxidatively damaged phospholipids and their derivatives, mainly from unsaturated lipids, directly lead to the formation of water wires. Water wires lead to proton uncoupling, which leads ultimately to apoptosis. The new aspect proposed here is that oxidative damage to unsaturated membrane phospholipids is proposed to directly create water wires due to changes in the molecular architecture of the membrane and diminish the proton or cation permeability barrier. Previous theories on why aging mitochondria become more porous to protons have focused on proton flux through inner membrane mitochondrial proteins that behave like proton gates, with the open gate conformation being modulated by various metabolic signals. There is convincing data that these proton gates are important as a driver of apoptosis. However, these data do not rule out the possibility of other mechanisms, such as oxidatively produced water wires, modulating proton permeability.

An abundant class of highly unsaturated phospholipids universally present in the inner mitochondrial membrane caught our attention as a possible weak link in membrane permeability against protons. It is well known that the most highly unsaturated class of phospholipids found almost exclusively in the inner membranes of mammalian, including human, mitochondria is cardiolipin (CL; see Chapters 12 and 21). The focus here is limited to CL as a possible primary target of lipoxidation. We suggest that lipoxidation of CL yields derivatives that may directly increase proton energy uncoupling via a water wire mechanism.

Unlike other mitochondrial phospholipids that have a single glycerolphosphate backbone and two fatty acyl side chains, CL has a double glycerolphosphate backbone and four fatty acyl side chains. In humans, CL is biosynthesized by joining of phosphotidylglycerol and cytidinediphosphate-diacylglycerol by CL synthase located on the inner face of the inner mitochondrial membrane. CL generally has high unsaturation, which renders these molecules vulnerable to lipoxidation, with possible negative impact on longevity. CL may be particularly susceptible to peroxidation not only because of the abundance of double bonds in its structure but also its close association with the proteins of the respiratory chain. The rates of lipoxidation of 18:2 often found in CL are roughly 15-fold higher than 18:1, but this may not be the complete story. For example, the total number of double bonds in CL may double in certain animals, with rates of oxidation of DHA chains being greater than 10-fold higher compared to 18:2. Also the close packing of four consecutive 18:2 chains, a common CL structure, might create oxidative "hot spots" in the membrane. Several studies indicate that peroxidized CL is unable to support the reconstitution of delipidated mitochondrial respiratory enzymes. Also, the reported loss of CL during aging and disease is consistent with a mechanism involving lipoxidation. Net losses of CL have been implicated in a number of diseases, but we're not aware that CL depletion and production of lipophilic oxidation by-products have been linked to energy stress. There is a series of papers linking CL loss to decreasing activity of a series of critical mitochondrial enzymes.

We hypothesize that oxidatively damaged CL increases the probability of formation of water wires in the inner mitochondrial membrane, subjecting the cell to energy stress at a time when aging makes it most vulnerable. Two mechanisms for creating water wires in aging mitochondria are being considered, as follows. (1) Loss of CL is predicted to increase water wires by a mechanism involving hypofluidity of the bilayer. At the same time, when putative proton leakage is increasing, hypofluidity is expected to reduce rates of respiration. (2) Damaged phospholipids such as oxidatively truncated derivatives as well as incomplete structures generated by repair lipases are predicted to directly create water wires and exacerbate other stresses.

## 19.5 INTEGRATING ENERGY STRESS CAUSED BY THE PLASMA MEMBRANE INTO THE AGING CASCADE(S)

According to the mitochondrial theory of aging, proton leakage across mitochondrial membranes increases with age. Less attention has been given to permeability properties of aging plasma membranes. The best evidence that oxidatively damaged phospholipid structures are important in the plasma membrane comes from studies of apoptosis. The main finding derives from molecular genetic studies of the effects of overexpression of repair phospholipase required to clear the plasma membrane of oxidatively damaged phospholipids, which are thought to be universally present in aerobic cells. This important experiment shows that oxidatively damaged phospholipids can drive apoptosis. In other words, these studies expand the current oxidative stress pathway by indicating that oxidative damage to unsaturated phospholipids of the plasma membrane leads to energy stress, which ultimately leads to cellular death. The importance of energy stress in aging is discussed in more detail next.

We suggest that pro-apoptotic properties of oxidatively damaged phospholipids in the plasma membrane can be explained from an energy or bioenergetics perspective. The critical assertion we make concerns how signals caused by lipoxidation of the plasma membrane are sensed and transmitted to mitochondria, which are at the heart of the oxidative stress cascade. Recall that the plasma membrane in human cells is responsible for preventing energy uncoupling caused by unwanted flow of $Na^+$ from the high outside to low inside concentrations. We propose that oxidative damage to unsaturated phospholipids in the plasma membrane in essence loosens the permeability barrier against $Na^+$, opening water wires for conductance of $Na^+$, a form of energy uncoupling. Thus, the pathological events in aging cells are hypothesized to involve energy stress caused in part by lipoxidation of the plasma membrane as well as the mitochondrial membrane. It is proposed that aging cells gradually run out of options for maintaining a healthy energy status. As a result, some critical processes will receive an insufficient energy supply, essentially activating a time clock of aging. It is searching for the proverbial "needle in a haystack" to attempt at this time to single out specific bioenergetic weak points. However, if key ROS defenses and membrane repair mechanisms begin to falter due to lack of energy availability, a chain reaction of great destructive power seems sure to follow. This discussion is continued in Chapter 20 dealing with neurodegenerative diseases where energy stress seems especially pertinent.

## 19.6 SUMMARY

The membrane unsaturation concept of aging has now expanded beyond a simple derivative of the mitochondrial-ROS theories of aging for several reasons, as follows. First, oxidatively damaged phospholipids are hypothesized to directly lead to defects in membrane permeability. Second, a mechanism based on water wires is proposed to explain how both mitochondria, and the plasma membrane become leaky for $H^+$ and $Na^+$, respectively. Third, the putative cross-talk between the plasma membrane and mitochondria might involve both ROS and energy stress caused by uncoupling of cation gradients. Lastly, this molecular model is consistent with unsaturated phospholipids serving as proaging molecules.

The introduction of the naked mole rat as a model for studying aging raises questions that go beyond the dramatic differences seen with membrane fatty acid composition. What evolutionary driving forces are behind the unexpectedly long life spans of these animals compared to similar-sized mice? Are differences in membrane composition the main story, or is there a more complex underlying physiology? The idea that these rodents are chronically starved of nutrients and thus burn less oxygen, a well-known mechanism to gain longevity, has already been considered and discounted by researchers in this field. However, because these animals live in $O_2$-starved and carbon dioxide–rich burrows, a condition that is difficult to reproduce with laboratory animals, the question of net $O_2$ uptake as a factor in aging remains of interest. Nevertheless, the correlation between longevity and DHA content of membranes in these two test animals is striking and consistent with the molecular perspective of DHA's benefits and risks, as suggested in this book. We believe that the era of research on membrane unsaturation and aging is destined to be an exciting one.

## SELECTED BIBLIOGRAPHY

Haddad, L. S., L. Kelbert, and A. J. Hulbert. 2007. Extended longevity of queen honey bees compared to workers is associated with peroxidation-resistant membranes. *Exp. Gerontol.* 42:601–609.

Hulbert, A. J. 2005. On the importance of fatty acid composition of membranes for aging. *J. Theor. Biol.* 234:277–288.

Hulbert, A. J., S. C. Faulks, and R. Buffenstein. 2006. Oxidation-resistant membrane phospholipids can explain longevity differences among the longest-lived rodents and similarly-sized mice. *J. Gerontol. A Biol. Sci. Med. Sci.* 61:1009–1018.

Hulbert, A. J., R. Pamplona, R. Buffenstein, and W. A. Buttemer. 2007. Life and death: Metabolic rate, membrane composition, and life span of animals. *Physiol. Rev.* 87:1175–1213.

Pamplona, R., G. Barja, and M. Portero-Otín. 2002. Membrane fatty acid unsaturation, protection against oxidative stress, and maximum life span: A homeoviscous-longevity adaptation? *Ann. NY Acad. Sci.* 959:475–490.

Wang, Y., D. S. Lin, L. Bolewicz, and W. E. Connor. 2006. The predominance of polyunsaturated fatty acids in the butterfly *Morpho peleides* before and after metamorphosis. *J. Lipid Res.* 47:530–536.

# 20 DHA Principle Applied to Neurodegenerative Diseases

Hypothesis: DHA-laden membranes play cardinal roles in modulating the health and causing the death of neurons.

Dramatic advances have been made in understanding the molecular pathology of age-related neurodegenerative diseases such as Alzheimer's disease (AD; Figure 20.1). Four of the most widely accepted unifying factors linking brain diseases in general are (1) oxidative stress, (2) energy stress, (3) toxic peptides, and (4) apoptosis. We present evidence in this chapter suggesting DHA membranes serve as a fifth unifying factor of age-related neurodegenerative diseases. Indeed, classifying these diseases as membrane diseases subject to the DHA principle provides a framework for possible integration of all five of these factors. According to this membrane hypothesis oxidative stress, energy stress, and toxic peptides all work together to modulate what we term membrane structural homeostasis. Membrane structural homeostasis is envisioned as comprising multiple parameters, including DHA levels, saturated fatty acid levels, cholesterol levels, and so forth, all aimed at maintaining a healthy long-term environment to facilitate membrane biochemistry. A critical corollary of the DHA principle is that large deviations in the physical state of the bilayers of neurons, such as caused by depletion of DHA, will gradually lead to catastrophic consequences (i.e., neuron death), a time clock that can be accelerated or slowed down by both internal and external signals (e.g., Parkinsonian chemicals and toxic prion peptides). Recent data even implicate DHA-enriched membranes in the final stages of suicide or destruction of sensory axons.

## 20.1 DHA AS A RISK FACTOR IN AGING NEURONS

According to the DHA principle, DHA is more prone to oxidation than any other common fatty acid in nature, and in some respects enrichment of the vast network of neural membranes in the brain is a gamble. Indeed, from an ecological perspective, a DHA membrane working for decades in the highly oxygenated environment of the brain is a disaster waiting to happen. What is even more amazing is that neurons, as fragile as their membranes are, last a human lifetime. A number of mechanisms have evolved to protect DHA membranes of neurons from oxidative damage. We will discuss a few among those listed below:

- Enzymatic detoxification of reactive oxygen species (ROS)

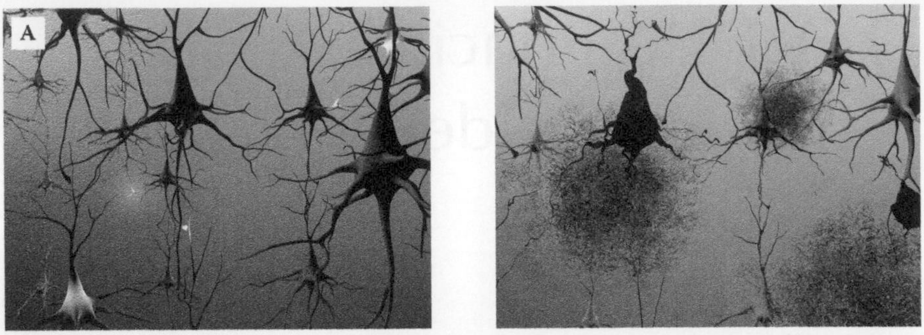


**FIGURE 20.1A** (A color version of this figure follows p. 140.) Loss of DHA from axonal membranes during Alzheimer's disease might be a cause rather than effect. (a) At left: healthy neurons maintain their DHA-enriched axonal membranes through an active and sophisticated repair and remodeling system for removing damaged DHA. DHA is replenished through the diet and via biosynthesis. With age the delicate balance needed for maintaining membrane structural homeostasis might slowly shift such that net DHA levels begin to drop. As discussed in the text a steep drop of DHA is characteristic of AD and might be a cause rather than an effect. At right: loss of DHA in neurons has been implicated in defective protein processing and generation of toxic peptides such as β-amyloid shown in this diagram. However, a direct linkage between β-amyloid and AD has not been established. [(A) Courtesy of Alzheimer's Association. (B) and (C) Courtesy of M. Tessier-Lavigne and Genentech Inc. Reprinted with permission of MacMillan Publishers Ltd: Nikolaev, A., McLaughlin, T., O'Leary, D. D., and M. Tessler-Lavigne. *Nature* 457:981–989, 2009.]

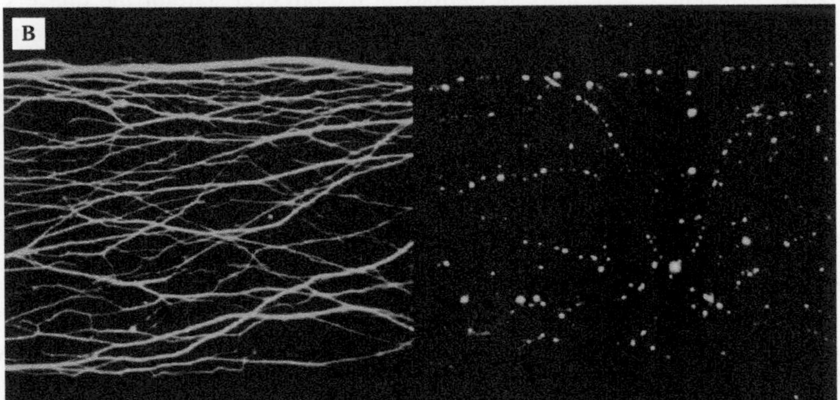


**FIGURE 20.1B** (A color version of this figure follows p. 140.) A different toxic peptide generated by defective protease processing of the same amyloid precursor protein (APP) acts as a suicide molecule destroying sensory connections. At left: fluorescent microscopy image of healthy sensory connection of neurons. At right: fluorescent microscopy image of neurons after treatment with N-APP, displaying destruction of neural connections.

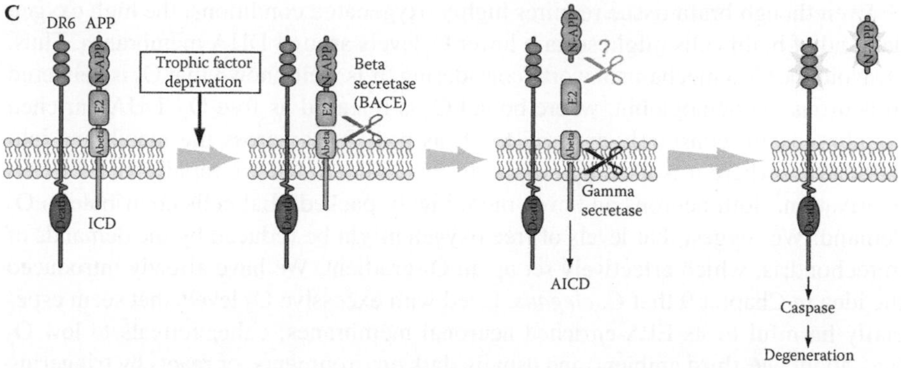


**FIGURE 20.1C** (A color version of this figure follows p. 140.) Model proposed for role of N-APP in AD by scientists at Genentech. We propose that depletion of DHA in neural membranes might result in increased production of N-APP by a mechanism similar to that described in the text for β-amyloid.

- Repair
- Shunting DHA away from mitochondrial membranes to reduce the ROS pool
- Antioxidants
- Careful regulation of levels of metals such as iron that accelerate lipoxidation
- Myelination (an avoidance mechanism?)
- Normal prion protein as antioxidant (controversial)
- DHA plasmalogens as antioxidants (theoretical)

Two additional global mechanisms worth considering are

- Avoidance via modulation of free $O_2$ levels surrounding neural membranes by concerted mitochondrial respiration
- Maintaining energy supply above a threshold needed to support continuous membrane repair, for an entire lifetime

Perhaps an eleventh condition can be added as a joke—don't expose your brain to light (see Chapters 9 and 17)! One of the first points to be made about this list is that conventional defenses against oxidative stress, which work well in other parts of the body, appear to be insufficient for the brain. For example, the body has evolved sophisticated protection mechanisms that direct or target DHA into specific neuronal membranes where these chains are most needed. This specificity likely applies to membrane classes within neurons such as plasma membranes enriched with DHA versus mitochondrial membranes, which are not. We interpret these targeting mechanisms as being defensive or protective in nature. Thus, we suggest that Darwinian selection of the fittest neuronal membranes involves targeting DHA to specific classes of membranes to maximize benefit while avoiding lipoxidation (see Chapter 13). Wrapping of axons in a protective sheath of myelin also might help shield against $O_2$ damage, although there is little information on this point.

Even though brain tissue requires highly oxygenated conditions, the high oxygen demand of brain cells might sustain lower $O_2$ levels around DHA membranes. Thus, $O_2$ avoidance is a mechanism worth considering. It is well known that $O_2$ is delivered to neurons via hemoglobin, where bound $O_2$ is released as free $O_2$. DHA-enriched membranes are constantly exposed to $O_2$ as it diffuses across the membrane into the neuron, where it is used by mitochondria, and neurons are rapidly killed by $O_2$ deprivation. Both neurons and even more highly packed glial cells contribute to $O_2$ demand. We suggest that levels of free oxygen might be reduced by the demands of mitochondria, which effectively set up an $O_2$ gradient. We have already introduced the idea in Chapter 9 that *C. elegans*, faced with excessive $O_2$ levels that seem especially harmful to its EPA-enriched neuronal membranes, either retreats to low $O_2$ (i.e., about one-third ambient) and usually dark environments, or reacts by triggering a social feeding frenzy, which lowers $O_2$ levels. The phenomenon of capitalizing on respiratory chains as a means of modulating cellular $O_2$ levels to protect sensitive cellular enzymes, such as nitrogenase, is also well known in bacteria. We are not claiming that the high $O_2$ demand characteristic of brain tissue evolved primarily as a defense against lipoxidation of its DHA-enriched membranes. However, it makes sense that once available, this mechanism might have enabled or benefited development of DHA membranes, and thus provides a critical source of protection.

The need for constant repair of DHA membranes first demonstrated in the case of rhodopsin disk membranes (see Chapter 9) also applies to neuronal membranes. Neuronal membrane lipids are not subject to photooxidation, which might explain the longer half-life of DHA membranes in neurons compared to rhodopsin disks. However, it is clear that the relatively rapid turnover of DHA membranes of neurons makes them, along with rhodopsin disks, examples of bilayers requiring constant repair. See Chapter 13 for an explanation of how sperm cells avoid lipoxidation of their DHA membranes.

We suggest that the factor most enabling long-lived neurons is likely linked to the net activity of all mechanisms for protecting DHA, both known and undiscovered. We hypothesize that as these processes begin to fail DHA will become an oxidative risk, perhaps even contributing catalytically to its own net loss. That is, we consider DHA levels as a dynamic steady-state system, and an increasing sink term will lead to a drop in the steady-state DHA content of the membrane. We suggest that this vicious cycle partially explains the risks of DHA in aging neurons.

If these arguments hold true, DHA should be added to a growing list of molecules contributing to oxidative stress of aging neurons, several of which are listed as follows:

- Chronic or acute exposure to pesticides (e.g., free radical-generating herbicides, rotenone)
- Excess iron and manganese in the diet or workplace
- Use of antipsychotic drugs (e.g., contaminated with the toxin MPTP)
- Carbon monoxide poisoning
- Trauma and swelling in the brain caused by hard blows to the head (e.g., the case of Mohammad Ali)
- Viral infection

- Prion infection
- Other diseases of the brain, including brain tumors and meningitis
- Aging
- DHA membranes in neurons?

## 20.2 DROP IN DHA LEVELS AS A DISEASE MARKER AND RELATIONSHIP TO ENERGY STRESS

AD is a slowly progressing disease that starts by killing neurons at a focal point, eventually orbiting the brain and leaving a deathly landscape after many years. As this wave of neuronal death spreads, DHA levels plummet. Is this a cause or effect relationship? We believe that the drop in DHA is an important signpost or marker of AD and have developed the following roadmap of events to explain this relationship. We suggest that aging leads to energy stress in neurons, which leads to unacceptable levels of DHA oxidation. The hyperoxidation of DHA triggers a net loss of DHA from neuronal membranes and facilitates defective processing of amyloid precursor protein (APP), leading ultimately to apoptosis of axons. DHA levels plummet as a result of this scheme.

We do not attempt to cover this entire scheme in this section and instead focus primarily on the roles of energy stress introduced in Chapter 19. Note that DHA oxidation and the chain reaction it can cause are covered in the previous section and discussed in Chapters 8, 17, and 21. The relationship between DHA levels and defective processing of APP is described in the next section.

The importance of energy supply in neurons is well illustrated by the interesting statistic that, whereas the brain makes up only about 2% of average body weight, this organ consumes an estimated 20% of total energy. In other words, the brain consumes about ten times more energy than other tissues. The tight linkage between neural health and bioenergetics is also apparent at the level of neurons making up specialized regions of the brain. For example, there is increasing evidence that energy stress is pivotal in the chain of events leading to death of Parkinsonian neurons. For example, a well-studied chemical known to induce Parkinson's disease (PD) was found to inhibit the electron transport chain of mitochondria at the level of Complex I. This inhibition results in rapidly decreasing levels of tissue ATP content in regions of the brain most sensitive to this toxic molecule. Studies of mitochondria isolated from distinct neuronal populations distinguish two classes, one in which inhibition of Complex I by as little as 25% significantly impairs ATP production, and a second class where much stronger inhibition of Complex I is needed to cause a similar drop in ATP production. This implies that mitochondria localized in different regions within neurons (e.g., synaptic regions) or from neurons populating different parts of the brain might be more or less sensitive to energy stress.

Neuronal bioenergetics is exceptional not just because of the fast pace but also because of the decades of consistent high-energy output by these cells. This brings us back to the necessity for continuous and energized repair of neuronal machinery, specifically DHA membranes, in order to maintain cellular function. In other words, neuronal bioenergetics has evolved for high-energy use over long periods of time. However, certain neurons may be running so close to full capacity that adding relatively small amounts of additional stresses (e.g., environmental, biotic, or genetic) can kill the neuron, sometimes after a long latent period when no disease symptoms are apparent.

One case history illustrating this point involves Leber's hereditary optic neuropathy (LHON). This is a mitochondrial genetic disease that kills neurons in the optic nerve. This is a bioenergetic disease attributable to a mutation in NADH-ubiquinone reductase (i.e., Complex I) of the electron transport chain, which functions in the inner mitochondrial membrane. This sophisticated enzyme behaves as a proton pump initiating the electron transport chain and thus sits at the heart of bioenergetics in mitochondria of these apparently vulnerable neurons. What is remarkable is that this defect, like AD, is age related where the "energy-stress" cascade is somehow suddenly amplified catastrophically, killing neurons specifically in the optic nerve and leading to blindness. Neurons in patients with LHON attempt to overcome their energy deficit by elevating levels of Complex II (i.e., succinic dehydrogenase), which only partially circumvents or compensates for loss of Complex I activity (i.e., this enzyme can shuttle more electrons into the electron transport chain, but is not itself an energy-yielding proton pump). It is not known whether ATP levels in the mitochondria in the optic nerve are lowered, but this possibility is supported by the finding of decreased ATP levels in certain neurodegenerative diseases, such as PD, as discussed above. One explanation for the late onset of LHON is that mitochondria in neurons making up the optic nerve are required to operate at near full capacity and thus are predisposed to being sensitive to any further stresses. Any number of stresses, such as oxidative stress, may eventually accelerate as a chain reaction and might trigger a cellular death cascade causing blindness later in life. Several other mechanisms likely feed into the energy stress cycle and become almost inseparable. These include excess ROS production such as triggered by hyperoxidation of DHA membranes, as discussed in a later section.

A recent study focuses attention on respiration and energy stress as a major risk factor in death of neurons during Huntington's disease (HD). This involves the case history of a young runner specializing in marathons, who years before any symptoms of HD were detected complained of muscle pain and severe weakness. Baffled at first, a specialist finally noticed major changes in mitochondrial structure and function during the course of his ailment. The take-home message is the finding that intense physical exercise may trigger mitochondrial myopathy in early stages of HD. Mitochondrial dysfunctions have been previously reported in muscle and brain tissues of some HD patients. Mitochondrial respiratory deficits have also been found in muscles of HD transgenic mice. However, mitochondrial myopathy as unmasked in this study as a first symptom of HD was never recognized before. It is known that huntingtin protein is expressed in all cells of the body, but it has been puzzling why HD affects preferentially the brain, usually beyond age 45. Medium spiny neurons of the striatum (a brain structure responsible for movement control), the most severely affected in HD, show one of the highest metabolic rates of the brain. It has been hypothesized for some time that this feature makes neurons of the striatum highly susceptible to energy stress leading to HD.

In summarizing this section, neurons appear to have evolved to operate near their maximum threshold of energy availability. That is, energy production systems in these cells seem to operate at near maximum rates to meet the intensive energy demands of neurosensory biochemistry. We propose that with aging of their mitochondria certain neurons may reach a critical stage of energy deficit. The extensive processes required to constantly repair and remodel oxidatively damaged DHA membranes might be exceptionally sensitive to energy stress. That is, failure to replace and manage these defective membranes might not only increase energy stress but also lead to a chain reaction of oxidative destruction. We envision energy stress and oxidative stress reinforcing one another and preferentially affecting the most oxidatively vulnerable membrane structure, which is DHA itself. At the beginning of this section we suggested that the hyperoxidation of DHA triggers a net loss of DHA from neuronal membranes. As discussed in the last section, a drop in DHA levels is known to alter the physical state of the membrane, which in turn is known to modulate processing of membrane proteins such as APP. Recall that defective processing of APP is believed to generate pro-apoptotic toxic peptides, considered next.

## 20.3 TOXIC PEPTIDES AS ENERGY UNCOUPLERS?

Stanley Prusiner, Nobelist for studies of the molecular biology of prion diseases, showed that a toxic peptide is responsible for this fatal neurodegenerative disease (Prusiner 1991, 2001). That prion protein behaves as an agent of disease is revolutionary and led Prusiner to propose that toxic peptides in general are at the heart of other age-related brain diseases such as AD. We suggest that toxic peptides such as prion peptide and β-amyloid (Figure 20.2) represent the third corner of a membrane-based stress triangle that also includes oxidative stress and energy stress.

Just as oxidative and energy stresses are most investigated in the case of PD, the mechanism of action of toxic peptides has been studied in greatest detail in AD. However, the mode of action of a toxic peptide that is unrelated to neurodegenerative diseases is even better understood, and it serves as a model for β-amyloid, characterized as a toxic peptide in AD.

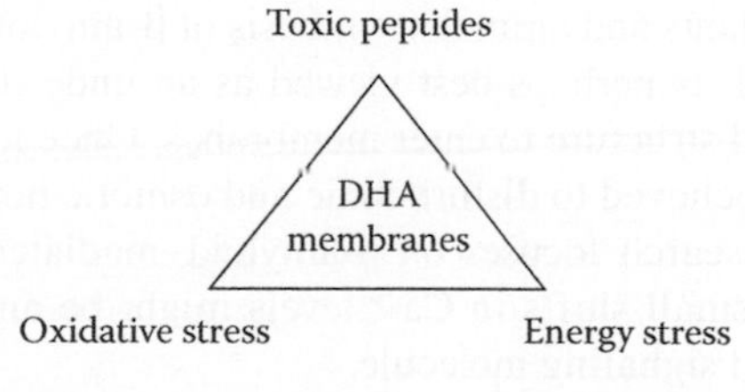


**FIGURE 20.2** Unified concept of neurodegenerative diseases. There is now ample evidence that implicates oxidative stress, energy stress, and toxic peptides as being important in the molecular pathology of AD and other age-related neurodegenerative diseases. We hypothesize that DHA membranes are a common thread connecting these well-known parameters. This is essentially an extension of a unified concept of the roles of toxic peptides in neurodegenerative diseases proposed by Stanley Prusiner.

The bioactive peptide protegrin-1 (PG-1) shares properties with β-amyloid. PG-1 is well known as a potent, broad-spectrum antibiotic. It prevents infection by attacking the membranes of invading microorganisms. This class of peptides is also known to be harmful to mammalian cells. In the membrane, PG-1 forms a pore or channel about 6 to 7 Å wide through polymerization of multiple subunits. Water and ions pass through the pore, but a clustering of arginine residues at the pore entrance seems to bias ionic specificity towards $Cl^-$ over $Na^+$. Amino acids forming the 15-residue peptide chain have been resynthesized, making PG-1 structure-function one of the best understood of all toxic peptides. However, there are a number of unexplained aspects of PG-1 that might contribute to its bioactivity. For example, structures of PG-1 pores in membranes generated by chemical simulation show a marked thinning of the bilayer, which might create a generalized ionic leakage in the pore region. Thus, whereas PG-1 is the best peptide available for comparison with peptides harmful to neurons, there is still plenty of room for surprises in this field.

Next, several assumptions derived from the literature regarding the mode of action of β-amyloid are listed as follows:

- This peptide is directly involved in AD.
- An oligomer rather than clumps or the monomer is responsible for bioactivity.
- The oligomer targets the membrane, altering permeability.
- A pore about 6 to 7 Å in width is generated, large enough for water and ions as large as $Ca^{2+}$ to pass.
- β-amyloid like PG-1 causes membrane thinning, which might help account for its putative harmful effects on $Ca^{2+}$ homeostasis.

Even though our mechanistic understanding of amyloid toxicity is incomplete, any breaching of the permeability barrier of neurons is likely to exacerbate stress, especially energy stress. β-amyloid is effluxed out of the cell into the space surrounding neurons, ultimately showing up as plaque, an insoluble structure. However, to create membrane pores, soluble oligomers would likely enter the membrane of neurons either from the outside or concurrently with synthesis. Unlike antimicrobial peptides produced by plants and animals, synthesis of β-amyloid is likely accidental. β-amyloid, unlike PG-1, is perhaps best viewed as an undesirable by-product with an appropriate chemical structure to enter membranes. Once formed, amyloid pores or thinned regions are believed to disturb ionic and osmotic homeostasis. One of the most active areas of research focuses on β-amyloid–mediated uncoupling of $Ca^{2+}$ gradients, where even small shifts in $Ca^{2+}$ levels might be amplified via pathways triggered by this critical signaling molecule.

The main point here is that increasing but incomplete evidence implicates toxic β-amyloid peptides generated in AD as playing important roles in premature death of neurons. It is interesting that toxic peptides might work by altering membrane properties, perhaps creating harmful effects that converge with other stresses to determine the fate of neurons.

## 20.4 CONSIDERATION OF NEURODEGENERATIVE DISEASES AS MEMBRANE DISEASES

Alzheimer's disease is by far the most prevalent brain disease, and the major drop of DHA in diseased brain tissue is considered here both as an accepted disease marker and as a clue regarding the importance of DHA. Researchers have met with success in exploiting this finding in treating AD mice with increased dietary levels of DHA as a means to slow progression of the disease. There are promising hints coming from early clinical trials that dietary DHA might have similar effects in humans.

The concept that loss of DHA in membranes favors synthesis of toxic peptides is considered in this section (Figure 20.1). This concept is based on studies linking defective processing of amyloid precursor protein (APP) with the physical state of the membrane. For example, the molecular biology of protease processing of β-amyloid precursor protein, which leads to formation of senile plaque, is understood in some detail and believed to be dependent on the physical state of the membrane, which is strongly influenced by DHA. APP is a large transmembrane protein, and it is cleared out sequentially by competing proteases also present in the membrane. The critical cleavage step, which prevents amyloid formation, appears to occur in the liquid region of the DHA-enriched membranes where the protease has ready access to its substrate, amyloid precursor protein. In contrast, the minor protease called β-secretase has been identified as a novel membrane-bound aspartyl-protease and cleaves amyloid precursor protein in its luminal domain, leading to formation of β-amyloid.

Recent data support the concept that β-secretase is compartmentalized along with its substrate within lipid rafts. In other words, this appears to be an example of a reaction that requires entrapment of both enzyme and substrate within gel-phase lipids. According to one model, disruption of lipid rafts by DHA phospholipids in normal membranes would favor the dominant protease in the liquid phase and slow amyloid plaque formation occurring in the gel phase. A drop in DHA levels in the brains of patients with AD is well known, and this would favor lipid raft formation and increased production of plaque. However, according to an exciting new model, production of an excess of yet another peptide fragment called NAPP from APP by β-secretase (BACE) is the actual culprit in triggering apoptosis of sensory axons. NAPP derived by defective cleavage near the N-terminal region of APP is believed to bind directly to a suicide protein, triggering destruction of sensory axons. Is the defective cleavage step leading to NAPP also accelerated in DHA-deficient membranes and vice versa?

Thus, there are three key findings that help define the importance and possible benefits of dietary DHA in AD, as follows.

1. DHA levels in neuronal membranes drop sharply in AD patients; the mechanism is unknown but might involve the oxidative instability of DHA. There is also evidence that a drop in serum levels of DHA is a possible predictor of neurodegenerative diseases. Thus, a drop in the pool of DHA might contribute to DHA depletion in neurons.
2. Feeding DHA to transgenic mice with human Alzheimer's genes slows amyloid plaque formation and disease progression (i.e., does not cure!).

3. High cholesterol levels seem to serve as a predictor of increased likelihood of AD, raising the question: does cholesterol replace dropping levels of DHA chains in neural membranes?

DHA losses that exceed available supply of DHA might serve as a tipping point not only resulting in increased recruiting of cholesterol but also replacement of DHA by more saturated fatty acids. Might high levels of saturated fatty acids in the diet or as released from deposits of stored body fat also be a risk factor in AD? Some recent results concerned with stored belly fat as a risk factor in AD are consistent with this view.

DHA-enriched membranes can also be visualized as key players in the development of the prion pathological state. Prion proteins from all species of animals contain a C-terminal glycerolphosphoinositol (GPI) anchor, a posttranslational modification that adds the GPI molecule to the outside of the cytoplasmic membrane, while allowing for later release by phospholipase. Recent studies show that folding-misfolding of prion protein is also a membrane-mediated process. For example, membrane anchoring above a threshold level induces misfolding of prion protein. Thus, excessive membrane anchoring seems to be a crucial event in the development of prion diseases. At least two separate mechanisms can account for excessive anchoring of prion proteins. One involves excessive posttranslational modifications involving the natural GPI anchor. The second involves bulk changes in membrane lipid architecture favoring aberrant proteolysis and leading to excessive production of toxic prion peptides, perhaps by a mechanism similar to that described for AD.

Since the pathology of neurodegenerative diseases seems to be tightly linked to protein processing at the membrane level, one of the most interesting approaches to treatment involves molecular management of activities of critical protease enzymes themselves. Using recombinant mice as surrogates for humans, it has recently been reported that knockout mutations of a proteolytic enzyme Caspase 6 slows disease progression in HD. These studies are consistent with the critical role of protein processing in the HD cascade. It is also interesting to note that the pathway of triggering of apoptosis by external signals involves a critical membrane stage that might also be modulated by the physical state of the bilayer.

We have emphasized the importance of energy stress in neurodegenerative diseases such as Alzheimer's disease, Parkinson's disease, and Huntington's disease. Thus, DHA-enriched plasma membranes of neurons, thought to cross-talk with mitochondria (see Chapter 19), as well as polyunsaturated membranes of mitochondria, become targets for the design of drugs that slow progression of neural diseases. Recently, a class of steroid derivatives has been identified where cytoprotection is independent of hormonal potency. Studies of structure-activity relationships, glutathione interactions, and mitochondrial function have led to a mechanistic model in which these steroidal phenols intercalate into cell membranes where they block lipid peroxidation reactions and are in turn recycled via glutathione. Such a mechanism would be particularly relevant in mitochondria where function is directly dependent on the permeability barrier of the inner membrane against futile cycling of protons and where glutathione levels are maintained at extraordinarily high (8 to 10 mM) concentrations. Indeed, these steroids stabilize mitochondria under $Ca^{2+}$ loading otherwise sufficient to collapse membrane potential. The cytoprotective and mitoprotective potencies for 14 of these analogs are significantly correlated, suggesting that these compounds prevent cell death in large measure by maintaining functionality of mitochondrial membranes. It is interesting to speculate that lipophilic antioxidants targeting mitochondria work by decreasing formation of oxidatively derived water wires. This class of lipophilic antioxidants would also be predicted to enter DHA-enriched plasma membranes and block membrane damage.

The strong linkage between DHA and cholesterol in neural membranes provides yet another strategy for drug design against neurodegenerative diseases. For example, according to a new review of 4.5 million medical records, a popular anticholesterol drug has recently been found to cut older adults' chances of developing dementia by more than half. Of three statin drugs on the market today, simvastatin (Zocor) offered the greatest protection against dementia. Patients 65 and older who took it for at least 7 months during the 3-year study period were 54% less likely to develop clinical dementia than patients taking nonstatin heart medicines. Simvastatin also cut the risk of a Parkinson's disease diagnosis by 49%.

## 20.5 SUMMARY

Comparative ecological studies involving a range of organisms show that DHA membranes have properties capable of both providing great benefit and causing great harm. These studies also show that a fine line often separates benefits from risks and that even a slight change in the cellular environment can rapidly transition DHA membranes from being beneficial to posing a great risk. This principle, when applied to neurons, opens up a new way of thinking about the roles of DHA membranes in the molecular pathology of neurodegenerative diseases.

Thus, we hypothesize that DHA membranes represent an important missing biochemical link in understanding age-related brain diseases such as AD, as well as prion diseases. As neurons age the sophisticated umbrella of defenses preventing oxidative damage of DHA membranes is proposed to weaken. Oxidatively damaged membranes might affect aging neurons in two ways. Damaged DHA phospholipids are proposed to work indirectly in increasing the level of the ROS pool, which if not kept below a critical threshold can trigger a chain reaction of cellular destruction, including damaging the membrane itself and causing a reduction to the cation permeability barrier of the membrane. Damaged membranes are also proposed to act directly by facilitating the formation of molecular threads of water that disrupt $Na^+$/$K^+$ homeostasis. Cation uncoupling drains energy from neurons already in a weakened bioenergetic state and could be the final insult leading to cellular dysfunction and death. Interestingly, certain toxic peptides such as β-amyloid have also been proposed to act at the membrane level as uncouplers of ion gradients and thus might further increase energy stress. Energy stress might also lead to a decrease in the pool size of bioavailable DHA required to replace damaged chains, perhaps reducing DHA levels in neural membranes.

Perhaps the best evidence for the important roles played by DHA membranes in brain diseases comes from molecular and biochemical studies of defective membrane protease processing of amyloid precursor protein in AD. In this case loss of

DHA from diseased membranes is believed to favor over-production of toxic peptides caused by aberrant protease processing associated with a build up of lipid rafts. These studies focus attention on mechanisms that trigger initial loss of DHA, an area that deserves more attention.

It is encouraging that 2009 has already seen two exciting new models regarding the molecular pathology of AD. Both are revolutionary in nature and show the intense and penetrating research being conducted with AD and related neurodegenerative diseases. Interestingly, the Genentech model might be explained by a DHA membrane scenario such as discussed here.

Finally, understanding the molecular pathology of neurodegenerative diseases is gaining momentum, yet remains one of the greatest challenges for health science in the twenty-first century. We suggest that oxidative stress, energy stress, and toxic peptides might be intimately linked at the level of DHA membranes, essentially feeding on each other and leading to premature death of neurons.

## SELECTED BIBLIOGRAPHY


Calon, F., G. P. Lim, F. Yang, et al. 2004. Docosahexaenoic acid protects from dendritic pathology in an Alzheimer's disease mouse model. *Neuron* 43:633–645.

Dauer, W., and S. Przedborski. 2003. Parkinson's disease: Mechanisms and models. *Neuron* 39:889–909.

Dawson, T. M., and V. L. Dawson. 2003. Molecular pathways of neurodegeneration in Parkinson's disease. *Science* 302:819–822.

Ehehalt, R., P. Keller, C. Haass, C. Thiele, and K. Simons. 2003. Amyloidogenic processing of the Alzheimer beta-amyloid precursor protein depends on lipid rafts. *J. Cell Biol.* 160:113–123.

Elfrink, K., J. Ollesch, J. Stöhr, D. Willbold, D. Riesner, and K. Gerwert. 2008. Structural changes of membrane-anchored native PrP(C). *Proc. Natl. Acad. Sci. U.S.A.* 105:10815–10819.

Graham, R. K., Y. Deng, E. J. Slow, et al. 2006. Cleavage at the caspase-6 site is required for neuronal dysfunction and degeneration due to mutant huntingtin. *Cell* 125:1179–1191.

Hall, E. D. 1992. Novel inhibitors of iron-dependent lipid peroxidation for neurodegenerative disorders. *Ann. Neurol.* 32 (Suppl.):S137–S142.

Hedge, R. S., J. A. Mastrianni, M. R. Scott, et al. 1998. A transmembrane form of the prion protein in neurodegenerative disease. *Science* 279: 827–834.

Jang, H., J. Zheng, R. Lal, and R. Nussinov. 2008. New structures help modeling of toxic amyloid beta ion channels. *Trends Biochem. Sci.* 33:91–100.

Langham, A. A., A. S. Ahmad, and Y. N. Kaznessis. 2008. On the nature of antimicrobial activity: A model for protegrin-1 pores. *J. Am. Chem. Soc.* 130:4338–4346.

Langston, J. W., P. Ballard, J. W. Tetrud, and I. Irwin. 1983. Chronic Parkinsonism in humans due to a product of meperidine-analog synthesis. *Science* 219:979–980.

Morillas, M., W. Swietnicki, P. Gambetti, and W. K. Surewicz. 1999. Membrane environment alters the conformational structure of the recombinant human prion protein. *J. Biol. Chem.* 274:36859–36865.

Nikolaev, A., T. McLaughlin, D. D. O'Leary, and M. Tessier-Lavigne. 2009. APP binds DR6 to trigger axon pruning and neuron death via distinct caspases. *Nature* 457:981–989.

Praticò, D., K. Uryu, S. Leight, J. Q. Trojanoswki, and V. M. Lee. 2001. Increased lipid peroxidation precedes amyloid plaque formation in an animal model of Alzheimer's amyloidosis. *J. Neurosci.* 21:4183–4187.

Prusiner, S. B. 1991. Molecular biology of prion diseases. *Science* 252:1515–1522.

Prusiner, S. B. 2001. Shattuck Lecture: Neurodegenerative diseases and prions. *N. Engl. J. Med.* 344:1516–1526.

Simons, M., P. Keller, J. Dichgans, and J. B. Schulz. 2001. Cholesterol and Alzheimer's disease: Is there a link? *Neurology* 57:1089–1093.

Trushina, E., R. D. Singh, R. B. Dyer, et al. 2006. Mutant huntingtin inhibits clathrin-independent endocytosis and causes accumulation of cholesterol in vitro and in vivo. *Hum. Mol. Genet.* 15:3578–3591.

Wang, X., F. Wang, L. Arterburn, R. Wollmann, and J. Ma. 2006. The interaction between cytoplasmic prion protein and the hydrophobic lipid core of membrane correlates with neurotoxicity. *J. Biol. Chem.* 281:13559–13565.

Whitmer, R. A., D. R. Gustafson, E. Barrett-Connor, M. N. Haan, E. P. Gunderson, and K. Yaffe. 2008. Central obesity and increased risk of dementia more than three decades later. *Neurology* 71:1057–1064.

Wolozin, B., S. W. Wang, N.-C. Li, A. Lee, T. A. Lee, and L. E. Kazis. 2007. Simvastatin is associated with a reduced incidence of dementia and Parkinson's disease. *BMC Med.* 5:20.

# 21 Dietary DHA in Prevention of Colon Cancer: *How a Risk to the Cell Benefits the Organism*

Hypothesis: Colonic cells incorporate dietary DHA to help exterminate precancerous mutants.

Billions of colonic epithelial cells (colonic cells or colonocytes) join together to form the surface or lining of the colon, an organ that is essential for our survival. As in the case of sperm cells, colonic cells have evolved to inhabit a location contacting the external environment. This location allows them to obtain nutrients directly from the gut. Colonic cells are also exposed to a rapidly changing and potentially lethal environment determined by conditions and contents of the colon, largely an anaerobic world. In a strange quirk of nature, colonocytes have evolved to benefit from DHA present in the diet and transported via the circulatory system, or alternatively, transported via the gastrointestinal tract and incorporated into membranes of these cells. Uptake of DHA is also believed literally to save lives by preventing colon cancer.

It is estimated that 70% of colorectal cancers are preventable by moderate changes in diet. Epidemiological studies indicate that populations ingesting higher amounts of fish are at lower risk for colon cancer and a lower mortality rate of colorectal cancer, compared to those with diets high in saturated fat. In addition, other dietary components also appear to favorably impact colonic epithelial cell biology. Most notably dietary fiber, which enhances butyrate production by microbes in the colon, can decrease the incidence of cancers of the digestive tract. Experts in this field suggest that the combination of dietary fish oil and fermentable fiber may provide the basis for a food-based strategy for maintaining a healthy colon.

## 21.1 NATURE OF COLON CANCER

Colorectal cancer is the third most common form of cancer and the second leading cause of cancer-related death in the Western world. Based on current rates, 5.2% of men and women born today will be diagnosed with cancer of the colon and rectum at some point during their lifetime. In 2004, in the United States there were more than 1,000,000 men and women who had a history of colon cancer, with more than 100,000 new cases added per year. Many cases of colon cancer begin as small, benign, clumps of cells or polyps that originate from cells lining the walls of the

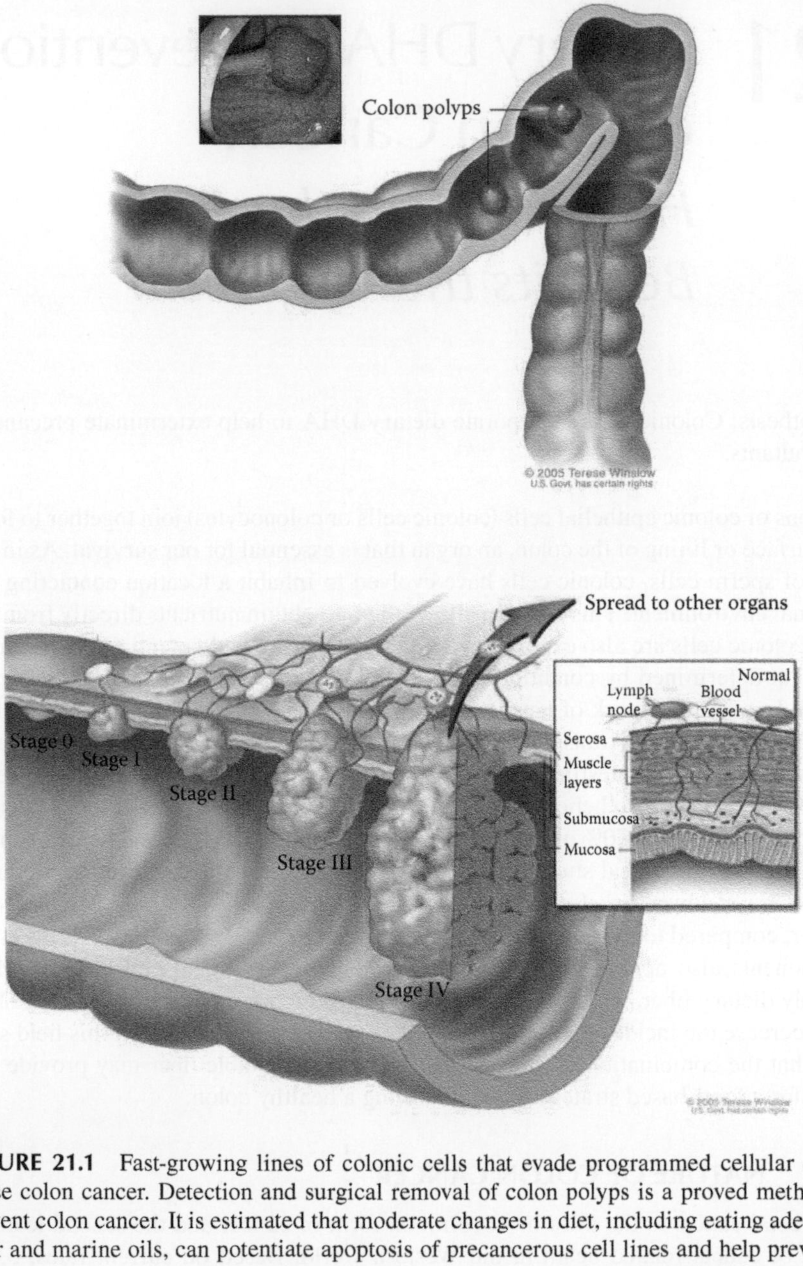


**FIGURE 21.1** Fast-growing lines of colonic cells that evade programmed cellular death cause colon cancer. Detection and surgical removal of colon polyps is a proved method to prevent colon cancer. It is estimated that moderate changes in diet, including eating adequate fiber and marine oils, can potentiate apoptosis of precancerous cell lines and help prevent a majority of colon cancer. This is one of the most dramatic examples of a crucial link between diet and disease discovered to date. (Courtesy of the National Cancer Institute.)

colon (Figure 21.1). Recent data also show that "pancake-like" clumps of cells that are much more difficult to detect than polyps can also become cancerous, opening up a second cellular-based pathway toward colon cancer. Like other cancers, colon cancer requires a series of genetic changes.

The life cycle of colonocytes is one of continuous cellular death and rapid growth of new cells. This constant recycling might be explained on the basis of their unique ecological niche. In essence colonic cells are subject to numerous environmental and chemical insults not encountered by cells in other parts of the body. The list of stresses likely includes the detergent effects of cholic acid, energy uncoupling by short-chain fatty acids such as butyrate produced by colon bacteria, variable levels of dietary DHA and other unsaturated chains, a rapidly changing ionic milieu, increased solubility and toxicity of ferrous iron favored by the anaerobic environment of the colon, toxins produced by colon or foodborne bacteria, low $O_2$ levels, and so forth. This helps explain why the life cycle of colonocytes is one of rapid aging and continuous renewal. Unfortunately, the unique life cycle of these rapidly regenerating cells favors the selection of fast growing mutant cell lines, increasing the probability of precancerous mutants.

## 21.2 DIETARY DHA TARGETED TO MITOCHONDRIAL CARDIOLIPIN OF COLON CELLS

Immortalized human colonic, adenocarcinoma (HT-29) cells, as well as "normal colon cells," fed exogenous supplies of DHA have been found to target DHA to cardiolipin of their mitochondria. DHA in cultured cells reaches levels as high as 48% of the total acyl chains in cardiolipin. Cardiolipin is present primarily in the inner mitochondrial membrane, and in tissues with high respiration rates, such as heart cells, CL can account for 25% of the phospholipids. The HT-29 cell line was nonspecific regarding incorporation of unsaturated fatty acids into phospholipids in general, with values of 38, 18, 25, 13, or 11 mol% for 18:1, 18:2, 20:4, 20:5, or 22:6 in cells enriched with these fatty acids, respectively. Cardiolipin from fatty acid–enriched HT-29 cells contained 26, 41, 6, 9, or 48 mol% for 18:3, 18:2, 20:4, 20:5, and 22:6, respectively. Thus, if HT-29 cells in culture are representative of colonic endothelial cells lining the colon, then one would expect to find selective incorporation of DHA along with linoleic acid from dietary sources into lipids of these cells.

As mentioned in Chapter 12, cardiolipin is often highly enriched with linoleic acid (18:2), with CL species composed of four 18:2 chains being common. The presence of two DHA chains replacing two of the four 18:2 chains, as occurs in cell cultures, generates one of the most unsaturated phospholipids found in humans, a total of 16 double bonds per CL molecule. Synthesis of such highly unsaturated CL structures is likely carried out by a combination of conventional fatty acid pathways in concert with active remodeling steps in which DHA is substituted for more saturated chains. It seems unlikely that dietary sources of DHA alone will drive such high levels of 22:6 incorporation into CL in the intact colon. However, feeding studies with whole animals confirm that DHA is targeted to CL, in agreement with studies of cells in culture.

## 21.3 OXIDATION OF DHA CARDIOLIPIN AS A TRIGGER OF APOPTOSIS

Animals have evolved three different apoptotic mechanisms to kill unwanted cells, such as precancerous cells forming in the colon. Two of the three are mediated by oxidative stress, and this is likely where DHA enters the picture. As summarized in Figure 21.2, DHA cardiolipin is highlighted as a target of lipoxidation resulting in oxidative stress. According to this model, incorporation of DHA into cardiolipin increases levels of oxidative stress and induces apoptosis. This is consistent with studies correlating ROS production to the unsaturation index of cardiolipin acyl chains. In addition, the oxidation of the cardiolipin acyl chains is associated with the release of cytochrome C from mitochondria, triggering caspase activation. Thus, it is possible that a DHA-containing cardiolipin strategically located in the inner mitochondrial membrane may act as an essential signal for the execution of an apoptotic cascade targeting colon cancer cells.

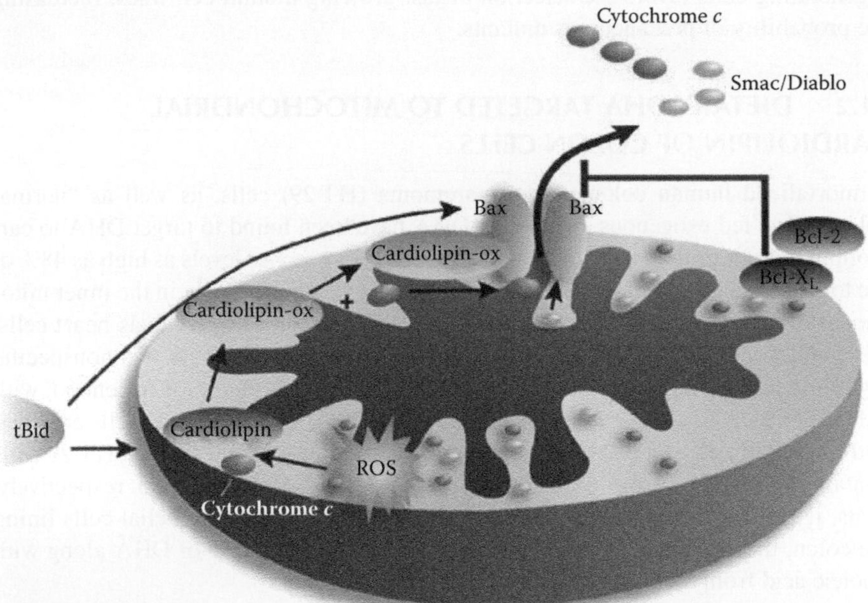


**FIGURE 21.2** DHA from the diet may exert anticancer effects following incorporation into colonic cell membranes by enhancing naturally occurring cascades of programmed cellular death. In this model DHA incorporated into cardiolipin and targeted to mitochondrial membranes acts to intensify apoptosis, effectively ridding the colon of potentially precancerous cells. Cardiolipin ox, oxidized cardiolipin; tBid, 16 kDa truncated peptide of Bid, which is a proapoptotic effector protein. Bax, a pore forming protein in the outer mitochondrial membrane. Bcl or Bclx are proteins that inhibit pore formation. Smac and Diablo, like cytochrome C, are proapoptotic proteins that are released into the cytoplasm through Bax. (Reproduced from Orrenius, S. and Zhivotovsky, B. *Nature Chem. Biol.,* 1:188–189, 2005. With permission.)

For maximum effectiveness, DHA might require molecular cross-talk with other molecules present in the colon, such as butyrate made by colon bacteria. According to a leading researcher in this field, DHA in concert with butyrate potentiate mitochondrial lipid oxidation and dissipation of membrane potential (Chapkin et al., 2008). It is proposed that dietary DHA, along with butyrate produced in the colon, prime colonic cells for apoptosis via a mechanism involving the mitochondrial permeability transition (MPT) pore. This has led to the proposal that DHA be added to the list of physiological regulators of the MPT pore. On the other hand, damaged DHA might directly cause proton leakiness of mitochondria, as already discussed in Chapters 9, 19, and 20, providing an alternative mechanism for modulating apoptosis.

## 21.4 SUMMARY

There is growing evidence that DHA incorporated directly from dietary sources alters mitochondrial membrane composition and function in colonocytes, thereby creating a permissive environment for apoptosis. This is proposed to widen the spectrum of potential precancerous cells targeted for apoptosis and serves as a defense against colon cancer. One interesting aspect of DHA's action in colon cells that needs more attention concerns the possible differential targeting of DHA to the fastest growing classes of precancerous cells. According to this idea, fast growing, precancerous, mutant cell lines might be simultaneously programmed to incorporate more DHA compared to normal colonocytes. This idea is based on the data described above in which colon cancer cell lines incorporate extremely high levels of DHA, when available, into their mitochondria. These faster growing cells might have evolved this pattern as a means of increasing respiration rates and energy output, such as occurs in mitochondria of fast muscles. In this case, the benefit of enhanced energy production may also increase the risk of apoptosis, in essence targeting these fast growing cells for death in the presence of sufficient dietary DHA. Finally, it is ironic to note that one of the riskiest properties of DHA molecules, rapid lipoxidation, is proposed to reduce the risk of colon cancer to humans by selectively increasing the risk to colon cells (i.e., risk → benefit). The molecular basis of DHA and other dietary constituents as anticancer molecules is being actively researched in several laboratories. These studies continue to shed light on the fundamental nature of cancer and highlight how dietary preferences can dramatically lower risks of colon cancer.

## SELECTED BIBLIOGRAPHY


Chapkin, R. S., J. Seo, D. N. McMurray, and J. R. Lupton. 2008. Mechanisms by which docosahexaenoic acid and related fatty acids reduce colon cancer risk and inflammatory disorders of the intestine. *Chem. Phys. Lipids* 153:14–23.

Chicco, A. J., and G. C. Sparagna. 2007. Role of cardiolipin alterations in mitochondrial dysfunction and disease. *Am. J. Physiol. Cell Physiol.* 292:33–44.

Gonzalvez, F., and E. Gottlieb. 2007. Cardiolipin: Setting the beat of apoptosis. *Apoptosis* 12:877–885.

Kato, T., R. L. Hancock, H. Mohammadpour, et al. 2002. Influence of omega-3 fatty acids on the growth of human colon carcinoma in nude mice. *Cancer Lett.* 187:169–177.

McMillin, J. B., and W. Dowhan. 2002. Cardiolipin and apoptosis. *Biochim. Biophys. Acta* 1585:97–107.

Nakagawa, Y. 2004. Initiation of apoptotic signal by the peroxidation of cardiolipin of mitochondria. *Ann. NY Acad. Sci.* 1011:177–184.

Ng, Y., R. Barhoumi, R. B. Tjalkens, et al. 2005. The role of docosahexaenoic acid in mediating mitochondrial membrane lipid oxidation and apoptosis in colonocytes. *Carcinogenesis* 26:1914–1921.

Orrenius, S., and B. Zhivotovsky. 2005. Cardiolipin oxidation sets cytochrome c free. *Nat. Chem. Biol.* 1:188–189.

Watkins, S. M., L. C. Carter, and J. B. German. 1998. Docosahexaenoic acid accumulates in cardiolipin and enhances HT-29 and cell oxidant production. *J. Lipid Res.* 39:1583–1588.

Watkins, S. M., T. Y. Lin, R. M. Davis, et al. 2001. Unique phospholipid metabolism in mouse heart in response to dietary docosahexaenoic or alpha-linolenic acids. *Lipids* 36:247–254.

# Index

Printed and bound by CPI Group (UK) Ltd, Croydon, CR0 4YY
21/10/2024
01777111-0002